CW00709279

Contributions from Science Education Research

Volume 3

Series Editor
Robin Millar, University of York, UK

Editorial Board
Constantinos P. Constantinou, University of Cyprus, Nicosia, Greece
Justin Dillon, King's College, London, UK
Reinders Duit, University of Kiel, Germany
Doris Jorde, University of Oslo, Norway
Dimitris Psillos, Aristotle University of Thessaloniki, Greece
Andrée Tiberghien, University of Lyon II, France

More information about this series at http://www.springer.com/series/11180

Kaisa Hahl • Kalle Juuti • Jarkko Lampiselkä
Anna Uitto • Jari Lavonen

Editors

Cognitive and Affective Aspects in Science Education Research

Selected Papers from the ESERA 2015 Conference

 Springer

Editors
Kaisa Hahl
University of Helsinki
Helsinki, Finland

Kalle Juuti
University of Helsinki
Helsinki, Finland

Jarkko Lampiselkä
University of Helsinki
Helsinki, Finland

Anna Uitto
University of Helsinki
Helsinki, Finland

Jari Lavonen
University of Helsinki
Helsinki, Finland

ISSN 2213-3623   ISSN 2213-3631   (electronic)
Contributions from Science Education Research
ISBN 978-3-319-58684-7   ISBN 978-3-319-58685-4   (eBook)
DOI 10.1007/978-3-319-58685-4

Library of Congress Control Number: 2017944438

© Springer International Publishing AG 2017
This work is subject to copyright. All rights are reserved by the Publisher, whether the whole or part of the material is concerned, specifically the rights of translation, reprinting, reuse of illustrations, recitation, broadcasting, reproduction on microfilms or in any other physical way, and transmission or information storage and retrieval, electronic adaptation, computer software, or by similar or dissimilar methodology now known or hereafter developed.
The use of general descriptive names, registered names, trademarks, service marks, etc. in this publication does not imply, even in the absence of a specific statement, that such names are exempt from the relevant protective laws and regulations and therefore free for general use.
The publisher, the authors and the editors are safe to assume that the advice and information in this book are believed to be true and accurate at the date of publication. Neither the publisher nor the authors or the editors give a warranty, express or implied, with respect to the material contained herein or for any errors or omissions that may have been made. The publisher remains neutral with regard to jurisdictional claims in published maps and institutional affiliations.

Printed on acid-free paper

This Springer imprint is published by Springer Nature
The registered company is Springer International Publishing AG
The registered company address is: Gewerbestrasse 11, 6330 Cham, Switzerland

Contents

About the Editors and Contributors

Editors

Kaisa Hahl PhD, is a postdoctoral researcher and teaches in an English-medium subject teacher education program at the University of Helsinki in Finland. Her research interests relate to the internationalization of teacher education, education for diversities, English as a lingua franca, and teacher development. Kaisa has experience from several years of working in different international projects that aim at enhancing teacher competences and teacher education. Kaisa was the secretary of the ESERA 2015 Conference in Helsinki, Finland.

Kalle Juuti PhD, is university lecturer in physics education in the Faculty of Educational Sciences at the University of Helsinki. He has been a researcher in national, EU-level, and inter-Atlantic research projects, which have focused on several aspects of science education including gender issues, learning environment design, and interest. Recently, his research has focused on students' engagement in school science. His research on learning environments has been engaged with a design-based research approach. Kalle Juuti is the author of several scientific articles in journals and compilations. He is also an editor of books and journal theme issues.

Jarkko Lampiselkä PhD, is university lecturer in chemistry and physics education and has worked at the University of Helsinki since 2004. His research interests are teaching and learning science. He has published several scientific papers in journals, in refereed conference proceedings, and in books and has been an editor of scientific proceedings. He was co-opted member in the ESERA (European Science Education Research Association) executive board (2012–2015), and he was the president of the ESERA 2015 Conference. He is also the head of the Päijät-Häme Region Centre for Mathematics and Science Education.

Jari Lavonen is professor of physics and chemistry education and vice-dean of the Faculty of Educational Sciences, University of Helsinki, Finland. Previously he was head of the Department of Teacher Education, director of the Finnish Graduate School for Research in Science and Mathematics Education, and president of the Finnish Association for Research on Teaching of Mathematics and Science. He has developed and taught numerous courses on teaching and assessment methods in science and technology education. His main research interests lie in science and technology teaching and learning, teacher education, and ICT use in science and technology education. He has participated in several European Union research projects as a principal investigator. He has published hundreds of scientific papers in various journals and books for science (teacher) education.

Anna Uitto is professor in biology education in the Faculty of Educational Sciences, University of Helsinki. She is the head of the research unit of biology and sustainability education. Her research interests include conceptual change, motivation and interest, socio-scientific issues, sustainability education, and curricular aspects in biology education. Her current projects deal with class and subject teacher training, inquiry approach in biology education, education for sustainability development, and out-of-school learning environments in biology and science education.

More information about the editors can be found online: https://tuhat.halvi.helsinki.fi/portal/

Contributors

Nadja Belova is a postdoctoral researcher in the chemistry education group at the Institute of Science Education of the University of Bremen, Germany. Her current research interests comprise societal-oriented science education, critical scientific media literacy, and advertising in science education. She was part of the EU-funded project PROFILES and now works in the EU project *ARTIST – Action Research to Innovate Science Teaching*.

Ángel Blanco-López is associate professor of science education at the University of Málaga, Spain. He heads the Research Group in Science Education and Competence (ENCIC). His research interests include the development of competencies, context-based and modeling approaches in science teaching, and science teacher training.

Markus Bliersbach is a staff member of the Institute of Chemistry Education at the University of Cologne in Germany. He studied chemistry, history, and mathematics for the secondary higher level at the University of Cologne and finished his studies in 2012. After that, he worked as a supply teacher in chemistry and mathematics for one year. In 2013, he started his PhD project, in which he addresses the role of creativity in chemistry and chemistry education. He investigates in particular how preservice chemistry teachers can be supported in developing appropriate conceptions in this context, so that they can function as "multipliers" and convey their perceptions to their future students.

Richard Brock is a PhD student at the Faculty of Education at the University of Cambridge. Prior to this, he taught physics at a secondary school in England. His PhD thesis is a microgenetic multiple-case study examining how students develop coherent conceptual structures related to concepts in physics. He has published articles on the role of insight and intuition in learning about science and on the use of the microgenetic method as a research approach in science education.

Juliana Bueno is manager of biology curriculum of the state government of São Paulo in Brazil. She graduated in biology and pedagogy. She has a master's in education and is a member of the Study and Research Group on Non-formal Education and Science Communication (GEENF; www.geenf.fe.usp.br).

Aurelio Cabello-Garrido is a teacher of science and health education in secondary schools in Málaga, Spain, and is a member of the group ENCIC. His research interests are food competence and nutrition education.

Giuliana Capasso graduated in 2015 in mathematics at the University of Naples Federico II. She currently works at "Ettore Pancini" Physics Department as fellowship owner to implement laboratory activities for secondary school students. Her research interests include inquiry-based teaching and the development of teaching-learning activities to introduce quantum mechanics at secondary school level.

Lyn Carter began her professional career teaching junior science and senior biology in secondary colleges and held the position of school curriculum coordinator. She moved to the tertiary sector and for many years taught science education to preservice teachers and future studies and research methods at master's and doctoral level. At present, she has the role of education doctorate degree coordinator at Australian Catholic University. She is well known in research circles, having served on various editorial boards of high-profile international research journals and written seminal articles on globalization and postcolonialism and their impacts on science education as well as on the environment.

Philip Clarkson is emeritus professor from Australian Catholic University. Before joining Australian Catholic University, he was for nearly 5 years director of a research center at the Papua New Guinea University of Technology and prior to that was at Monash University and tertiary colleges in Melbourne. He began his professional life as a teacher of chemistry, environmental science, mathematics, and physical education in secondary schools. Clarkson has led major consultancies and ARC research projects and was president, secretary, and vice president (publications) of the Mathematics Education Research Group of Australasia and founding editor of the *Mathematics Education Research Journal*. He continues to speak at various international conferences, gives teacher workshops, and publishes regularly.

Digna Couso is reader in *Science Education* and director of the research center CRECIM at Universitat Autònoma de Barcelona, Spain. Her research interests are models and modeling for the teaching of science, participatory design-research approaches to teacher education, and equity issues in STEM education, with special interest in ICTs. She has led and participated in a wide range of national and European projects and is referee of some prestigious journals in the field.

Bette Davidowitz is an associate professor and has been working at the University of Cape Town (UCT), South Africa, for the last 30 years where she specializes in teaching chemistry to underprepared students registered for the extended degree program in the science faculty. She leads research projects in the areas of pedagogical content knowledge as well as teaching and learning in the classroom and laboratory. She has a master's and PhD in organic chemistry from UCT and is the recipient of a Distinguished Teacher Award from UCT and the South African Chemical Institute Medal for Chemistry Education.

Johanna Dittmar is an upper secondary school teacher for chemistry and mathematics education in Bremen, Germany. She is also a scientific coworker in the chemistry education group at the Institute of Science Education of the University of Bremen, Germany. Her current research interests are societal-oriented science education, critical scientific media literacy, and science-related discussions in Internet forums. She was part of the EU project *TEMI – Teaching Enquiry with Mysteries Incorporated*.

Ingo Eilks is a full professor of chemistry education at the Institute of Science Education at the University of Bremen, Germany. His current research areas are socio-scientific issues-based science education, education for sustainable development, action research, research-based curriculum development, and science teacher education. He has been part of many EU projects like ECTN, SALiS, PROFILES, or TEMI. Currently he leads the EU-funded project *ARTIST – Action Research to Innovate Science Teaching*. He has received numerous awards for his research and teaching, the most recent being the Award for Incorporation of Sustainability into Chemical Education by the American Chemical Society.

Yrjö Engeström is professor emeritus of adult education at the University of Helsinki and professor emeritus of communication at the University of California, San Diego. He is director of the Center for Research on Activity, Development and Learning (CRADLE) at the University of Helsinki. He applies cultural-historical activity theory in studies of learning and transformations in organizations, communities, and social movements. He is known for his theory of expansive learning and for the formative intervention method called the Change Laboratory. His most recent books are *Studies in Expansive Learning* (2016) and *Learning by Expanding,* 2nd edition (2015), both published by Cambridge University Press.

Sibel Erduran is professor of science education at the University of Oxford, UK. She also holds a distinguished chair professorship at National Normal Taiwan University, Taiwan, and adjunct professorship at the University of Limerick, Ireland. She is an editor for the *International Journal of Science Education* and section editor for *Science Education* and serves as a member of the executive board of the European Science Education Research Association. Her research interests focus on nature of science in science education.

Enrique España-Ramos is a lecturer of science education at the University of Málaga, Spain, and is a member of the group ENCIC. His research interests are the development of competencies, context-based approach in science teaching, and preservice science teacher training.

Mariona Espinet is a professor of science education at Universitat Autònoma de Barcelona in Catalonia, Spain. She earned a PhD degree in science education at the University of Georgia, Athens, GA, in 1990. She now teaches science education courses in preschool and primary teacher education and science education research courses at the master's and doctoral levels. She is the coordinator of the Official Master of Research in Education Specialty in Science Education at the UAB and the coordinator of two research groups. Her research and innovation interests are strongly interdisciplinary and focus on classroom discourse and critical literacy in multilingual science learning environments and education for sustainability at the interphase between schools and communities.

Wilmo Ernesto Francisco Junior graduated in chemistry and has a PhD in chemistry education from the Institute of Chemistry at São Paulo State University (UNESP/Brazil). Currently, Wilmo is an adjunct professor of chemistry education at the Federal University of Alagoas in Brazil in undergraduate and postgraduate courses. His main research interests lie in science and technology teaching and learning, science history and philosophy, science teacher education, and ICT use in science education.

Silvia Galano is a PhD student at the International School of Advanced Studies of the University of Camerino in collaboration with the University of Naples Federico II. Her research work focuses on the development of a learning progression about basic astronomy phenomena from elementary to secondary school level. She is also interested in investigating children's representations of astronomical phenomena.

Anna Garrido-Espeja is a researcher at the research center CRECIM at Universitat Autònoma de Barcelona, and she has worked as a teacher educator at Universitat de Barcelona, Spain. She has participated in several national and European projects regarding models and modeling in science education, preservice teacher education and socio-scientific issues in science teaching. Her research work is focused on the learning of models and the participation of preservice teachers in the scientific practice of modeling, identifying teaching mechanisms to promote so, and suggesting a modeling cycle to guide the design of lesson plans.

Lena Hansson is an associate professor in science education at Kristianstad University, Sweden. Lena has a broad interest in science education research, but her main research interest lies in the intersection between nature of science perspectives and cultural perspectives on science teaching. One of her interests regards how different images of science, communicated in science class, includes and excludes different groups of students. Related to this, she is involved in research on messages communicated in recruitment initiatives from different actors. Lena has also been engaged in research on the inclusion of socio-scientific issues in the teaching of science.

Anu Hartikainen-Ahia is a senior lecturer in biology and geography education. She educates primary and secondary school teacher students and gives lectures focusing on science education, especially on biology and geography education. She has MSc in geography and PhD in education. Her research areas have primarily focused on inquiry-based and collaborative learning. She has participated in European science teacher education projects, i.e., JoCiTE and PROFILES, and currently in the research project MultiCo.

Christine Heidinger is psychologist and research assistant at the Austrian Educational Competence Centre for Biology Education since 2009. Her expertise is in quantitative and qualitative research methods, philosophy of science, and learning theories. From 2004 to 2010, Christine Heidinger worked at the SCHUHFRIED Company, where she developed computer-based psychological tests and cognitive training programs. At the Austrian Educational Competence Centre, Christine Heidinger is currently working in the fields of inquiry-based science education (IBSE) in authentic learning environments and learning based on socio-scientific issues (SSIs). She developed and implemented various student-scientist partnerships together with science educators and science teachers, and she teaches IBSE and learning based on SSIs in preservice teacher training programs.

Avi Hofstein is a professor (emeritus) of science education at the Weizmann Institute of Science in Israel. He was head of the chemistry education R&D group, head of the National Center for Chemistry Teachers, and head of department. His research is mainly focusing on issues related to teaching and learning in chemistry laboratories, relevance of learning science, attitude and interest in learning science (mainly chemistry), and development of models for professional development of chemistry teachers.

Martin Hopf is full professor for physics education research at the Austrian Educational Competence Centre Physics at the University of Vienna, Austria. His research foci are design-based research, competence orientation in middle and high school, and teachers' and teacher students' pedagogical content knowledge. He acts as editor for two teacher journals: *Plus Lucis* and *Praxis der Naturwissenschaften – Physik in der Schule*. A recent project on teaching Newtonian mechanics to seventh graders was awarded the Polytechnik Prize in 2011.

Maria Irakleous is a research associate of the Research in Science and Technology Education Group at the University of Cyprus. She holds a bachelor's degree in primary education and a master's degree in learning in natural sciences from the University of Cyprus, and she is currently a PhD candidate in learning in natural sciences at the University of Cyprus. Her research interests involve the exploration of teachers' participation roles in the design and implementation of professional development programs in promoting pre- and in-service teachers' competence in teaching science through inquiry.

Lana Ivanjek is a senior lecturer at the University of Vienna, Austria. She obtained a doctorate in physics in 2012 from the University of Zagreb, Croatia, with the dissertation "An Investigation of Conceptual Understanding of Atomic Spectra Among University Students." Her main research interest is physics education research, especially development and validation of new teaching materials, as well as development of different inquiries to assess student learning at university level. She combines the development of evidence-based teaching materials with substantial empirical research on students' understanding of physics. As a senior lecturer at the University of Vienna, she teaches laboratory on experiments in physics education.

Ilpo Jäppinen is a PhD student in science education at the University of Eastern Finland. He obtained his MSc degree from the University of Eastern Finland in 2014. Jäppinen's general research area is science teaching processes and in-service teacher education. His research interests include developing and testing study modules, which enhance students' motivation. His present study is about investigating students' motivation by gender, age, and subject during science lessons. He has worked in several EU projects that focus on students' motivation. He also has experience working as a teacher trainer and has funding from the Finnish Cultural Foundation for his thesis.

Jennifer Johnston is a senior lecturer in the School of Education, University of Lincoln, UK. Dr. Johnston's research interests include science and mathematics integration, peer learning, conceptual understanding of science, and teacher education. Dr. Johnston holds a PhD in science education from the University of Limerick, Ireland, and previous roles include head of education at St. Patrick's College, Thurles, Ireland, and senior project officer in science teaching and learning at EPI-STEM, the National Centre for STEM Education, at the University of Limerick, Ireland.

Alexander F. Koch has been a researcher at the University of Applied Sciences and Arts Northwestern Switzerland, School of Education, Centre for Science and Technology Education, in Basel, Switzerland, since 2012. Alexander studied educational science at the University of Bochum, Germany, and received a PhD from the University of Basel, Switzerland. His main research interests include competence assessment and development, adult education, and methodological issues in educational research.

Irina Kudenko MA, PhD, is an education research specialist, currently working as research and evaluation lead in STEM Learning Ltd., one of the UK's leading providers of professional development and resources for STEM (science, technology, engineering, and mathematics) educators. Irina's research interests include effective professional in-service learning of STEM teachers and measuring impact of educational interventions on student outcomes. In the project *ECB inGenious*, which has findings that are presented in this volume, Irina worked as a principal evaluator and work package leader with responsibilities for evaluation design, implementation, analysis, and reporting.

Martha Marandino is associate professor in the Faculty of Education, University of São Paulo, Brazil. She graduated in biology and has MSc in education and PhD in education. Marandino is coordinator of the Study and Research Group on Nonformal Education and Science Communication (GEENF; www.geenf.fe.usp.br). She has experience in teaching and research in the areas of science education and museum education. Her research interests include science and particularly biology dissemination in out-of-school contexts such as museums, science centers, zoos, aquaria, and botanical gardens.

Irene Marzoli is researcher at the School of Science and Technology (physics division) of the University of Camerino, Italy. Her research interests span from theoretical quantum optics and atomic physics to physics education. She is a member of the Optical Society of America (OSA) and of the Italian Physical Society (SIF). In July 2013, she was one of the directors of the course "Ion Traps for Tomorrow's Applications" of the International School of Physics "Enrico Fermi" (Varenna, Italy). Since 2005, she is local coordinator of *Piano Nazionale per le Lauree Scientifiche*, an initiative funded by the Italian Ministry of Education, University and Research to improve and promote science education in secondary schools.

George McClelland PhD, is a senior lecturer (emeritus) in the Department of Physics at the University of Limerick. He was head of the department on two occasions and was also a director of EPI-STEM, the National Centre for STEM Education, at the University of Limerick. He has taught a variety of undergraduate physics courses, with a particular interest in physics education for preservice physics teachers. His research interests have been primarily in the area of physics education, spanning second and third levels.

Robin Millar is emeritus professor of science education at the University of York. Following a degree in physics and a PhD in medical physics, he taught science for 8 years in secondary schools before moving to a lectureship at York in 1982. He taught on initial and in-service teacher education programs and supervised PhD students. His main research interests are teaching and learning in science (especially physics), science curriculum design and development, and the assessment of science learning. He has codirected several large research and curriculum development projects, including *Evidence-Based Practice in Science Education* and *Twenty-First Century Science*.

Jan Alexis Nielsen is associate professor and head of section at the Department of Science Education, University of Copenhagen, Denmark. His primary research interest is formative and summative assessment of complex skills and competences in science education. Nielsen has participated in several EU projects on socio--scientific issues and inquiry teaching and learning. His most recent work has focused on teachers' assessment practice vis-à-vis innovation competence and inquiry skills.

Jonathan Osborne is the Kamalachari chair in science education at the Graduate School of Education, Stanford University. Previously he held the chair in science education at King's College London. He was a member of the US National Academies Panel that produced the *Framework for K-12 Science Education* that is the basis for the Next Generation Science Standards. Currently he is chair of the expert group that produced the framework for the PISA science assessments in 2015 and 2018. His research interests are in the role of argumentation in science and improving the teaching of literacy in science.

Magnus Oskarsson has a PhD in science education at Mid Sweden University in Sundsvall and graduated on thesis about student attitudes and interest in science. His main research is focused on affective components in student learning. He has strong experience and expertise in large-scale assessment and in cooperation in different research networks around the world. He served as national project manager in PISA in Sweden between 2010 and 2015 and as TIMSS science expert in Sweden between 2009 and 2015. He is implementing and doing research on both local and international programs for school improvement with aims to increase social, emotional, and intercultural skills among students.

Peter Pany is a biology and chemistry teacher in Vienna, Austria, and he also works as research assistant at the Austrian Educational Competence Centre for Biology Education since 2006 and cooperates with the Botanical Garden of the University of Vienna since 2005. His expertise is botany and plant science as well as in didactical research and teaching science. For his PhD thesis, he studied students' interests in plants. At the Austrian Educational Competence Centre, Peter Pany is currently working in developing preservice teacher training programs and in-service teacher professional development courses. In addition to his academic activities, he is teaching science at the Wiedner Gymnasium/Sir Karl Popper School in Vienna.

Marios Papaevripidou is a specialist teaching fellow of science education at the University of Cyprus and senior researcher of the Research in Science and Technology Education Group at the University of Cyprus. He completed a BA in education, an MA in learning in natural sciences, and a PhD in learning in natural sciences at the University of Cyprus. His research interests entail the use of modeling-centered inquiry as both a learning tool and an instructional approach in science teaching and learning and also the design and research validation of computer-supported curriculum materials to foster teachers' professional development in modeling and inquiry-based learning.

Verena Pietzner is full professor of chemistry education at the Carl von Ossietzky University in Oldenburg (Germany). After studying mathematics and chemistry for teaching profession at upper secondary schools including teacher training, she started her PhD thesis at TU Braunschweig (Braunschweig, Germany) and then became a postdoc at TU Braunschweig in 2002. Since 2009, she was associate professor at the University of Hildesheim (Germany) before she went to Oldenburg. Her research interests are computer-based teaching and learning, career education, creativity, and food chemistry as part of chemistry education.

Roser Pintó is a doctor in physics and expert researcher in Science Education, and she is tenured university professor and former director of CRECIM and of the Thematic Network of Research in STEM Education of Catalonia (REMIC), which includes more than 15 research groups from the seven Catalan universities. Internationally, Dr. Pintó is well known for her publications and talks and her role in science education institutions and networks, having been member of the board of the prestigious association in the science education field ESERA (European Science Education Research Association). She has also organized the GIREP 2000 and the ESERA 2005 International Conferences in the field.

Maja Planinic is a senior lecturer and head of the physics education group at the Faculty of Science, University of Zagreb, in Croatia, where she graduated in physics in 1988 and obtained a master's degree in atomic and molecular physics in 1994 and doctorate in physics (in physics education research) in 2005. She has served on committees for introduction of state matriculation exam in Croatia (2005–2010) and the reform of the national physics curriculum (2015–2016). She was the principal investigator on a physics and mathematics education research project (2007–2014), funded by the Croatian Ministry of Science, Technology and Sports. Her main research interests are investigation of student conceptual understanding of physics and educational applications of Rasch modeling.

Marietjie Potgieter is an associate professor and has been working at the University of Pretoria, South Africa, for the last 20 years where she teaches organic chemistry to undergraduate students and leads research projects in science education. Her research interests include effective instruction and student success in tertiary mathematics and science. She has a PhD in bio-organic chemistry from the University of

Illinois, USA, and a master's degree in carbohydrate chemistry from the University of South Africa. She is a recipient of the Chemical Education Medal from the South African Chemical Institute.

Christiane S. Reiners studied chemistry and English in a degree of education program for teaching at secondary and lower level at the University of Cologne in Germany. After in-service teacher training, she finished her PhD in 1988 and her habilitation in 1993 at the University of Cologne. In 1996, she became full professor for chemistry education at the FU Berlin, and since 1997, she is full professor at the Institute of Chemistry Education at the University of Cologne. From 2003 to 2005, she was vice rector of the University of Cologne. From 2008 to 2015, she served as national representative from Germany on the IUPAC Committee on Chemistry Education.

Marissa Rollnick is a professor at Wits University in South Africa. She has taught at school level in South Africa and at teachers' college and university level in Swaziland before returning to South Africa in 1990. Since then, she has worked at Wits University and has been chair of science education since 2005 in the Marang Centre, Wits School of Education. She is engaged in research into subject matter for teaching or pedagogical content knowledge.

Luzie Semmler is a PhD student at the Department of Chemistry Education at the Carl von Ossietzky University in Oldenburg (Germany). Her research interest focuses on the investigation of teachers' and student teachers' views and conceptions about creativity in general and creativity in chemistry class. She studied chemistry and German language for teaching profession at upper secondary schools. She received her bachelor of arts degree in 2012 at the University of Hildesheim (Hildesheim, Germany) and her master of education degree in 2014 at TU Braunschweig (Braunschweig, Germany).

Margareta Serder has a PhD in science and mathematics education from Malmö University, where she has her affiliation as researcher. Serder has a position as research and school development leader at Skåne Association of Local Authorities in Sweden. Her research concerns the sociocultural and sociomaterial aspects of education and of assessment. In her publications, Serder has specifically explored these aspects in relation to the Programme for International Student Assessment (PISA).

Cristina Simarro is an engineer, specialized in materials. She holds a master's degree in science education (2010) and a master's degree in science education research (2011). After her 6-year experience working on materials engineering and on purchasing management in enterprises of the information sector, she joined the TIREC research group, in which she has participated in several national and international projects (ECB-inGenious, TRACES, etc.). She is now a PhD candidate, working on new STEM (science, technology, engineering, and mathematics) educational methodologies, specifically in the context of making movement both in formal and informal education.

Svein Sjøberg is professor in science education at Oslo University, Norway. His current research interests are the social, ethical, and cultural aspects of science education, in particular the impacts and influence of large-scale assessment studies like PISA and TIMSS. He has worked extensively with international and comparative aspects of science education through, e.g., UNESCO, OECD, ICSU, and the EU. He has won several prizes and awards for his research, teaching, and promotion of science literacy and public understanding of science. He is an elected member of two Norwegian academies: Norwegian Academy of Science and Letters and Norwegian Technical Science Academy.

Jesper Sjöström is an associate professor of science and chemistry education at the Department of Science-Environment-Society, Malmö University, Sweden. Prior to joining Malmö University, he did postdoctoral research at the Research Policy Institute, Lund University. His current research interests encompass sustainability and socio-politically oriented science education, with a particular focus on the molecular sciences and their links to philosophy of education. Examples of journals, which have published his research, are *Science & Education* and *Chemistry Education Research and Practice*. Since 2015 Sjöström participates in the EU project PARRISE, which combines socio-scientific issues with inquiry-based learning and citizenship education.

Luigi Smaldone is a retired associate professor of astronomy at the University of Naples Federico II. His most research interests include astronomy education and the development of low-cost inquiry experiments for secondary school students in optics and electromagnetism.

Kari Sormunen PhD, is a senior lecturer on physics and chemistry education at the University of Eastern Finland. He has been educating primary, secondary, and special teachers on his field for over 24 years. He has coordinated nationally many EU teacher education projects. He was a Fulbright visiting scholar at Stanford University, CA, USA, in fall 2016. His research interests are in diverse learners in science education and science teacher education.

Annika Springub studied chemistry and mathematics at the Carl von Ossietzky University in Oldenburg (Germany) for teaching profession for upper secondary level. She got her master's degree in 2014 and then entered the teacher training in Bremen. After finishing teacher training with the Federal State Examination for Teachers, she is now working at the Hermann Tempel Comprehensive School in Ihlow (Germany).

Elke Sumfleth PhD, is full professor for chemistry education and has worked at the University of Duisburg-Essen, Germany, since 1990. Her research interests relate to academic teaching and learning of chemistry, competence measurement, improvement of experimental learning environments, and professional knowledge of chemistry teachers. Elke is an author of many scientific journal articles and an editor of different journals. She was head of the German Research Training Group 902 "Teaching and Learning of Science" (nwu-essen) and is now head of the German Research Group 2242 "Academic Learning and Academic Success During University Entry Phase in Natural and Engineering Sciences."

Ana Susac is a senior lecturer at the Department of Physics, Faculty of Science, University of Zagreb, in Croatia. Her main research interests are educational physics and educational neuroscience. She is particularly interested in student understanding of basic concepts in physics and the role of mathematics and formal logical reasoning in physics teaching and learning. Besides standard physics education research (PER) methods, she also uses eye tracking and functional neuroimaging techniques in exploration of student attention patterns while they are solving physics problems.

Keith S. Taber is the professor of science education at the University of Cambridge. He taught sciences (mainly physics and chemistry) in secondary schools and in further education, before joining the Faculty of Education at Cambridge, where he is currently the chair of the Science, Technology and Mathematics Education Academic Group. He works with graduate students, teaching research methods and supervising student projects. His research interests are largely related to aspects of learning in science. He is interested in student conceptual understanding (including student thinking about the nature of science) and how this develops, and potentially becomes better integrated, over time.

Oliver Tepner PhD, has been professor for chemistry education at the University of Regensburg in Bavaria, Germany, since 2012. His main research interests are in the field of chemistry teachers' professional knowledge, their acting in classroom situations, and students' conceptions and students' outcome. His research is focused on teachers' pedagogical content knowledge (PCK) and content knowledge (CK), use of technical language, and teacher students' development of PCK and CK. Oliver is an author of several scientific journal articles, refereed conference proceedings, and book chapters. He has organized several international joint researcher trainings and national conferences.

Italo Testa PhD, is assistant professor of physics education at the University of Naples Federico II. His current research interests include physics teacher training, astronomy education, students' interpretation of simulations and textbook images, and the design of inquiry-based teaching-learning sequences at secondary school level.

Holger Tröger studied chemistry and German language and literature at the University of Duisburg-Essen to become a secondary school teacher (German

Gymnasium). After his first state examination for teaching in secondary school, he started his research on teachers' professional knowledge. Holger developed video-based test instruments and paper-pencil tests in order to analyze the use of technical language and its meaning for science teaching. At present, he is in secondary school teacher training.

Shingo Uchinokura is associate professor in Kagoshima University, Japan. He finished his master's program in education at the University of Tsukuba (Japan). After he taught chemistry at high school for 4 years, he started to research students' conceptual and analogical understandings in the doctoral program in school education at the University of Tsukuba (Japan) as a JSPS research fellow. In 2008, he became assistant professor at Shizuoka University (Japan). Since 2013, he has engaged in the study of analogy, model, and creativity in science education at Kagoshima University.

Laura Valdés-Sánchez is an associate professor and a research assistant at Universitat Autònoma de Barcelona in Catalonia, Spain. She earned a degree in biology at Universitat de Barcelona (2008) and a PhD degree in science education at Universitat Autònoma de Barcelona (2016). She has worked as an educator at the Natural Science Museum of Barcelona and at the Botanic Garden of Barcelona, and nowadays, she teaches science education in preschool and primary teacher education. Her research focuses on classroom discourse in multilingual science learning environments and on co-teaching as a tool for professional development and as a strategy for the collaborative construction of content and language integrated learning projects.

Roald P. Verhoeff is assistant professor of science education at the Freudenthal Institute for Science and Mathematics Education at Utrecht University in the Netherlands. He has a doctorate in biology education and was involved in several projects which aimed to translate ethical quandaries on emerging science and technologies to science education. He has lectured on science communication and education for more than 10 years. His current research interest is in the integration of moral and societal implications of research in science education. As such he is involved in the EU project PARRISE, which promotes civic involvement in scientific research and innovation through activities in schools.

Gráinne Walshe PhD, is the coordinator of the Science Learning Centre at the University of Limerick, Ireland. She also works as a researcher in EPI-STEM, the National Centre for STEM Education, in the University of Limerick. Her research interests are in interdisciplinary science education, investigating mathematical practices in science, teacher-based curriculum development, physics education research, and student retention and progression in science subjects at third level.

Zacharias C. Zacharia is an associate professor of science education and director of the Research in Science and Technology Education Group at the University of

Cyprus. He completed his BA in education at the University of Cyprus (Cyprus), BA in physics at Rutgers University – New Brunswick (USA), and graduate studies (MA, MSc, MPhil, and PhD) in science education at Columbia University, New York (USA). He was the (co-)coordinator of several research projects concerning science and technology education that received continuous financial support over the years from the European Commission. His research interests include, among others, the design and development of inquiry learning environments in science and the development and assessment of science curriculum.

Alessandro Zappia graduated in physics at the University of Naples Federico II in Italy in 2011 and obtained the teaching certification in math and physics for secondary school in 2013. He was employed in the period 2011–2016 as fellowship owner at the Department of Physics of the University of Naples Federico II in European funded projects involving students at secondary school level and implementing science teachers' training activities. His interests span from modern physics to implementations of teaching-learning activities using low-cost laboratory and inquiry methodology. He is currently a PhD student of the International School of Advanced Studies at the University of Camerino (Italy) in collaboration with the University of Naples Federico II.

Introduction

Kaisa Hahl, Kalle Juuti, Jarkko Lampiselkä, Anna Uitto, and Jari Lavonen

This edited volume is comprised of selected studies that were presented at the 11th European Science Education Research Association (ESERA) Conference, held in Helsinki, Finland, from 31 August to 4 September, 2015. The ESERA science education research community consists of professionals with diverse disciplinary backgrounds from the natural sciences to social sciences. This diversity provides a rich understanding of cognitive and affective aspects of science teaching and learning in this volume.

The European Science Education Research Association (ESERA) is an international organization for science researchers and science educators and its aims are to: (i) enhance the range and quality of research and research training in science education; (ii) provide a forum for collaboration in science education research; (iii) represent the professional interests of science education researchers in Europe; (iv) seek to relate research to the policy and practice of science education in Europe; and, (v) foster links between science education researchers in Europe and elsewhere in the world (www.esera.org). The biennial ESERA conference is the main forum for direct scientific discourse within the community, for the exchange of insightful practices, and for extending networks among the researchers and educators.

The chapters in this book showcase current orientations of research in science education and are of interest to science teachers, teacher educators and science education researchers around the world with a commitment to evidence-based and forward-looking science teaching and learning.

K. Hahl (✉) • K. Juuti • J. Lampiselkä • A. Uitto • J. Lavonen
University of Helsinki, Helsinki, Finland
e-mail: kaisa.hahl@helsinki.fi; kalle.juuti@helsinki.fi; jarkko.lampiselka@helsinki.fi; anna.
uitto@helsinki.fi; jari.lavonen@helsinki.fi

© Springer International Publishing AG 2017   1
K. Hahl et al. (eds.), *Cognitive and Affective Aspects in Science Education
Research*, Contributions from Science Education Research 3,
DOI 10.1007/978-3-319-58685-4_1

A Look Back at the ESERA 2015 Conference

The ESERA 2015 Conference theme was *Engaging Learners for a Sustainable Future*, thus underlining aspects that are of great relevance in contemporary science education research. The conference theme called on researchers to reflect on different current approaches to science education in order to enhance our shared knowledge of learning and instruction in various educational contexts.

In total almost 1150 single and multi-paper proposals were submitted to the conference in early 2015. Eventually, for example, out of 810 proposals for single oral presentations initially submitted for review, 580 were presented as such at the conference. Out of the 230 interactive posters presented at the conference, 35 were presented by young researchers who had attended the ESERA Summer School in 2014. The workshops or ICT demos were less popular as a presentation format as just 12 of them were presented or carried out in the conference. In total 72 symposia (each with four papers) were held at the conference, with 16 of them being invited symposia. In general, it was the symposia that received the highest average session ratings from the participants. Each symposium was organized by an expert around a specific topic and each of the papers, presented by scholars from different countries, addressed the topic from different perspectives.

The conference week was thus tightly scheduled with hundreds of presentations, either single oral presentations, symposia, interactive posters or ICT demos or workshops, divided into 19 different strands based on their topic (see www.esera2015.org). In addition, the conference had also four invited plenary talks by prominent scholars. Education for sustainability or sustainability related issues were considered in several symposia and other types of presentations in many strands of the conference, especially in Strand 9 (Environmental, health and outdoor science education). However, most presentations dealt with other current topics of science education. Many conference presentations dealt with research on students' and teachers' knowledge, engagement, experiences, competences and social interaction in different learning environments, reflecting the importance of various cognitive and affective aspects in science education. After the conference, all presenters were invited to submit revised and extended papers of their conference presentation to the electronic Proceedings of the ESERA 2015 Conference, titled *Science Education Research: Engaging Learners for a Sustainable Future* (Lavonen et al. 2016). The electronic publication comprises over 300 papers and, as a whole, represents the interests and areas of emphasis in the ESERA community at the end of 2015.

The ESERA 2015 Conference was attended by nearly 1300 science education researchers from around the world and thus the conference was indeed a very international meeting. About two thirds of the participants came from 29 different European countries and one third from 27 countries in Asia, Australia, Africa, North America and South America. While presenting one's own research and engaging others in discussion of it was naturally one of the most important aspects of the conference, having an opportunity to meet other members of the science community was just as valuable. The countless encounters with other professionals throughout the week enabled the participants to strengthen their existing networks, make new acquaintances and set seeds for future cooperation.

In education, it is important to "practice what you preach" and, for the first time in an ESERA conference, the organizers – with the help of the ESERA Board – encouraged the keynote speakers to plan and incorporate social interaction into their presentations. The participants enjoyed the opportunity to send questions beforehand, follow the interaction of two keynote speakers sharing the stage on the opening day of the conference, and use Twitter and Flinga (a digital infinite whiteboard) online during some of the keynote presentations. Besides the formal conference programme, the participants had an opportunity to attend pre-conference workshops and different receptions, and take part in other excursions such as school visits, industrial site visit, national park trip or, for example, the first ever ESERA football game. Student helpers were incremental to the daily operations throughout the conference and were also present and ready to help during the conference sessions, outings and at evening events.

The general atmosphere at the conference exuded positive energy that all the participants brought together. The participants were delighted to be in Helsinki and share their contributions at the conference. The excitement and commitment of being part of the ESERA science education community culminated at the traditional conference gala dinner on the evening before the final conference day. The gala dinner was organized at Finlandia Hall which is one of the most iconic buildings in Finland and designed by Alvar Aalto, a world-renowned Finnish architect. This historic location provided a perfect setting for the three course dinner and entertainment before and after. The evening ended in dancing to the tunes of a wonderful dance band – all too soon for many of the conference participants still present.

Highlights of the Chapters

This volume brings forth intriguing research that is both novel and innovative in the field of science education. We are certain that will continue to spark interest and discussion within the community of science education researchers and science educators and extend beyond to new contexts and new actors. We, as the editors, are grateful for all the work carried out by the strand chairs and reviewers who made it possible to include these selected papers to this compilation. In the first step, the strand chairs recommended the most interesting original conference synopses as possible papers for the book. We invited the recommended scholars to submit full manuscripts. Based on at least two reviewer reports, we determined the papers to be selected in this book. Thus, the chapters underwent a rigorous scientific process, guided by the editors, before being accepted into this volume in their final form. Initially 35 invitations to submit manuscripts were sent to different authors and 34 manuscripts were received.

This book contains 25 research chapters that each take a specific perspective into a relevant aspect of contemporary science education. The chapters are multifaceted and examine different science education phenomena in a wide range and thus they could be grouped in various ways and many could fit under more than one of the following titled parts. To help the reader, the chapters are loosely organized into six parts, each of which revolves around certain cognitive and/or affective aspects in science education research. The parts are *Teacher Knowledge*, *Student Engagement*, *Student learning and assessment*, *Language in science classrooms*, *Professional development*, and *Expanding science teaching and learning*.

In what follows, we will bring up the main topic or theme of each particular study. The reader will get a quick overview of the variety of different subjects, contexts and countries, and research approaches.

Teacher Knowledge

The volume starts off with different views of *Teacher Knowledge* that address the notions of Nature of Science (NOS), formative assessment, Pedagogical Content Knowledge (PCK) and creativity in science education. **Sibel Erduran**, one of the keynote speakers at the ESERA 2015 Conference, tackles the fundamental question of what science is and argues that visualizing NOS with images can help bring coherence to the ways NOS is conceptualized and enacted in science education. In her chapter, Erduran gives examples of holistic images in science that can inform curricular activities and she discusses a framework that can be used as a tool to identify and address gaps and missing links within the science curriculum.

In his chapter, **Robin Millar** discusses a curriculum development project in England that has its starting point in assessment instead of content or pedagogy. The project aims to support teachers in embedding formative assessment in their science teaching at lower secondary school level. Millar uses data from questionnaires and interviews to find out how teachers' practices and thinking are impacted by carefully-designed and research-informed assessment materials.

The two following chapters deal with creativity in chemistry education from different perspectives and thus aim to fill the gap in science education research as related to creativity. Both studies are carried out in the German education system, the first among practicing teachers and the second in pre-service teacher education. **Annika Springub**, **Luzie Semmler**, **Shingo Uchinokura**, and **Verena Pietzner** used online questionnaires to gain insight into chemistry teachers' perceptions and attitudes to creativity and to find out what creative methods, if any, the teachers themselves use. The authors discuss the results based on, for example, how the teachers rated their own personal traits in relation to creativity.

The study by **Markus Bliersbach** and **Christiane S. Reiners**, aims at supporting prospective teachers in developing their conceptions about creativity in chemistry education practice. The authors designed two different interventions (one approach with historical case studies and the other with student-generated analogies) that they integrated into teacher education courses. The authors discuss the differences between these approaches and conclude that while one approach can help students to recognize the importance of creativity in chemistry, the other approach can help students to realize how creative processes can be implemented in school chemistry.

The two last chapters in the first part examine pedagogical content knowledge (PCK) from different angles. Although there are different models that make up teachers' expertise, recent research often focuses on three distinct dimensions of teachers' professional knowledge which are content knowledge, pedagogical knowledge and pedagogical content knowledge. While PCK is considered to be a construct unique to teachers, its definition and understanding among researchers may vary. **Marissa Rollnick**, **Bette Davidowitz** and **Marietjie Potgieter** introduce the notion of teachers' specialised knowledge at topic level, calling it topic-specific PCK, and in their study they attempt to discover whether topic-specific PCK is unique to teachers. The authors gathered both qualitative and quantitative data with two different previously validated instruments that included paper-and-pencil tests. In the study they compare the performance of chemistry subject matter specialists to chemistry teachers in both content knowledge and topic-specific PCK in organic chemistry in South Africa.

In his study, **Alexander F. Koch** addresses PCK from the question of its predictive power on constructivist teaching habits in science classrooms. Data were gathered from experienced science teachers from kindergarten up to lower-secondary school levels and evaluated with multiple measures for 3 years. The study is part of a science education project in Switzerland which is focused on competence development in in-service teacher education and aims at fostering activating, inquiry-based teaching practice.

Student Engagement

The second part is *Student Engagement* which includes studies that address student interest and study ways which can enhance student engagement in science classrooms or build connections between science and students' lives. **Kari Sormunen**, **Anu Hartikainen-Ahia** and **Ilpo Jäppinen** tackle the problem of students' lack of interest and motivation in science studies from the perspective of teachers' pedagogical practices with socio-scientific issues (SSI). The authors examine scenarios (problems or cases that relate to real-life situations) that were designed by Finnish teachers for their science classes and analyze the quality of the designs. The scenarios are compared against the components of an ideal scenario and examined for their relevance, students' involvement and reflection, and stimulation of scientific questions.

Drama in science education is a relatively new phenomenon and little is yet known of its effectiveness, for example, for student learning of difficult concepts, forming of opinions, or understanding social problems. In the study by **Roald**

P. Verhoeff, university students in the Netherlands from various science disciplines were involved in a drama experiment and, instead of being merely spectators, performed multiple roles as audience, reviewers, authors and actors. In this chapter Verhoeff explores how the use of drama can enhance reflexivity in embracing and critiquing socio-scientific issues on emerging neuro technologies which involve various ethical and moral aspects.

The study by **Peter Pany** and **Christine Heidinger** is set within biology education and, more specifically, in botany education among secondary students in Austria. The authors describe the notion of 'plant blindness' that refers to the phenomenon that is currently taking place when people ignore plants in everyday life. Gaining knowledge about plants and plant life is important for students so that they understand the major global challenges that our societies are facing, such as climate change and global warming. The authors' approach to battle students' disinterest in plants is to examine their existing pre-interest in plants. The authors collected data with a questionnaire in order to find out differences between relevant groups (age and gender) and calculate interest profiles on an individual level. In their conclusions Pany and Heidinger discuss how useful plants can be used as 'flagship' species to trigger student interest in botany.

The inclusion of digital technologies in education is nowadays considered as an integral part of school work and explicitly called for in many curricula (e.g. Finnish National Board of Education 2016). The use of new technologies can also help enhance student interest and engagement in lesson activities. **Wilmo Ernesto Francisco Junior** examines the contribution of video production to chemistry education in a university course in Brazil. He analyses videos produced by students and information about the production process gathered through a questionnaire. The author discusses different demonstrations of student engagement and argues that the production of videos creates learning opportunities that can promote student reflections over a long time and help clarify remaining misconceptions.

Student Learning and Assessment

The third part, *Student learning and assessment,* includes chapters that each takes a different approach to issues that deal with an aspect of student learning in science education. **Richard Brock** and **Keith S. Taber** view learning as a learner's personal act of structuring multiple conceptual elements. In their chapter the authors use the term 'making sense' to describe student learning in science education and they define it as "the formation or modification of a conceptual structure in which concepts are related in a coherent system that may be applied to a range of contexts." In their microgenetic case study set in England, Brock and Taber use the notion of coherence to examine and discuss two secondary school students' organizations of resources in order to find out how the students' conceptual structures related to physics developed over a 5-week period.

The study by **Lana Ivanjek**, **Maja Planinic**, **Martin Hopf** and **Ana Susac** investigates university students' strategies and difficulties with graph interpretation in different domains related to mathematics and physics and compares the results between students in both Croatia and Austria. The authors conclude that student reasoning about problems seems to be bound by the contexts and the disciplines in which their knowledge was acquired. They also discuss the possibilities and potential of using mathematics or physics related problems in contexts other than standard mathematics or physics to better expose and develop student reasoning.

Aurelio Cabello-Garrido, **Enrique España-Ramos** and **Ángel Blanco-López** conduct a literature review of last three decades' publications of human nutrition in order to identify and describe students' mental models of human nutrition and to consider how these evolve over time. The authors define mental models as constructs that are built both by scientists and learners to interpret their experiences and to make sense of the physical world. Knowledge of the structures of students' mental model can be used for developing school curricula. Cabello-Garrido et al. explain that as overweight and obesity are becoming serious public health issues worldwide, it is important to raise students' awareness of this problem and encourage healthy eating habits.

The PISA studies have received considerable attention throughout the world since their inception in 2000. The Programme for International Student Assessment is an international survey conducted every 3 years which evaluates education systems worldwide by testing the skills and knowledge of 15-year-old students (e.g. OECD 2000, 2016). In their chapter **Jonathan Osborne**, **Magnus Oskarsson**, **Margareta Serder** and **Svein Sjøberg** seek to explore the positive and negative aspects of the PISA assessment. The authors explain the changes and improvements in the assessment framework between the 2006 and 2015 science assessment and examine the major social and political impact that PISA has on education systems, schools and the learning of science. Each of the authors takes a specific role to consider PISA from different perspectives, and, for example, to argue that it can be seen as an instrument of power that leads to a global race, or as an external indicator that can help measure the performance of an education system in relation to others. The chapter also presents an interpretation of PISA results from a science classroom perspective from Sweden and describes a study that seeks to understand the interaction between the students and the items used for PISA testing.

Language in Science Classrooms

The fourth part in this volume is titled *Language in science classrooms* and it contains three chapters that each take a look at language(s) used in science teaching and learning, either by teachers or students. Language is of central importance in learning as it is the primary medium of interpersonal communication to convey information such as thoughts, beliefs and knowledge, and it is closely connected to cognition and thinking. The focus in the study by **Holger Tröger**, **Elke Sumfleth** and **Oliver**

Tepner is on pedagogical content knowledge about technical language used by teachers at secondary schools in Germany. The authors examine the relation between teachers' content knowledge, pedagogical content knowledge regarding technical language, teachers' acting in class and students' learning achievement in chemistry. Tröger et al. used multiple methods to collect data at different points in time, through paper-and-pencil tests, questionnaires and video recordings. The authors discuss the relevance and influence of technical language used by teachers for students' improved learning outcomes.

The aim of language education in Europe is not only for each student to learn two foreign languages besides their first language, but to develop a linguistic repertoire, in which all linguistic abilities have a place (Council of Europe 2001). In order to improve foreign language learning, new and more effective approaches to learning languages are promoted by the Council of Europe, such as Content and Language Integrated Learning (CLIL). The study by **Laura Valdés-Sánchez** and **Mariona Espinet** explores the potential of co-teaching as a strategy to help build a Science-and-English CLIL project in Catalonia, Spain. The authors explain how the collaboration in a CLIL class between teachers from different specialized disciplines can be challenging to achieve genuine integration between the learning objectives of different subjects. Valdés-Sánchez and Espinet collected video data of three pairs of co-teachers and developed an analytical tool used for the analysis of the discursive collaboration of co-teacher pairs. The authors discuss the different discursive participation strategies and their development and transfer between the co-teachers as they progress toward more co-constructive partnerships.

In the past decades societies all over the world have become increasingly multicultural and multilingual. The diversification of societies is also seen and felt in the school systems. **Philip Clarkson** and **Lyn Carter** provide an overview of issues related to research in multilingual and multicultural STEM (Science, Technology, Engineering, and Mathematics) teaching and learning environments. The authors discuss the diverse scenarios of mixed language groups and mixed language abilities (including STEM languages) that challenge teachers and students in STEM classes – not only in the authors' own Australian context but around the world. Clarkson and Carter sketch a list of different practical and theoretical issues that arise from the complexity of multilingualism and that they consider should be taken into account in STEM research and teacher education.

Professional Development

The fifth part of this book consists of chapters that address the topic of teachers' professional development either in pre-service or in-service teacher education, and it is titled *Professional development*. In order for teachers to be able to involve their students in scientific practices, they must first have a chance to practice these themselves and reflect on their experiences (Russell and Martin 2014). **Digna Couso** and **Anna Garrido-Espeja** design and investigate a teacher education course for pre-service primary school teachers in Spain who construct scientific models while participating in modelling practices. The authors' view of modelling includes the assumption that participation in scientific practices is not only to learn to engage in the practices or to learn about the practices but also to learn the conceptual knowledge in which to frame them. In order to analyze how the pre-service teachers' modelling practices and their versions of the school-based scientific model evolved, the authors gathered data through audio and video recordings. Couso and Garrido discuss the results and the participants' development and consider ways that can support pre-service teachers' conceptual learning and construction of more sophisticated scientific models.

The chapter by **Marios Papaevripidou, Maria Irakleous**, and **Zacharias C. Zacharia** presents a study that investigates the effect of a professional development (PD) programme designed for pre-service elementary school teachers within a science methods course in Cyprus to develop their inquiry competence. During the programme the teachers acted in different roles at three different phases: they were engaged as learners in multiple inquiry-cycles, thinkers who studied the curriculum from its pedagogical rationale, and as reflective practitioners who designed and implemented lesson plans and curriculum materials. The authors collected data from multiple sources (teachers' definitions of inquiry, reflective diaries, pre-and-post-assessment of teachers' inquiry skills, project work, and individual interviews). Papaevripidou et al. discuss the features of the course and the three distinct participatory roles and their contribution to the effectiveness of the PD programme.

In the next chapter, **Alessandro Zappia, Giuliana Capasso, Silvia Galano, Irene Marzoli, Luigi Smaldone** and **Italo Testa** also address the topic of scientific inquiry when they set out to identify the aspects of inquiry teaching that teachers mostly accept or transform in classroom practice. Their study was carried out among secondary school science teachers in Italy who first participated in a PD course and were familiarized with inquiry principles. The teachers later implemented a teaching-learning sequence (TLS) in their classroom. The authors collected data through audio and video recordings and used a knowledge transfer framework to examine teachers' transfer of TLSs in classroom practice. Zappia et al. discuss the results of their study by separating them into core and non core aspects that help to see what aspects are essential features of inquiry teaching and what can be modified to a specific educational context.

Expanding Science Teaching and Learning

The last part in this volume is titled *Expanding science teaching and learning* and it comprises five chapters that each take steps to expand science education beyond the regular science lesson and, for example, integrate teaching of a science subject with another subject or cross-curricular goals, engage schools with industry, introduce an out-of-school learning environment, or bring societal practices from outside of school into the classroom. In an effort to enhance student learning and understanding of issues and concepts related to complex global phenomena or transdisciplinary problems, students need exposure to such learning environments that integrate different subjects and create a multidisciplinary context where transversal competences can be practiced (e.g. Finnish National Board of Education 2016).

In their chapter **Nadja Belova, Johanna Dittmar, Lena Hansson, Avi Hofstein, Jan Alexis Nielsen, Jesper Sjöström**, and **Ingo Eilks** use the concept of cross-curricular goals that they describe as generally accepted educational demands across all school subjects and all educational levels. The authors focus the chapter on a set of cross-curricular goals (education for sustainability, critical media literacy, innovation competence, vocational orientation, and employability) and investigate their challenges in raising the relevance of chemistry and science education, i.e., making them more meaningful in relation to a learner's life and future. Belova et al. discuss the importance and the multifaceted nature of each goal in turn and provide examples of topics and issues that deal with the particular cross-curricular goal and offer guidance and strategies how these can be integrated with science education.

Although science and mathematics integration has long been recommended as a way to make meaningful connections between these subjects, teachers have lacked clear guidelines and access to suitable materials. In their chapter **Gráinne Walshe, Jennifer Johnston**, and **George McClelland** approach this issue and design and develop a curriculum model for assisting teachers to integrate mathematics into science in second-level education in Ireland. In the chapter the authors examine two major themes that they discovered in their analysis as relating to teachers' perceptions regarding disciplinary boundaries of subject communities. When examining their results, Walshe et al. discuss aspects beyond the subjects or artefacts of the model that affect the implementation of subject integration, such as school structure and teacher identity.

The next chapter by **Irina Kudenko, Cristina Simarro**, and **Roser Pintó** addresses the issue of bringing the world of work closer to education by the cooperation of schools with STEM-related industries. The authors assess a variety of contemporary initiatives developed by European industries for STEM learning and career education and analyze their impact on teachers and students. Although Kudenko et al. show that school-industry partnerships can have a positive role to play in addressing the needs for STEM enrichment and career learning in school, they also point out challenges that hinder effective implementation and sustainability of such projects.

Juliana Bueno and **Martha Marandino** set their study in a science museum in Brazil and analyze the production of a diorama (an exhibition object) in order to find out how the teaching process occurs in this out-of-school context. The authors use the concept of praxeology, a theoretical framework, to show how it can be used as a tool to identify the intentions related to what and how to teach in museums, and reveal the didactic potential of teaching ideas about ecology and biodiversity. The authors also discuss the potential and limitations of dioramas as teaching objects and consider the role of the designers in constructing dioramas.

The final chapter in this volume is authored by **Yrjö Engeström**, a keynote speaker at the ESERA 2015 Conference. Engeström draws on a range published studies that are based on or inspired by cultural-historical activity theory to discuss ways of expanding the focus of science education beyond the traditional textbooks confined in a classroom. Engeström argues that there are two major forces that demand this expansion of science education, namely students' increasing involvement in knowledge-related practices outside formal education, and the problematic character of natural phenomena. The author introduces and examines five layers of expansion and identifies their potentials and challenges of expansion. He argues that the layers represent opportunities that are dependent on specific cultural and historical circumstances and that should be considered as a repertoire of possibilities that can be combined and hybridized in various ways. Engeström also offers new theoretical and methodological openings to pursue the dimension of expansion in research.

Concluding Remarks

As the reader can see, this book deals with a wide variety of topics and research approaches, conducted in various contexts and settings, all contributing to our shared knowledge of science education. As the editors, we trust this volume will invoke discussion and ignite further interest in finding new ways of doing and researching science education for the future and looking for international partners for both science education and science education research. We also encourage the readers to take different aspects of sustainability into account in developing the future science education and science education research.

The Internet and other digital applications and media make it possible, feasible and attractive to organize collaborative international research groups that can jointly carry out science education research from physically distant locations. The ESERA biennial conferences provide an outstanding forum for science education researchers and practitioners to present their research and expose it for discussion and examination, and further build their networks – not only within Europe but all over the world. We want to extend a sincere thank you to the ESERA Board for the opportunity and for the confidence bestowed on us to enable us to make the ESERA 2015 Conference in Helsinki, Finland a great success.

References

Council of Europe. (2001). *Common European framework of reference for languages: Learning, teaching, assessment*. Cambridge: Cambridge University Press. Available at http://www.coe. int/t/dg4/linguistic/Source/Framework_EN.pdf

Finnish National Board of Education (FNBE). (2016). *National core curriculum for basic education 2014*. Helsinki: FNBE.

Lavonen, J., Juuti, K., Lampiselkä, J., Uitto, A., & Hahl, K. (Eds.). (2016). Electronic proceedings of the ESERA 2015 Conference. Science education research: Engaging learners for a sustainable future. Helsinki: University of Helsinki. Available at: https://www.esera.org/conferenceproceedings/19-esera-2015/270-science-education-research-engaging-learners-for-a-sustainable-future-proceedings-of-esera-2015

OECD. (2000). *Measuring student knowledge and skills: The PISA 2000 assessment of reading, mathematical and scientific literacy*. Paris: OECD Publishing. doi:http://dx.doi.org/10.1787/9789264181564-en

OECD. (2016). *PISA 2015 results (Volume I): Excellence and equity in education*. Paris: OECD Publishing. doi:http://dx.doi.org/10.1787/9789264266490-en

Russell, T., & Martin, A. K. (2014). Learning to teach science. In N. G. Lederman & S. K. Abell (Eds.), *Handbook of research on science education* (Vol. II, pp. 871–910). New York: Routledge.

Part I
Teacher Knowledge

Visualizing the Nature of Science: Beyond Textual Pieces to Holistic Images in Science Education

Sibel Erduran

Introduction

"Birds do it, bees do it. Even educated fleas do it. Let's do it, let's fall in love," goes the Cole Porter song popularized by Ella Fitzgerald. As science educators, we can extend an analogy to raise a fundamental question about science: physicists do it; chemists do it; even educated biologists do it. But what is this thing called "science"? We can proceed with further related questions: how do we know what this thing of science is? Where do we turn to answer the question of what science is? Previously (Justi and Erduran 2015), we likened science to a great landscape that is to be explored and understood, such as a major city like London. As vast and complicated a city as London is, we can get a glimpse of its various aspects through the giant Ferris wheel, the London Eye. In using the London Eye analogy, we developed an approach that we called a "Model of Science for Science Education" (Fig. 1) that aims to develop understanding of the various facets of science from different perspectives.

For example, one can have a view of science from a historical, a philosophical, a sociological or an economical perspective. Depending on the place of the individual disciplinary "capsule", the landscape will be understood in various ways. Furthermore, depending on the theoretical orientation and the diversity of orientations from each disciplinary framework, the view will be different from the *Science Eye* (Fig. 2).

S. Erduran (✉)
University of Oxford, Oxford, UK

National Taiwan Normal University, Taipei City, Taiwan

University of Limerick, Limerick, Ireland
e-mail: sibel.erduran@education.ox.ac.uk

© Springer International Publishing AG 2017
K. Hahl et al. (eds.), *Cognitive and Affective Aspects in Science Education Research*, Contributions from Science Education Research 3,
DOI 10.1007/978-3-319-58685-4_2

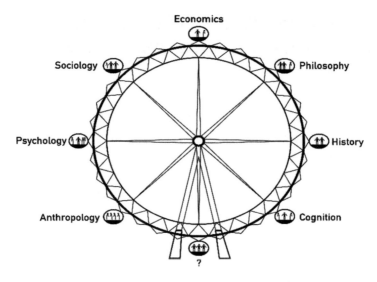

Fig. 1 Model of Science for Science Education (Justi and Erduran 2015)

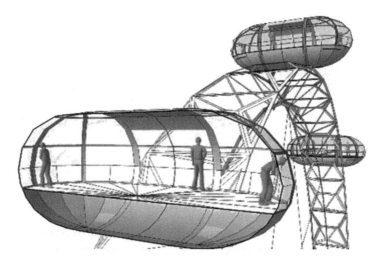

Fig. 2 "Science Eye" and disciplinary variations in understanding science (From Justi and Erduran 2015)

The visual representations in Figs. 1 and 2 capture the disciplinary perspectives that as science educators, we can appeal to in addressing the question "*What is this thing called 'science'?*" We can appeal to the anthropological studies on science to gain understanding of how scientific cultures and norms operate. Understanding of such issues might then provide some insight into how classroom learning cultures of science can be designed to have scientific authenticity. The representations are dynamic in nature communicating the ever-changing accounts of science. They also illustrate a collective and connected account on science, providing an overview of how we get to know what science is about.

As in the case of the preceding analogy, often we have appealed to contemporary philosophy of science in order to understand the nature of science (NOS). NOS is a significant body of work that has been of interest to science educators for at least the 1960s (e.g. Ackerson and Donnelly 2008; Abd-El-Khalick et al. 1998; Allchin 2013; Clough 2007; Duschl and Grandy 2013; Irzik and Nola 2011; 1968; Klopfer 1969; Lederman et al. 2002; Matthews 2014; McComas et al. 1998; Schwartz et al. 2004). NOS has been promoted in science curricula from around the world because it can help in supporting the development of scientific literacy (DfES/QCA 2006; CDC 1998). The contemporary arguments for the inclusion of NOS in science curriculum policy mirror earlier initiatives. For example, a crucial forerunner of science curriculum reform in the USA, Project 2061: *Science For All Americans*, a report prepared by the American Association for the Advancement of Science (1989), had articulated the view that an understanding of the nature of science is one of four categories considered essential for all citizens in a scientifically literate society.

The chapter aims to highlight that research on NOS in science education has primarily focused on textual representations of NOS and has not paid sufficient attention to the visualization of NOS. The "Science Eye" presented earlier is an example that highlights how a complex idea such as how we get to know about NOS can be communicated through visual analogies. Various aspects of NOS (e.g. scientific method, scientific knowledge) can also be represented and communicated visually. The chapter provides an overview of such visual tools that can be adapted for science education.

Nature of Science Research in Science Education

Definitions of the nature of scientific knowledge presented in the science education literature are diverse. The work in the 1960s included seminal pieces by Conant (1961) and Klopfer (1969). According to Klopfer (1969), the processes of scientific inquiry and the developmental nature of knowledge acquisition in science depict the nature of science. Klopfer identifies the understanding of how scientific ideas are developed as one of the three important components of scientific literacy. In this view, students must learn how scientific ideas are formulated, tested and, inevitably, revised, and he/she must learn what motivates scientists to engage in this activity. Kimball (1968) developed a model of the nature of science following an extensive review of literature on the nature and philosophy of science. The main statements guiding his model were the following:

1. The fundamental driving force in science is curiosity concerning the physical universe. It has no connection with outcomes, applications or uses aside from the generation of new knowledge.
2. In the search for knowledge, science is process-oriented; it is a dynamic, ongoing activity rather than a static accumulation of information.

3. In dealing with knowledge as it is developed and manipulated, science aims at ever-increasing comprehensiveness and simplification, emphasizing mathematical language as the most precise and simplest means of stating relationships.
4. There is no one "scientific method" as often described in school science textbooks. Rather, there are as many methods of science as there are practitioners.
5. The methods of science are characterized by a few attributes which are more in the realm of values than techniques. Among these traits of science are dependence upon sense experience, insistence on operational definitions and the evaluation of scientific work in terms of reproducibility and of usefulness in furthering scientific inquiry.
6. A basic characteristic of science is a faith in the susceptibility of the physical universe to human ordering and understanding.
7. Science has a unique attribute of openness, both openness of mind, allowing for willingness to change opinion in the face of evidence, and openness of the realm of investigation, unlimited by such factors as religion, politics or geography.
8. Tentativeness and uncertainty mark all of science. Nothing is ever completely proven in science, and recognition of this fact is a guiding consideration of the discipline (Kimball 1968: 111–112).

Some of the work conducted in the 1970s included that of Showalter (1974) who used the concepts tentative, public, replicable, probabilistic, humanistic, historic, unique, holistic and empirical to characterize the nature of scientific knowledge. After conducting a review of literature on the nature of scientific knowledge, Rubba and Anderson (1978) consolidated the nine concepts identified by Showalter into a six-factor model called "A Model of the Nature of Scientific Knowledge". The six factors included by Rubba and Anderson are defined as amoral (scientific knowledge itself cannot be judged as morally good or bad), creative (scientific knowledge is partially a product of human creativity), developmental (scientific knowledge is tentative), parsimonious (scientific knowledge attempts to achieve simplicity of explanation as opposed to complexity), testable (scientific knowledge is capable of empirical test) and unified (the specialized sciences contribute to an interrelated network of laws, theories and concepts).

Other researchers such as Cotham and Smith (1981) use the terms "tentative" and "revisionary" to describe the nature of scientific theories. The tentative component of this conception highlights the inconclusiveness of all knowledge claims in science. The revisionary component indicates the revision of existing scientific knowledge in response to changing theoretical frameworks. While NOS has been used as terminology in the literature to represent the same facets as scientific knowledge, it is usually presented in a broader context. This broader context includes not only the nature of scientific knowledge but the nature of the scientific enterprise and the nature of scientists as well (Cooley and Klopfer 1963).

More contemporary accounts of NOS in the science education research literature have been reviewed by Chang et al. (2010) who traced the literature between 1990 and 2007. The key proponents during this period in science education (Abd-El-Khalick 2012; Lederman et al. 2002; McComas and Olson 1998) have outlined a set of statements that characterize what has been referred to as a "consensus view" of the nature of science. The key aspects of this approach are as follows:

1. Tentativeness of scientific knowledge: Scientific knowledge is both tentative and durable.
2. Observations and inferences: Science is based on both observations and inferences. Both observations and inferences are guided by scientists' prior knowledge and perspectives of current science.
3. Subjectivity and objectivity in science: Science aims to be objective and precise, but subjectivity in science is unavoidable.
4. Creativity and rationality in science: Scientific knowledge is created from human imagination and logical reasoning. This creation is based on observations and inferences of the natural world.
5. Social and cultural embeddedness in science: Science is part of social and cultural traditions. As a human endeavour, science is influenced by the society and culture in which it is practiced.
6. Scientific theories and laws: Both scientific laws and theories are subject to change. Scientific laws describe generalized relationships, observed or perceived, of natural phenomena under certain conditions.
7. Scientific methods: There is no single universal step-by-step scientific method that all scientists follow. Scientists investigate research questions with prior knowledge, perseverance and creativity (Lederman et al. 2002: 500–502).

The "consensus view" of NOS has led to a major body of empirical studies in science education (Ackerson and Donnelly 2008; Abd-El-Khalick and Lederman 2000). While many science educators agree with the key tenets of this definition of NOS, several points of debate have been prevalent in the community. For example, some authors (e.g. Lederman 2007) have advised that while NOS and scientific inquiry are related, they should be differentiated. The main premise of this argument is that "inquiry" can be specified as the methods and procedures of science, while the NOS concerns more the epistemological features of scientific processes and knowledge.

Grandy and Duschl (2007) have disputed such distinctions on the basis that they "greatly oversimplify the nature of observation and theory and almost entirely ignores the role of models in the conceptual structure of science" (2007: 144). Although Lederman (2007) advocates using the phrase "nature of scientific knowledge" (rather than NOS) to avoid the conflation issue, scientific inquiry (especially "scientific methods") has been considered an important aspect of NOS in other researchers' work (e.g. Ryder et al. 1999). A related set of research studies highlight the epistemological goal of inquiry (e.g. Sandoval 2005) and epistemological enactment through inquiry (e.g. Ford 2008).

The Missing Pieces in NOS Research in Science Education

The literature on NOS in science education has focused our attention on an important aspect of science to promote in science teaching and learning. It has provided an overview of some key ideas and has resulted in considerable empirical research. Yet there are still some questions that remain to be addressed as follows:

- Nature of *which* science is meant by NOS.

  - *How can we account for domain specificity as well as domain generality of science?*

- What's the big picture in terms of how the various components of NOS are related to each other?

  - *How can we move from disconnected fragments that are about declarative statements about NOS to holistic accounts of science in school science that can have some pedagogical utility?*

In order to address the first question about NOS, let's take one often-cited misunderstanding that concerns scientific laws. Classified as the number one NOS myth by McComas and Olson (1998: 54), many individuals tend to believe "...that with increased evidence there is a developmental sequence through which scientific ideas pass on their way to final acceptance as mature laws". Involved in this belief is the thought that science starts out with facts and progresses to hypotheses, then to theories then, when confirmed, to laws. Another myth pertains to the idea that scientific laws are absolute (McComas and Olson 1998). A "law" is typically defined as "a regularity that holds throughout the universe at all places and at all times" (Salmon et al. 1992: 17). Some laws in chemistry like Avogadro's law (i.e. equal volumes of gases under identical temperature and pressure conditions will contain equal numbers of particles) are quantitative in nature, while others are not. For example, laws of stoichiometry are quantitative in nature and count as laws in a strong sense. Others rely more on approximations and are difficult to specify in an algebraic fashion. Scerri (2000) takes the position that some laws of chemistry are fundamentally different from laws in physics. Further contrasts of the nature of domain specificity of laws in chemistry and biology have been examined in the context of science education (Dagher and Erduran 2017).

In addressing the second question, I want to highlight a typical activity that is carried out in science lessons. We referred to classification in school science as a sorting activity or a tool for organizing observations with little or no attention given to its explanatory and predictive power or to how it fits within a broader theoretical framework (Erduran and Dagher 2014a: 71). For instance, students might be asked to classify objects for which there is no broader theoretical significance, such as sorting out buttons and pencils. This sense of classification could be considered as an activity. This is in sharp contrast to how scientists use classification not only to organize existing relationships but also predict new ones all the while operating within a broader theoretical framework. Classification serves an epistemic purpose in explaining phenomena through scientific knowledge in the form of models and theories. Another example from chemistry is how Mendeleev's classification of elements on the basis of periodicity led to the prediction of gallium, hence highlighting the role that classification can play in predictions. Conceiving of classification as practice in science education lifts the level of engagement with it from being an isolated activity to one that is situated in the broader epistemic, cognitive and social-institutional practices of the discipline. Hence, our discussion brings us now to three major questions:

- How can we produce holistic accounts of NOS in school science for meaningful learning?
- How can we account for disciplinary variations as well as similarities in NOS?
- What visual tools can we produce to facilitate the teaching and learning of NOS?

In our work, we have taken an approach to NOS that can account for domain-general as well as domain-specific aspects of science (Erduran and Dagher 2014a). For this purpose, we found the so-called family resemblance approach (FRA) (Irzik and Nola 2014) useful as will be described in the following sections. This approach has also helped us to think about NOS in a unified manner where declarative and disconnected fragments of verbal statements could be unified into meaningful wholes. This is because the FRA is based on a set of categories such as the aims and values, knowledge, practices, methods, social interactions and institutional aspects of science that lead to a coherent narrative about science.

Rationale for the Family Resemblance Approach to NOS

In our rationalization of FRA for science education (Erduran and Dagher 2014a; Dagher and Erduran 2016, 2017), we have appealed to the work of philosophers of science Irzik and Nola (2014). The advantage of using FRA to characterize a scientific field of study is that it allows a set of broad categories to address a diverse set of features that are common to all the sciences and the activities carried out within them. This is particularly useful in science, where all disciplines share common characteristics but not all of these can define science or demarcate it from other disciplines. Irzik and Nola (2014) present the example of observation (i.e. human or artificial through the use of detecting devices) and argue that even though observing is common to all the sciences, the very act of observing is not exclusive to science and therefore does not necessarily allow grant family membership. The same applies to other practices such as inferring and data collection, whereby these are shared by the sciences but their use is not necessarily limited to science disciplines.

The discovery of the structure of DNA can provide an example to illustrate the broad categories that underlie the FRA framework. James Watson and Francis Crick published the double-helix model of DNA in Nature in 1953 (Olby 1994). Their account was based on the X-ray diffraction image generated by Rosalind Franklin and Raymond Gosling a year earlier as well as information from Erwin Chargaff on the pairing of bases in DNA. Maurice Wilkins and his colleagues had also published results based on X-ray patterns of DNA which provided evidence for the double-helix model proposed by Watson and Crick. Watson, Crick and Wilkins were acknowledged jointly for the discovery of the structure of DNA following the death of Franklin. The extent to which Franklin's contribution has been acknowledged has emerged as a contentious issue. In particular, there is widespread recognition that Franklin experienced sexism from Watson, Crick and Wilkins (Sayre 2000/1975) (Table 1).

The DNA example illustrates how the FRA framework can be applied to a scientific topic with implications for science education. Clearly, the argument for the inclusion of these various features of science is not new. Numerous science education researchers have already made this argument. However, what is novel about this approach in relation to NOS literature is that when covered together, in a collective and inclusive manner, NOS is presented to learners in a more authentic and coherent

Table 1 Application of FRA categories to the context of DNA discovery

FRA	DNA example
Aims and values	Although the base, sugar and phosphate unit within the DNA was known prior to the modelling carried out by Watson and Crick, the correct structure of DNA was not known. Their quest in establishing the structure of DNA relied on the use of such existing data objectively and accurately to generate a model for the structure. Hence the values exercised included objectivity and accuracy
Practices	In their 1953 paper in Nature, Watson and Crick provide an illustration of the model of DNA as a drawing. Hence they engaged in providing representations of the model that they built. They also included the original X-ray diffraction image generated by Franklin on which their observations were based. The scientific practices of representation and observation were thus used
Methodology	The methods that Watson and Crick used were Franklin's X-ray diffraction data which relied on non-manipulative observation. Hence, the methodology involved particular techniques such as X-ray crystallography and observations
Knowledge	The main contribution in this episode of science is that a model of the structure of DNA as a double helix was generated. This model became part of scientific knowledge on DNA and contributed to a wide range of scientific disciplines including chemistry, molecular biology and biochemistry
Social and institutional context	This episode illustrates some of the gender and power relations that can exist between scientists. There is widespread acknowledgment in the literature and also by Crick himself, for instance, that Franklin was subjected to sexism and that there was institutional sexism at King's College London where Franklin worked (Sayre 2000/1975, p. 97). The DNA case also illustrates that science is both a cooperative and a competitive enterprise. Without Franklin's X-rays, Watson and Crick would not be able to discover the correct structure of DNA. This is the cooperative aspect. However there was also competition within and across teams of researchers

Erduran and Dagher (2014a: 30)

fashion. When students confront this and other examples positioned in a similar fashion (where now comparative aspects across examples can be pursued as well), the "family resemblance" element can also be drawn in. For instance, the precise nature of observation in terms of it being a "scientific practice" in the DNA example can be contrasted with another instance, say, an example from astronomy to draw out the similarities and differences of observation in different branches of science. The domain-specificity aspects of the FRA approach is illustrated through the examples from different topics from biology, chemistry and physics (Table 2).

Table 2 Articulation of FRA components across science topics in Key Stage 4 in the National Curriculum for England and Wales (2013)

Science topic	Cell biology	Periodic table	Energy
Subtopic	*The importance of stem cells in embryonic and adult animals and of meristems in plants*	*Predicting chemical properties, reactivity and type of reaction of elements from their position in the periodic table*	*National and global fuel resources, renewable energy sources*
Aims and values E.g. *empirical adequacy*	Use data on stem cells to determine how they influence embryo development	Use data on the physical and chemical properties of elements to conclude which elements they belong to	Use data on fuel resources and how they provide energy
Practices	Discuss similarities and differences between experiments and simulations performed in class and those done in academic or industrial labs	Generate classifications of elements on the basis of their physical and chemical properties; consider how different classifications and arrangements of the elements in the periodic table illustrate different trends in properties	Generate classifications on the pros and cons of different energy sources and their risks to environment. Generate representations of data produced by scientists noting aspects of practices that explain differences between the two communities
Methods	Compare the different methods scientists use to conduct stem cell research. Discuss manipulative methods, compared to non-manipulative methods	Conduct experiments to compare chemical reactions of different elements, e.g. oxidation and solubility in water	Discussion and comparison of energy production techniques based on a range of energy sources like solar, wind and nuclear energy
Knowledge	Consider how stem cell theory fits in with other theories and how new explanatory models in this area revised our understanding about cell growth and development	Consider the variation between the columns and periods of the periodic table and what they indicate about chemical and physical properties of elements	Consider the nature of different sources of energy and compare their efficiency in generating energy

(continued)

Table 2 (continued)

Science topic	Cell biology	Periodic table	Energy
Social-institutional E.g. economic, ethical	Discuss impact of stem cell research on the health sector, medical field and personal decisions; ethical issues arising from stem cell research; funding issues (public *v* private) and knowledge ownership	Predict the personal and environmental safety of chemicals and hold institutions responsible for ethical disposal of chemical waste	Consider the political and economic interests governing the use of national and global energy resources, investment in researching green energy sources
		Consider the economic impact of some chemicals (e.g. in food processing industry, in air) on personal and public health	

From Erduran and Dagher (2014a: 172)

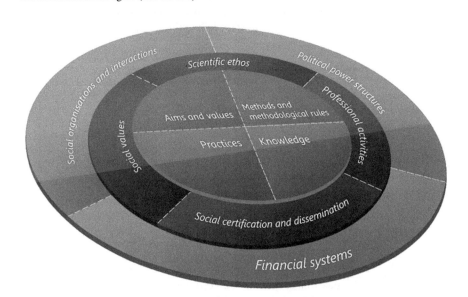

Fig. 3 FRA wheel: science as cognitive-epistemic and social-institutional system (Erduran and Dagher 2014a: 28)

We have extended Irzik and Nola's (2014) original set of categories in the FRA framework and added further categories to "social organizations and interactions", "political power structures" and "financial systems". Furthermore, we have transformed their list of categories to a visual representation in the form of a wheel where the categories are projected in an interactive manner (Fig. 3).

The FRA wheel hence provides us with a visual tool that is a summary of some major aspects of NOS. It is holistic and dynamic in that the various categories are conceptualized together, whereby they are related to each other. The FRA wheel is thus a "meta" tool in organizing some key concepts. It is also generative because as

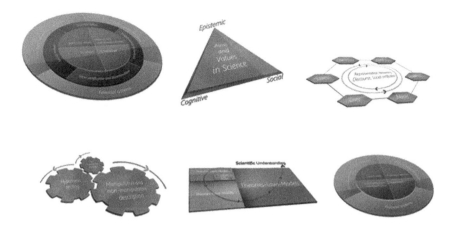

Fig. 4 Generative Images of Science (Erduran and Dagher 2014a: 164)

science educators, we can use it to generate some guidelines for how the various aspects of NOS can be considered for pedagogical, curricular and other educational purposes. Each category has further been articulated with a separate visual tool that helps unpack that particular category. Collectively, we called these visual tools *Generative Images of Science* (GIS) because they help generate educational ideas about NOS (Fig. 4).

Educational Applications of the FRA Framework

In a conventional science curriculum, science concepts are articulated vertically by ensuring that basic exposure to these concepts is implemented early in the primary grades and is developed as students progress from kindergarten to high school. This progression can be noted in many curriculum guides. In many curricula from around the world (e.g. Achieve, Inc., 2013 in the USA), basic understandings about a topic such as heredity are developed across the years along a developmental pathway where a more sophisticated understanding is targeted at secondary schooling. The FRA wheel can help structure curricular thinking and planning so that the various aspects of NOS can be covered in unison and in a consistent fashion across years of schooling (Fig. 5). FRA may increase in sophistication as science concepts get more complex moving from primary to secondary school (Erduran and Dagher 2014a).

The FRA categories can also be targeted across science topics taught in the same grade level. A similar process can be followed for outlining how the FRA categories can be connected to the content. This shows how the FRA can help maintain a continuity of coverage of NOS themes throughout the school year, a term or a sequence of lessons (Fig. 6). This is a matter of great concern to science educators who have often complained about the typical NOS coverage in an introductory textbook chapter that never gets to be revisited again in successive lessons.

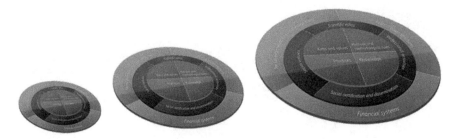

Fig. 5 FRA categories across schooling (Erduran and Dagher 2014a: 167)

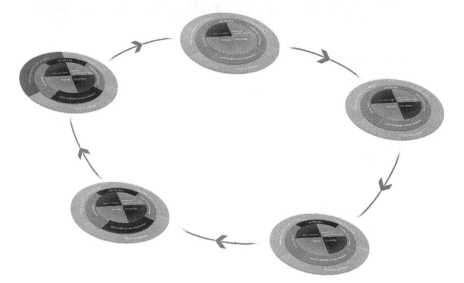

Fig. 6 Rotating emphases on FRA categories (Erduran and Dagher 2014a: 173)

The FRA framework can thus serve as a tool for thinking about what content the science curriculum should have and how it should be structured. In this vein, Kaya and Erduran (2016) have done a recent curriculum analysis study where they have contrasted two Turkish curricula using the FRA framework. The results of this analysis are consistent with previous research (Erduran and Dagher 2014b) in terms of the presence of some categories such as aims and values, knowledge, practices and methods. In order to investigate the potential of FRA for comparative international curriculum analysis, we focused on those categories that were not well represented in our analysis as well as those of other researchers. In the work of those researchers as well as ours, there is limited reference to the categories of professional activities, financial systems and political power structures. Hence we focused on how these categories compare across curriculum documents from three example countries: Turkey, the USA and Ireland.

With respect to the "social organizations and interactions" category, only the Turkish curriculum includes a statement of "The students investigate and present the studies conducted by public/private institutions and civil society organizations that contribute to the development of chemical industry in our country" (MEB 2013: 34). Related to "scientific ethos" category, there is a statement as follows: "Conduct research relevant to a scientific issue, evaluate different sources of information, understanding that a source may lack detail or show bias" (MEB 2013, 17). This example of "scientific ethos" is present only in the Irish curriculum, while the US and Turkish curriculum statements did not include any instances of this category. The lack of reference to the "professional activities" category is consistent with the curriculum analysis study by Erduran and Dagher (2014b) who reported the FRA categories in the Irish science curriculum. The "scientific ethos" category is referred to by only the NCCA in Ireland, while "social organizations and interactions" category is referred to by only MEB in Turkey. Overall, the NGSS in the USA referred to only one, whereas the NCCA in Ireland referred to two and MEB in Turkey referred to three out of the seven categories.

What the preceding discussion illustrates is that the FRA framework can be adapted as an analytical tool to investigate the science curriculum and to carry out international comparative curriculum analysis. This aspect of the work has far broader and more significant implications for science education than just NOS in science education as a research because it concerns the fundamental problem of what is included in the science curriculum in the first place. Similar concerns are raised in the context of the science curricula in Taiwan (Yeh et al. 2017).

Conclusions and Implications

The chapter is broadly related to the science education research literature on NOS. However, within the historical progression of NOS (e.g. Abd-el-Khalick and Lederman 2000; Lederman 1992, 2007; Schwartz et al. 2004), research has been limited in providing a holistic and visual account of NOS. The holistic aspect relates to the coordination of the cognitive, epistemic and social-institutional dimensions of science, while the visual aspect refers to the transformation of such dimensions to visual representations that can be effectively used in application to science education. In particular, the GIS (Generative Images of Science) provide some practical heuristics with which researchers, curriculum reformers and science teachers can articulate the complexity of NOS in science education. Initial empirical validation of GIS in science teacher education are encouraging (e.g. Kaya and Erduran 2016; Saribas and Ceyhan 2015).

Recent curriculum analysis studies (Kaya and Erduran 2016; Erduran and Dagher 2014b; Yeh et al. 2017) point out that FRA (family resemblance approach) categories about the epistemic and cognitive context such as aims and values, scientific practices and scientific knowledge were included in curriculum documents. However, the inclusion of FRA categories related to social-institutional context was limited in the curricula of various curriculum documents. Even in the case of those positive instances of FRA categories being present in the curriculum documents, there seems to be a trend in presenting these categories in a rather fragmented set of statements that do not add to a coherent overall vision for that category. For example, regarding scientific knowledge, model as a type of scientific knowledge is mentioned in the curriculum, but the relationship and coherence among theories, laws and models as types of scientific knowledge were not addressed in the curricula (Kaya and Erduran 2016). FRA is a framework for articulating NOS in a comprehensive manner such that gaps and missing links within the science curriculum can be identified and addressed. We have illustrated that representations like the "Science Eye" (Justi and Erduran 2015) and GIS (Erduran and Dagher 2014a) can provide some visual tools to conceptualize and communicate aspects of NOS. Without a comprehensive, holistic and inclusive approach to the content of the science curriculum, it is dubious how we as science educators can address the fundamental question of science education: *What is this thing called "science"?*

References

AAAS. (1989). *Science for all Americans*. Washington, DC: American Association for the Advancement of Science.

Abd-El-Khalick, F. (2012). Examining the sources for our understandings about science: Enduring conflations and critical issues in research on nature of science in science education. *International Journal of Science Education, 34*(3), 353–374.

Abd-El-Khalick, F., & Lederman, N. G. (2000). Improving science teachers' conceptions of nature of science: A critical review of the literature. *International Journal of Science Education, 22*(7), 665–701.

Abd-El-Khalick, F., Bell, R. L., & Lederman, N. G. (1998). The nature of science and instructional practice: Making the unnatural natural. *Science Education, 82*(4), 417–436.

Ackerson, V., & Donnelly, L. A. (2008). Relationships among learner characteristics and preservice teachers' views of the nature of science. *Journal of Elementary Science Education, 20*(1), 45–58.

Allchin, D. (2013). *Teaching the nature of science: Perspectives and resources*. St. Paul: SHiPs.

CDC [Curriculum Development Council]. (1998). *Science syllabus for secondary 1–3*. Hong Kong: CDC.

Chang, Y., Chang, C., & Tseng, Y. (2010). Trends of science education research: An automatic content analysis. *Journal of Science Education and Technology, 19*, 315–332.

Clough, M. P. (2007, January). Teaching the nature of science to secondary and post-secondary students: Questions rather than tenets. The Pantaneto forum, issue 25, http://www.pantaneto.co.uk/issue25/front25.htm

Cooley, W., & Klopfer, L. (1963). The evaluation of specific educational innovations. *Journal of Research in Science Teaching, 1*, 73–80.

Conant, J. (1961). *Science and common sense*. New Haven: Yale University Press.

Cotham, J., & Smith, E. (1981). Development and validation of the conceptions of scientific theories test. *Journal of Research in Science Teaching, 18*(5), 387–396.

Dagher, Z., & Erduran, S. (2016). Reconceptualizing the nature of science: Why does it matter? *Science & Education, 25*(1 & 2), 147–164. doi:10.1007/s11191-015-9800-8.

Dagher, Z., & Erduran, S. (2017). Abandoning patchwork approaches to nature of science in science education. *Canadian Journal of Science, Mathematics, and Technology Education, 17*(1), 46–52.

Department for Education and Skills and Qualifications and Curriculum Authority. (2006) Science. The National Curriculum for England, HMSO.

Duschl, R., & Grandy, R. (2013). Two views about explicitly teaching nature of science. *Science & Education, 22,* 2109–2139.

Erduran, S., & Dagher, Z. (2014a). *Reconceptualizing the nature of science for science education: Scientific knowledge, practices and other family categories.* Dordrecht: Springer.

Erduran, S., & Dagher, Z. (2014b). Regaining focus in Irish junior cycle science: Potential new directions for curriculum development on nature of science. *Irish Educational Studies, 33*(4), 335–350.

Ford, M. (2008). 'Grasp of practice' as a reasoning resource for inquiry and nature of science understanding. *Science & Education, 17,* 147–177.

Grandy, R., & Duschl, R. (2007). Reconsidering the character and role of inquiry in school science: Analysis of a conference. *Science & Education, 16*(1), 141–166.

Irzik, G., & Nola, R. (2011). A family resemblance approach to the nature of science. *Science & Education, 20,* 591–607.

Irzik, G., & Nola, R. (2014). New directions for nature of science research. In M. Matthews (Ed.), *International handbook of research in history, philosophy and science teaching* (pp. 999–1021). Dordrecht: Springer.

Justi, R., & Erduran, S. (2015). *Characterizing nature of science: A supporting model for teachers.* Paper presented at the international history, philosophy and science teaching biennial conference, Rio de Janeiro, Brazil.

Kaya, E., & Erduran, S. (2016). From FRA to RFN, or How the family resemblance approach can be transformed for curriculum analysis on nature of science. *Science & Education, 25*(9–10), 1115–1133. doi:.

Kimball, M. (1968). Understanding the nature of science: A comparison of scientists and science teachers. Journal of Research in Science Teaching, 5, 110–120.

Klopfer, L. (1969). The teaching of science and the history of science. *Journal of Research in Science Teaching, 6,* 87–95.

Lederman, N. G. (1992). Students' and teachers' conceptions of the nature of science: A review of the research. *Journal of Research in Science Teaching, 29*(4), 331–359.

Lederman, N. G., Abd-El-Khalick, F., Bell, R. L., & Schwartz, R. S. (2002). Views of nature of science questionnaire (VNOS): Toward valid and meaningful assessment of learners conceptions of nature of science. *Journal of Research in Science Teaching, 39*(6), 497–521.

Lederman, N. (2007). Nature of science: Past, present, future. In S. Abell & N. Lederman (Eds.), *Handbook of research on science education* (pp. 831–879). Mahwah: Lawrence Erlbaum.

Matthews, M. (Ed.). (2014). *Handbook of research on history, philosophy and sociology of science.* Dordrecht: Springer.

McComas, W. F., & Olson, J. K. (1998). The nature of science in international science education standards documents. In W. F. McComas (Ed.), *The nature of science in science education: Rationales and strategies* (pp. 41–52). Dordrecht: Kluwer Academic Publishers.

McComas, W. F., Clough, M. P., & Almazroa, H. (1998). The role and character of the nature of science in science education. *Science & Education, 7*(6), 511–532.

Milli Egitim Bakanligi (MEB). (2013). *İlkogretim Fen Bilimleri Dersi (3., 4., 5., 6., 7. ve 8. Siniflar).* Ankara: Ogretim Programi.

Olby, R. C. (1994). *The path to the double helix: The discovery of DNA.* New York: Dover Publications.

Rubba, P., & Anderson, H. (1978). Development of an instrument to assess secondary students' understanding of the nature of scientific knowledge. *Science Education, 62*(4), 449–458.

Ryder, J., Leach, J., & Driver, R. (1999). Undergraduate science students' images of science. *Journal of Research in Science Teaching, 36*(2), 201–220.

Sandoval, W. A. (2005). Understanding students' practical epistemologies and their influence on learning through inquiry. *Science Education, 89*(4), 634–656.

Salmon, M. H., Earman, J., Glymour, C., Lennox, J. G., Machamer, P., McGuire, J. E., Norton, J. D., Salmon, W. C., & Schaffner, K. F. (1992). *Introduction to the philosophy of science.* Englewood Cliffs: Prentice Hall.

Saribas, D., & Ceyhan, G. (2015). Learning to teach scientific practices: Pedagogical decisions and reflections during a course for pre-service science teachers. *International Journal of STEM Education, 2*(7), 1–13. doi:10.1186/s40594-015-0023-y.

Sayre, A. (2000/1975). *Rosalind Franklin and DNA.* New York: W.W. Norton & Co.

Scerri, E. (2000). Philosophy of chemistry: A new interdisciplinary field? *Journal of Chemical Education, 77,* 522–526.

Schwartz, R. S., Lederman, N. G., & Crawford, B. A. (2004). Developing views of nature of science in an authentic context: An explicit approach to bridging the gap between nature of science and scientific inquiry. *Science Education, 88*(4), 610–645.

Showalter, V. (1974). What is unified science education? Program objectives and scientific literacy (Part 5). *Prisim II, 2*(3), 1–3.

Yeh, Y. F., Erduran, S., & Hsu, Y. S. (2017, April). *From fragments to wholes: Investigating the NOS in science curriculum in Taiwan.* Paper presented at annual conference of NARST: A worldwide association for improving science teaching and learning through research, San Antonio, TX.

Using Assessment Materials to Stimulate Improvements in Teaching and Learning

Robin Millar

Introduction

It is widely accepted that the quality of school science education depends on three interrelated elements: the curriculum (what we aim to teach), pedagogy (how we teach) and assessment (how we evaluate what students have learned). Major curriculum development projects in many countries have tended to focus on the first two. The third element, assessment, is often considered after the curriculum content, teaching approaches and materials have been developed. This can significantly undermine the overall success of an innovative development (for fuller discussion of an example of this, see Millar 2013). Several major innovations at national policy level in UK school science over the past few decades have failed to meet their designers' expectations because insufficient attention was given to the issue of how the intended student learning might be assessed. Two examples are the introduction of investigative practical work (Donnelly et al. 1996) and of a strand on 'how science works' in curricula for the 14–16 and 16–18 age range (Hunt 2010). This chapter discusses a current development and research (D&R) project in England (*York Science*) for lower secondary school (students aged 11–14) which starts from assessment.

R. Millar (✉)
University of York, York, UK
e-mail: robin.millar@york.ac.uk

© Springer International Publishing AG 2017
K. Hahl et al. (eds.), *Cognitive and Affective Aspects in Science Education Research*, Contributions from Science Education Research 3,
DOI 10.1007/978-3-319-58685-4_3

Context

Why did the *York Science* project choose to focus on the 11–14 age range? The reasons lie in the importance of this phase of students' science education in shaping their views and aspirations (Archer et al. 2013) and in providing the foundation for their future science learning. Successive changes to the curriculum framework in England for this age group, however, are widely seen as having led to a loss of coherence and clarity of purpose about learning goals (Oates 2010). The curriculum reforms initiated by the incoming government in 2010 sought to address this, by emphasising teaching and learning of the core ideas of science and raising attainment targets to match those of other leading jurisdictions worldwide. Consultations around the proposed changes were protracted, and the new national curriculum for students aged 11–16 and the associated assessment framework were not finalised and implemented until September 2016. A major concern for teachers is how to modify their programmes to address these changes.

Theoretical Rationale

Although assessment is often associated with tests and examinations, its role in education is much broader. Kellaghan and Greaney comment that 'The term "assessment" may be used in education to refer to any procedure or activity that is designed to collect information about the knowledge, attitudes or skills of a learner or group of learners' (Kellaghan and Greaney 2001: 19). Assessment is a crucial aspect of the educational process because there is a very large difference between what is taught and what is learned. As Wiliam (2010) puts it:

> If what students learned as a result of the instructional practices of teachers were predictable, then all forms of assessment would be unnecessary; student achievement could be determined simply by inventorying their educational experiences. However, because what is learned by students is not related in any simple way to what they have been taught, assessment is a central – perhaps even *the* central – process in education. (Wiliam 2010: 254)

In discussions of assessment, three main purposes are often distinguished. Assessment may be used for summative purposes (to measure students' attainment at a specific moment – e.g. the end of a year, or term, or course – in a form that can be reported to the student and to others), or for formative purposes (to collect evidence of students' learning and use it to guide and encourage the subsequent actions of students and teachers), or for accountability purposes (to provide evidence of the effectiveness of teachers, schools and education systems). But lying behind all of these is a more fundamental role of assessment: to clarify the intended learning objectives of a teaching episode. Clarification is necessary because much of what is said and written about intended learning is ambiguous and unclear. For example, Mulhall et al. (2001: 583) ask: 'What, in detail, do we expect students to learn when we talk of "conceptual understanding" in electricity'? They go on to argue that 'we do not have even the beginnings of systemic answers' but that 'some justified response to [this question] is a necessary, if not sufficient, condition for any helpful advances in the thinking about and practice of teaching electricity' (ibid.). This does not apply only to teaching and learning about electricity. The same could be said about any science topic.

Assessment is the tool that clarifies learning objectives; 'by its very nature assessment *reduces ambiguity*' (Wiliam 2010: 254, emphasis in original). A question or a task that we would expect students to be able to accomplish after instruction, if learning has been successful, provides the clearest indication of what the learning objective really means.

In addition to the key role of assessment in clarifying objectives, there is also a considerable body of research evidence showing that the formative use of assessment by teachers is associated with significant gains in student attainment (Black and Wiliam 1998; Hattie 2009). The impact of formative assessment on learning outcomes, however, depends crucially on how well the assessment is embedded in classroom practice and on the quality of the questions asked (Wiliam 2011). Wiliam concludes that 'sharing high quality questions may be the most significant thing we can do to improve the quality of student learning' (Wiliam 2011: 104).

Because assessment tasks provide the greatest clarity about learning objectives, Wiggins and McTighe (2006) advocate a 'backward design' approach to the planning of instruction. They argue that the first step in the development process is to write the questions or tasks that students should be able to tackle successfully at the end of a teaching episode and only then begin to think about how to teach to get them there. From a curriculum developer's perspective, specifying exactly how the intended learning outcomes of a course or module will be assessed is the best way to make clear to potential users what these outcomes are and mean. This then enables more focused and effective teaching and in the longer run enables a more focused evaluation of the effectiveness of the approach that underlies the development and of the materials developed to help implement it.

These lines of thinking provide the rationale for a curriculum project that centred on the development of a large, structured set of diagnostic questions and tasks as a resource that might both facilitate and, at the same time, shape teachers' classroom actions and their longer-term planning.

Research Question and Methods

The central research question which the project addressed was:

- In what ways are teachers' practices and views changed by providing access to structured banks of diagnostic assessment resources?

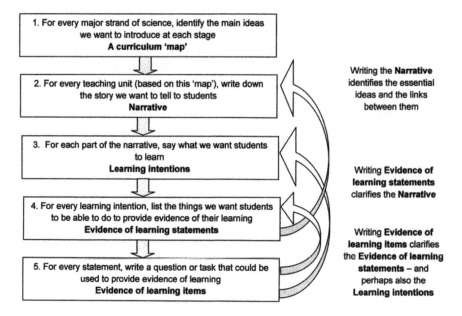

Fig. 1 The development process used in the York Science project

Development Phase

The development process which the *York Science* project adopted is shown in Fig. 1. As the project was dealing with a 3-year period within a 5–16 continuum, the first step was to develop a curriculum 'map' outlining how the major ideas in each of the main strands of science content might be expected to develop over the 5–16 age range. A 'main strand of science' here means a major topic like forces and motion, electricity and magnetism, chemical change or evolution. This 'map' in effect proposes an outline teaching sequence or learning progression (Corcoran et al. 2009).

The 'map' developed for *York Science* was influenced (though not totally constrained) by the requirements of the national curriculum but also informed by the available research evidence on students' learning (AAAS 2001; Driver et al. 1994; Duit 2009; Victoria State Government 2014) and by professional experience. A teaching sequence for the whole 5–16 age range enabled principled decisions to be made about the ideas to be introduced and developed in each strand within the 11–14 age range, which was the project's target, and made explicit what we assumed would have been taught by age 11 and what should be left until after age 14.

The second step was then, for each strand of science, to write down the story we want to tell to students at the 11–14 stage, as a continuous *narrative*. This is much more useful than a list of learning targets or objectives. Setting out the story briefly, but clearly, helps to identify the main ideas that have to be included and to sort out the order in which they need logically to come and the links between them. Although *narratives* were written with teachers (not students) in mind as the audience, they

Some objects can affect others at a distance by emitting *radiation* which travels from one object (the *source*) to another (the *receiver*), through the material or the space (the *medium*) between them. Light and sound are examples of radiation. Radiation travels out from a source in straight lines in all directions. When it strikes another object, it may go straight through (*transmission*), bounce off (*scattering* or *reflection*), or be stopped (*absorption*) – or a combination of these. When radiation is blocked by an *opaque* object, this causes a shadow region. The effects of radiation get steadily less the further it goes, because it is spread over an ever-increasing area, and because it may be gradually absorbed by the medium it is travelling through. When radiation is absorbed by an object, it has an effect on the object; this might be a chemical effect, an electrical effect or a heating effect.

Fig. 2 The first part of the York Science Narrative on Radiation (Light and Sound)

Table 1 Sample evidence of learning statements for the topic Radiation (Light and Sound)

Evidence of learning statements
Students should be able to:
Identify, in a given situation, the radiation source, the receiver of radiation and the medium through which radiation is travelling
Explain and predict the shapes and sizes of shadows cast by point sources
Explain why a source of light becomes dimmer (or the sound from a source fainter) the further you are away from it
Use the radiation model to explain familiar phenomena and events and to make predictions
Identify evidence for specific aspects of the radiation model
Offer plausible suggestions about what will happen in a given situation when radiation is absorbed: (i) to the absorber, (ii) to the radiation

use the language that would be used in 'telling the story' to students. An illustrative example of part of a *narrative* is shown in Fig. 2.

A *narrative* usually consists of a sequence of paragraphs (or sections). For each section, the next step (step 3 in Fig. 1) is to say briefly what the learning intention for that part of the story is: what we want students to learn. For the *narrative* section in Fig. 2, this might be that 'Students should understand and be able to use the source-radiation-receiver model'. This is then followed by two crucial steps. First (step 4 in Fig. 1), the learning intention is translated into a set of observable performances: a list of things we would expect students to be able to do if their learning has been successful. This step, in effect, involves operationalising the learning intentions. Words like 'know' and 'understand' disappear and are replaced by the observable actions that we would take as evidence of knowledge and understanding. To illustrate this, some *evidence of learning statements* for the *narrative* section shown in Fig. 2 are listed in Table 1.

Finally, and equally crucially, step 5 (in Fig. 1) is to write at least one question or task that a teacher could use in class to obtain reasonably good evidence of students' learning, quickly enough to be able to use this to inform the next actions of the students and/or the teacher. We called these *evidence of learning items*. Among the formats used were:

- Two-tier multiple-choice questions
- 'Talking heads' questions, where students are asked to evaluate a set of responses to a situation presented in speech bubbles and in terms that a student might use
- Predict-explain-observe-explain practical tasks
- Confidence grids: questions in which several statements are made about a given situation and students have to put each statement in one of the categories (I'm sure this is right/I think this is right/I think this is wrong/I'm sure this is wrong)
- Construct an explanation: where students have to select the correct option in each of a sequence of boxes to construct a correct explanation of a given event or phenomenon
- Evaluating a representation: where students have to identify aspects of a given representation (usually a textbook diagram) which they think are 'a good picture' of the real thing and aspects which they think are not

This list is not complete; *evidence of learning items* of other types and formats have also been developed.

As the right-hand side of Fig. 1 emphasises, this is an iterative process, not a linear one. Writing *evidence of learning items* often makes you question the way the corresponding *evidence of learning statement* has been expressed or helps you notice that a statement is missing and should be added. In some cases, this indicates a need to revise the stated *learning intention*, or even the *narrative*. The outcome of the development process is a large set of *evidence of learning items* for each of the main strands of science, linked clearly to (and consistent with) a *narrative*, a set of *learning intentions* and a list of *evidence of learning statements*.

At the time of writing, this development work has been completed for around half of the biology, chemistry and physics content required by the English national curriculum for the 11–14 age range. Work is continuing on the remaining strands.

Research Phase

To obtain evidence of the impacts of the project's approach and materials (and to obtain feedback to improve these), teachers in 45 schools were given a large set of *evidence of learning items* (ELIs) for one of the first three science strands developed. The three sets were allocated randomly to schools. They were accompanied by guidance material which encouraged teachers to use the ELIs for formative purposes, rather than summative ones, and suggested a range of ways of using ELIs that preliminary work had shown to be valuable. In particular, teachers were encouraged to use *evidence of learning items* as stimuli for small-group discussion rather than as individual written exercises or tests and to see the discussion these generated as a valuable source of evidence of students' thinking.

After the teachers had had the material for around 3 months, written question-naires (n = 45), augmented with interviews where this was feasible on grounds of availability and access (n = 13), were used to collect feedback including:

Descriptive data (What had they used? How had they used it?)

Evaluative data (What did they think of the materials? Any suggestions for improve-ment/addition?)

Reflective data (How might this change their teaching and planning?)

Responses were analysed, initially using predetermined categories implicit in the data collection instruments, modified by an inductive analysis using a grounded theory approach to pick up any unanticipated themes (see, e.g. Robson 2002).

Findings

The reception by teachers of the project materials and approach has been strongly and uniformly positive. Almost all saw the project as directly relevant to issues with which teachers are currently grappling as a result of policy-driven changes.

Many teachers said they were aware of common 'misconceptions' (the term they invariably chose to use) that some students are likely to hold, but several expressed surprise at their prevalence. One commented that 'without the questions, I might never have been aware how widespread particular misconceptions were' (T09). Others expressed surprise that many students did not understand things they expected them to have grasped. One wrote:

> When I was given the trial pack to try it out, I was in the middle of teaching light and I thought "Oh, I'll try some of these, they'll be able to do them, no problem for students." But they couldn't. (T02)

She followed this up by sending the response of a student group to a question designed to probe ideas about primary sources of light. It asked what you would see if you closed yourself inside a dark cupboard with a well-sealed door and no win-dow, in which there was a mirror and a cat. Students were given four statements to evaluate. Writing of her class, she said: 'One student got it right, the most common response by far was this' (Table 2).

Reflecting on this and other items on the same topic, this teacher commented that:

> I really like how I'm able to get down to the nitty-gritty of what the kids are thinking ... how are they actually thinking about it? There's an activity about light travelling in straight lines and where it travels from, and they all thought that light comes out of your eye. I really thought that they would know all of this, there'd be no problem with the science, and oh my goodness there were problems with the science. That was really eye opening and I really liked that. I thought, if this can tell me about things that I thought students would know then what could it tell me about the things that I'm actually teaching them? (T02)

Other respondents also replied that using items from the question bank not only showed them what many students thought but also gave them insights into the thinking

Table 2 Data from a teacher on the commonest response pattern in her class to one evidence of learning item

Statement		I am sure this is right	I think this is right	I think this is wrong	I am sure this is wrong
1	After a whilst, you will be able to see everything, but very dim		✓		
2	The only thing you will see is the cat's eyes shining		✓		
3	You will see the mirror shining, but everything else will be dark			✓	
4	You won't be able to see anything at all, no matter how long you wait				✓

that lay behind their answers. As one teacher put it, 'It makes you look at things from an understanding level and also informs you on an understanding level as well' (T01). He commented on some benefits of research-informed multiple-choice questions:

> The nice thing about this, it's multiple choice, you have various different answers, but there are some which if your thinking isn't quite right, that's the one you'll go for. And that's really really helpful and really useful. You can listen to the thought processes, they have discussions about it, what do you think, what's this, how does this work, and that really helps you into what they're thinking and how it works.' (T01)

This teacher went on to talk about how his use of a set of items on chemical substances and chemical change were changing the way he taught this topic:

> The YS materials pick up the misconceptions in such a way that it's clear what they don't understand and how they don't understand it. So it's better than simply getting a wrong answer on a test, you've actually got some sort of idea about what they don't understand and a potential way in to fix it. And it's mainly go back to the particular lessons where I knew there was a problem and take another look at them as well. When I've taught it again, I've approached it in a different way. (T01)

Another point made by several teachers was about the value of these questions in stimulating well-focused discussions in student groups. One remarked that 'they [the questions] were so interesting to use. The use for me is opening up the discussion, thinking about how they're actually perceiving things, that was the interesting bit' (T11). Another commented that 'so much of what is generated from this is discussion with the pupils … It's prompted more discussions than I would normally have had … which is good' (T07).

Whilst there are many challenges in designing good diagnostic assessment questions and tasks, teacher feedback has not reported any significant challenges or problems in using the materials produced by the *York Science* project.

Conclusions and Implications

This preliminary and small-scale evaluation of the *York Science* materials for three science strands, and of the embedded formative assessment approach that they promote, provides encouragement that carefully designed and research-informed assessment materials can have the intended impacts on teachers' practice and thinking. In general, the materials were used as intended, for assessment during lessons 'in real time' and as stimuli for 'on-task' discussion. This study provides 'proof of principle' that the strategy the project has adopted that can work indeed is quite likely to work. This strategy might be summarised as seeking to stimulate changes of specific and planned kinds in teachers' practices, by providing resources which make it easier for them to implement these changes and hence to influence their thinking about teaching and learning more generally and about the planning of lesson sequences.

The responses of teachers to the project materials and approach confirm the view from which we began that assessment items play a crucial role in communicating intended learning objectives clearly and that structured sets of items are particularly valuable in focusing teachers' attention on learning outcomes and facilitating the use of embedded formative assessment to monitor students' ideas and learning as the teaching proceeds and respond to this evidence 'in real time'.

Acknowledgements The *York Science* project involved a group of teachers and curriculum developers who have contributed to the ideas and the development work described in this chapter. I would particularly like to acknowledge the input of Mary Whitehouse, Anne Scott, Liz Swinbank and Alastair Moore. We are grateful for the financial support of the Salters' Institute for this development work.

References

AAAS (American Association for the Advancement of Science). (2001). *Atlas of science literacy*. Washington, DC: AAAS.

Archer, L., Osborne, J., DeWitt, J., Dillon, J., & Wong, B. (2013). *ASPIRES. Young people's science and career aspirations age 11–14* (Final report). London: King's College London.

Black, P., & Wiliam, D. (1998). Assessment and classroom learning. *Assessment in Education: Principles, Policy and Practice, 5*(1), 7–74.

Corcoran, T., Mosher, F., & Rogat, A. (2009). *Learning progressions in science: An evidence-based approach to reform* (CPRE research report#RR-63). Philadelphia: Consortium for Policy Research in Education.

Donnelly, J., Buchan, A., Jenkins, E., Laws, P., & Welford, G. (1996). *Investigations by Order. Policy, curriculum and science teachers' work under the Education Reform Act*. Nafferton: Studies in Education Ltd.

Driver, R., Rushworth, P., Squires, A., & Wood-Robinson, V. (1994). *Children's ideas in science*. London: Routledge.

Duit, R. (2009). *Bibliography STCSE. Students' and Teachers' conceptions and science education*. Kiel: IPN. Retrieved June 27, 2016 from archiv.ipn.uni-kiel.de/stcse/

Hattie, J. (2009). *Visible learning*. London: Routledge.

Hunt, A. (2010). *Ideas and evidence in science: Lessons from Assessment.* Report prepared for SCORE (Science Community Representing Education). Retrieved April 13, 2016 from www.score-education.org/media/7376/finalhsw.pdf

Kellaghan, T., & Greaney, V. (2001). *Using assessment to improve the quality of education.* Paris: UNESCO. International Institute for Educational Planning.

Millar, R. (2013). Improving science education: Why assessment matters. In D. Corrigan, R. Gunstone, & A. Jones (Eds.), *Valuing assessment in science education: Pedagogy, curriculum, policy* (pp. 55–68). Dordrecht: Springer.

Mulhall, P., McKittrick, B., & Gunstone, R. (2001). A perspective on the resolution of confusions in the teaching of electricity. *Research in Science Education, 31*(4), 575–587.

Oates, T. (2010). *Could do better: Using international comparisons to refine the National Curriculum in England.* Cambridge: Cambridge Assessment.

Robson, C. (2002). *Real world research* (2nd ed. Ch. 14: The analysis of qualitative data, pp. 455–499). Oxford: Blackwell.

Victoria State Government (2014). Science Continuum F-10. Retrieved April 13, 2016 from www.education.vic.gov.au/school/teachers/teachingresources/discipline/science/continuum/Pages/default.aspx

Wiggins, G., & McTighe, J. (2006). *Understanding by design* (2nd ed.). Upper Saddle River: Pearson.

Wiliam, D. (2010). What counts as evidence of educational achievement? The role of constructs in the pursuit of equity in assessment. *Review of Research in Education, 34*, 254–284.

Wiliam, D. (2011). *Embedded formative assessment.* Bloomington: Solution Tree Press.

Chemistry Teachers' Perceptions and Attitudes Towards Creativity in Chemistry Class

Annika Springub, Luzie Semmler, Shingo Uchinokura, and Verena Pietzner

Theoretical Background

Creativity is a phenomenon that lacks a uniform definition. It refers to something that is neither directly observable nor quantifiable without further notice. Therefore, the understanding of creativity is individually different, as well as its measurement and evaluation. What is ultimately called creative is highly dependent on the society, the development of society and the Zeitgeist of each era. Hence, a wealth of definitions can be found over a long period of time from the past; many of them are still relevant today.

The modern research on creativity was initiated in 1950 with a presentation by Guilford at the congress of the American Psychology Association (Guilford 1987). Since the 1980s, creativity has become an intense field of research (Runco 2004). In Europe, however, the interest in creativity has developed later and it has remained a relatively neglected area of research (Urban 2004: 5). Nevertheless, creativity gets new importance in today's rapidly evolving times as industry and technology are increasingly dependent on innovation (European Commission 2009). Thus, creativity is required of graduates, although it is not a determined part of their education. It finds its place, mostly in the artistic, musical and craft subjects. Especially in the STEM subjects, creativity seems not yet established in the classroom and is not described in the instructional plans and curricula.

For creativity has many facets, no overall definition of creativity exists. However, suitable definitions of several aspects of creativity can be found, like the one by Guilford (1987) dealing with divergent thinking as part of creativity. For Torrance

A. Springub • L. Semmler • V. Pietzner (✉)
Carl von Ossietzky University, Oldenburg, Germany
e-mail: verena.pietzner@uni-oldenburg.de

S. Uchinokura
Kagoshima University, Kagoshima, Japan

© Springer International Publishing AG 2017
K. Hahl et al. (eds.), *Cognitive and Affective Aspects in Science Education Research*, Contributions from Science Education Research 3,
DOI 10.1007/978-3-319-58685-4_4

(1970), the main aspects of creativity are the creative process as well as the creative product, not the creative person, though. Looking at the creative process and product, four aspects are of central importance: *fluidity* (drawing conclusions and making associations in short time), *flexibility* (in using methods and ways of thinking), *originality* (diversity in finding solutions for a certain problem) and *elaboration*, which means the ability to implement the solutions. Finally, the definition of Rhodes (1961) covers all main components of creativity. With his 4P model of creativity – "p" for person, product, process and place – he offers a definition that allows us to focus on different aspects of creativity, depending on the research question. Nevertheless, it is important that, especially for students and teachers, a common understanding of creativity is found on which creative teaching can be built. For this, the British government formed the National Advisory Committee on Creative and Cultural Education (NACCCE). The Committee defines creativity as:

> ... imaginative activity fashioned so as to produce outcomes that are both original and of value. (NACCCE 1999: 30)

This definition implies that creativity includes all human activities and that every human being carries creative potential that can be discovered, developed and promoted.

Regarding the creative person, plenty of research has been done. Barron and Harrington (1981) identified core characteristics of creative persons, like autonomy, intuition, self-confidence or being independent from judgement. In addition, creative persons show unsocial behaviour more often than noncreative persons (Cropley 2001). Creative students attract attention by showing dominant patterns (MacKinnon 1967), having a high self-belief (Drevdahl 1956) and a certain disregard for authorities (Hilgard 1959). Following the investigations of Mackinnon (1962) and Weisberg and Springer (1961), autonomy plays a large role in the development of creativity. Children should start early to build an autonomous behaviour and, in addition, learn to act responsibly. Altogether, a basis for the development of creativity is established. Regarding creative children, Westby and Dawson (1995) developed a ranking of character traits of creative children; they then let teachers rate the character traits. They could show that teachers, on the one hand, think that creative students are an enrichment in class, but, on the other hand, they prefer character traits that refer to a noncreative person.

It is important to remember that fostering creativity usually refers to a specific student population and not to the whole society. This is expressed in the term "little c creativity", which refers to everyday ideas that can be original and creative for one or more people, but it does not have to be for the society (Craft 2005: 19). This kind of creativity is mostly to be used in school. Whether anything can be called creative or not depends on the evaluation of the students and of the respective teacher. Creativity can be implemented, however, and integrated differently in school. The NACCCE report makes a distinction between *teaching creatively* and *teaching for creativity* (NACCCE 1999: 102f). The former says that teachers employ imaginative approaches and methods to make learning interesting and effective and to motivate students. Teaching for creativity includes this but is more related to the environment and the abilities of the teacher himself. Another term can be distinguished from those already mentioned: *teaching creativity*. Here it comes to convey creativity directly and watch it more as a separate subject instead of integrating it into a sub-directional fold (Simonton 2012). Furthermore, there is the concept of *creative learning*; student-centred tasks are understood to be part of it (Craft 2005: 23).

Natural sciences involve creative potentials that do not exist in this form in other subjects, for it is domain specific. Creativity in science means amongst other things to form hypotheses and to plan experiments to perform, reflect and revise, if necessary (Newton and Newton 2009). The results must be interpreted with caution; that too is a creative process. There is also the possibility to independently design and build scientific models. This promotes problem-solving skills, because knowledge must be applied to new situations. Therefore, it can stimulate creative thinking processes (Sawyer 2012: 401).

Creativity can only take place in an environment that also allows errors, makes room for fear and promotes independent work (Jeffrey and Craft 2004). Several publications describe how creative teaching should look like and how the creativity of students can be promoted (e.g. Craft 2005; Hallmann 1970; Cropley 1991; Urban 2004). However, these publications relate more to the general education. With regard to teaching chemistry, only a few publications can be found. Especially current studies dealing with the teachers' attitude towards concepts of creativity like the ones of Cachia and Ferrari (2010) or Tanggaard (2011) are rare.

Although teachers think positively about creativity, there is a discrepancy between the view on creativity and the fostering of creativity in class (Cachia and Ferrari 2010). In particular, scientific creativity at school has been investigated in various studies focusing on creative teaching, inquiring science and how to measure scientific creativity (Kind and Kind 2007). In general, it is possible to foster scientific creativity by implementing open tasks or cooperative, student-oriented work. However, teachers that use these tools often limit the openness by giving "recipes" or providing a lot of guidance (Kind and Kind 2007), but this contradicts the aim of a more open teaching approach. Teachers should keep in mind that they cannot force creativity; they can only stimulate it (Sawyer 2012: 390). Therefore, they should teach students to recognise their own creative potential and develop it accordingly.

An important prerequisite in order to implement creativity in school are creative teachers. Even if teachers have a positive attitude towards creativity, this can only be effective if they show themselves as a creative personality. The latter requires effort and will. To facilitate the entry of a teacher in creative teaching, Hallmann (1970) listed several aspects that are typical for creative teachers. Some of these aspects are the following:

- Teacher must first promote the independence of the pupils and thus educate them to take the initiative.
- Teacher must provide a certain amount of freedom to the students so as to promote creativity.

- Teacher should encourage the students to develop flexible problem-solving strategies and show them not only one scheme that leads to the desired solution.
- The focus is on the questions of students. They should not be ignored but lead to a discussion. On the other hand, teachers should ask questions in a way that they allow creative solution processes (Hallmann 1970).

Other studies have shown that creative students benefit when teachers show an overall positive attitude towards creativity because an atmosphere is created that promotes the creativity of the students. Cropley (1982) has investigated the promotion of creativity in class and derived recommendations how teachers should act in class in order to promote the creativity of their students:

> Teachers should therefore rate creative thinking positively, encourage playful dealing with things and ideas, show tolerance for new ideas, not impose fixed pattern, promote learning on their own initiative and rate this positively, provide support for the development of new ideas, criticize constructively and support self-assessment, promote multidisciplinary learning, be open for adventure. (Cropley 1982: 272)

Although creativity is in the focus of public debate, only little research regarding creativity and science can be found. Therefore, it is necessary to investigate today's chemistry teachers' views about creativity to get a current impression about today's situation. Within the study, we wanted to answer the following research questions:

1. How do chemistry teachers define creativity, and do they see themselves as creative persons?
2. How do gender, age, second subject or school type influence the attitude of chemistry teachers towards creativity in chemistry class?
3. What do today's chemistry teachers think about character traits typical for creative children following the study of Westby and Dawson (1995)?

Method

For the study, an online questionnaire was developed consisting of three parts. In the first part, teachers were asked in two open questions to give their own definition of creativity and to rate their own creativity (creative – little creative – noncreative) including an explanation for their self-assessment. Afterwards, the teachers had to rate statements showing their creativity-related attitude; the items had to be rated with a four-point Likert scale (from 1 = totally disagree up to 4 = totally agree). Both open questions as well as the items for the attitude scale were taken from the study of Williams and Wyburd (2006). In the second part, the teachers had to rate character traits of students on a four-point Likert scale (scaling: ++; +; −; −−) that were taken from the study of Westby and Dawson (1995). In our study, only the five most typical and the five most untypical character traits, according to the study of Westby and Dawson (1995), were used. The most typical character traits are determined, impulsive, emotional and nonconformist and make up rules as he/she goes along; the most untypical character traits are logical, sincere, good-natured, understanding and appreciative. The second part comprises questions about creativity in chemistry class. The teachers were asked to give information about creative methods they know and use in class. For our study, we chose creative methods described by Gärtner (1997): explorative teaching, project work, reconstructive teaching, imitative teaching and egg races. With this, we wanted to find out if teachers are aware of and use creative teaching methods to get an impression of the possibilities of the students to be creative in class. Finally, personal data like gender, age, school type and the subjects taught were collected.

For the analysis of open questions, methods for qualitative research were used. First, the answers of the open questions were categorised in several categories which were derived inductively from the data. This inductive method was taken from the method of evaluating qualitative data according to Mayring (2014). The validation of the derived categories was done in a group discussion of researchers working in the field of chemistry education. With this method, intersubjectivity and comprehensibility as well as semantic validity can be ensured (Flick et al. 2004: 187; Mayring 2014: 110). All mathematical evaluations were calculated with SPSS 23. The revalidation was done for the creativity attitude scale of Williams and Wyburd (2006). The scale had a Cronbach Alpha of $\alpha = 0.506$, which is good enough. All other items within the questionnaire do not form a scale. For all Likert-scaled questions, for example, the rating of character traits and descriptive and inferential statistics (chi-square test or Kruskal-Wallis test, both for non-parametric data) were done. Non-parametric tests were taken due to the number of teachers that took part in the study. For further analysis of the creativity attitude scale, a factor analysis was done. Here, we used the main component analysis with varimax rotation.

Results

In total, 64 questionnaires could be analysed. The relatively low size of the sample limits the generalisation of the study. However, it casts a spotlight on the perspectives and views of German teachers. This will have to be taken into account when the results are discussed.

Table 1 shows the composition of the teachers that participated in the study (also called participants) regarding age and gender.

Most teachers teach at schools of upper secondary level (Gymnasium 47.5%, Gesamtschule 20.3%), and 32.2% teach at schools of lower secondary level (Hauptschule, Realschule, Oberschule).

Due to small numbers of teachers in each group regarding age, the group was split into two for the analysis: teachers younger than 40 (37 teachers) and teachers equal or older than 40 (24 teachers; three cases are missing). The different second subjects of the teachers (in Germany, every teacher teaches two different subjects) were also grouped: 32 teachers had mathematics or another science as second

Table 1 Composition of the participants regarding age and gender (N = 61)

Age	Female		Male		Total	
	Freq.	%	Freq.	%	Freq.	%
Under 25	3	4.9	1	1.6	4	6.6
25–29	10	16.4	9	14.8	19	31.1
30–39	7	11.5	7	11.5	14	23.0
40–49	10	16.4	7	11.5	17	27.9
50–59	3	4.9	3	4.9	6	9.8
60 and above	0	0	1	1.6	1	1.6
Total	33	54.1	28	45.9	61	100

Table 2 Overview of the different categories that were derived from the individual definitions of creativity of the participants (N = 63). The frequency and the ratio of the derived categories are shown

	Frequency	%
Versatile ideas/problem-solving strategies	34	54.0
Create something new from known things	8	12.7
Artistic, musical talent	7	11.1
Divergent thinking	7	11.0
Solve artisan/technical tasks originally	4	6.4
Expression of personality	3	4.8

subject, six teachers had a language, ten had history or social sciences, five had arts or music and three had sports as their second subject. Thus, the group with maths/ sciences and the group with the remaining, previously mentioned, second subjects were analysed separately. In case of school types, also two groups were analysed: 19 teachers who work at a school without upper secondary level and 40 teachers who are working at upper secondary level or in a school with upper secondary level; five cases are missing.

In their definitions of creativity, the teachers emphasise versatile ideas as being typical for creativity (see Table 2). Other aspects, like elaboration, fluidity or personal aspects, are rarely taken into account. Therefore, creativity is not seen as a whole process to include fluidity, flexibility or elaboration; rather only the ability to find more or less unusual solutions or to use different problem-solving strategies is predominantly given credit. No significant differences of the definitions regarding gender, age, school type or second subject were found.

Regarding themselves, around 35.9% of the teachers think that they are not creative at all, and 28.1% consider themselves to be little creative or creative, which was partly motivated by music, art or craft activities. The overall rather low degree of creativity was often justified by the desire to follow a certain structure.

Nevertheless, the participants of the study give a high approval for statements that support fostering creativity, e.g. "It is possible to support and develop creativity" or "Creative teaching supports the creativity of the students" (see Fig. 1). But at

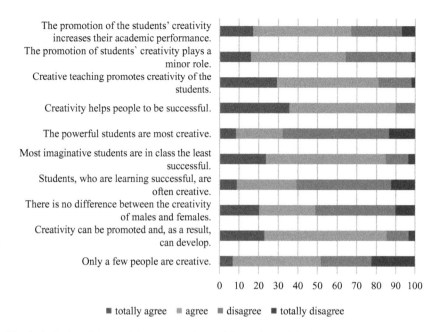

Fig. 1 Attitudes of the participants towards creativity and creativity in school; all data are in percent. The items were Likert-scaled from totally disagree up to totally agree

the same time, they do not really feel responsible for promoting creativity in their class (mean, 2.79). It is also interesting to see that the teachers rate the two terms "Creativity helps people to be successful" and "Most imaginative students in class are the least successful" as almost being the same. Possible reasons for this could be that for the teachers, being imaginative does not relate to being creative or that they differentiate between success in school and success in working life. Female teachers have a higher approval regarding the positive influence of creative teaching on the creativity of students (Chi^2 = 0.882; df = 1; p = 0.009). In addition, they are more convinced that fostering creativity leads to higher achievement (Chi^2 = 4.597; df = 1; p = 0.032). Regarding the other groups, no difference can be found regarding age, second subject or school type. Taking the attitude items as one scale, also no differences between the groups can be found.

For further evaluation, a factor analysis with main component analysis with varimax rotation has been done. The results are shown in Table 3.

In total, four factors could be identified to explain 66% of the variance observed. The factors can be described as follows:

- Factor 1: Creative people are more successful, and creative teaching can foster this.
- Factor 2: Creativity is independent from gender and therefore it is not important to support this.
- Factor 3: Creative students are successful, and because only a few students are really successful, there must be only a few creative students.

Table 3 Factor analysis of the creativity attitude items. Only factor loadings >||0.4|| are shown. The items can be grouped into four different factors

	1	2	3	4
Only a few people are creative			0.468	
Creativity can be promoted and, as a result, can develop				−0.732
There is no difference between the creativity of males and females		0.782		
Students, who are learning successful, are often creative			0.719	
Most imaginative students are in class the least successful				0.872
The powerful students are most creative			0.777	
Creativity helps people to be successful	0.772			
Creative teaching promotes creativity of the students	0.851			
The promotion of students' creativity plays a minor role		0.816		

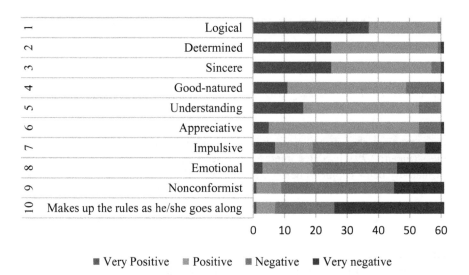

Fig. 2 Chemistry teachers' ranking of character traits. The ranking was done using a four-point Likert scale from 1 = very negative to 4 = very positive; the figure shows the descriptive results (N = 64)

- Factor 4: It is better not to support creativity, because creative students perform less and creativity cannot be promoted anyway.

Regarding school type and second subject taught, no differences could be observed, but regarding gender, female teachers show a significant higher belief that creativity promotes success (factor 1; U = 131.0, Z = −3.23, p = 0.001). Teachers who consider themselves noncreative are significantly different in factor 1: they do not think that creative people are more successful (Chi^2 = 6.362, df = 2, p = 0.042).

To compare the teachers' rating of the personal traits with the ones of Westby and Dawson (1995), a ranking was derived from the ratings of the teachers (see Fig. 2).

Table 4 Comparison of the rankings of Westby and Dawson (1995) with the rankings of the teachers. The ranking of Westby and Dawson ranks the character traits from typical for a creative child to more untypical for a creative child. The ranking of participants shows what are more positive and more negative character traits of children in their eyes

Ranking of Westby and Dawson	Character traits of creative children	Ranking of the teachers
Most typical character traits		
1	Makes up the rules as he/she goes along	10
2	Impulsive	7
3	Nonconformist	9
4	Emotional	8
5	Determined	2
Most untypical character traits		
6	Logical	1
7	Understanding	5
8	Appreciative	6
9	Good-natured	4
10	Sincere	3

For this, the personal traits were sorted following the rankings of the teachers. First, the personal traits were sorted in accordance with the highest "very positive" scores. If the scores were the same, the "positive" scores were additionally taken into account.

Comparing the results of the chemistry teachers with the ones of Westby and Dawson (see Table 4), it must be stated that the teachers still favour character traits that do not belong to a creative person, although most teachers favour creative students in the classroom. An exception is the property *determined*, a character trait that is rated as typical for a creative child but is assessed positively by the chemistry teachers. In addition, some differences in rating between different groups could be identified. Male teachers rate the character trait *logical* more positive than female teachers ($U = 334.5$, $Z = -2.189$, $p = 0.029$); teachers with math or a science subject as second school subject rated *logical* and *impulsive* more positive than other teachers ($U = 248.0$, $Z = -2.634$, $p = 0.008$; $U = 253.0$, $Z = -2.298$, $p = 0.022$), and teachers teaching at upper secondary level rated the character trait *emotional* higher than teachers teaching at lower secondary level ($U = 225.5$, $Z = -2.416$, $p = 0.016$). Looking at the different groups of teachers regarding their own creativity, no significant differences can be found.

Regarding the knowledge of creative teaching methods, it can be claimed that the teachers know several methods that have the potential to promote creative learning (see Fig. 3).

Those who knew the respective teaching method also specified whether they already used this method or not. Explorative teaching was already done by 74.1% of the teachers, followed by project work (61.0%) and reconstructive teaching (45.6%). One can clearly see that the teaching methods are not used by all teachers who know them well. No significant differences were found between the different groups

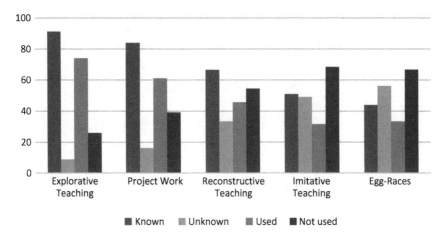

Fig. 3 Overview of typical creative methods; the participants had to tell whether they know the method or not, and in case they know the respective method, they were asked to specify if they have used it so far or not; all data in percent (N = 57)

except that younger teachers use project work more often than older teachers (U = 312.5, Z = −1.963, p = 0.050). Taking teachers' own creativity into account, some differences between the respective groups (creative, little creative, noncreative) can be found. Teachers who regard themselves as noncreative know the methods of reconstructive teaching (Ch^2 = 7.503, df = 2, p = 0.023) and project work (Ch^2 = 6.283, df = 2, p = 0.043) significantly less than the other two groups. Looking at the usage of creative methods, it is the group of creative teachers that shows significantly higher usage of imitative teaching (Ch^2 = 8.047, df = 2, p = 0.018), as well as reconstructive teaching (Ch^2 = 7.132, df = 2, p = 0.028).

Finally, the chemistry teachers rated different types of work in chemistry classes (see Fig. 4). On a four-point Likert scale, activities associated with concepts and dimensions are seen as less creative, whereas open tasks are rated very creative. In total, the teachers have seen all activities as more or less creative, except working with closed tasks. No differences regarding second subject, school type or age were found. Female chemistry teachers rated activities associated with models and laws as more creative than male teachers (U = 318.000, Z = −2.231, p = 0.026). Teachers who consider themselves creative show a significantly higher rating of activities associated with concepts and dimensions, such as defining (Chi^2 = 6.641, df = 2, p = 0.036), as well as activities associated with experimentation, such as fairs, logging and interpreting (U = 8.427, df = 2, p = 0.015).

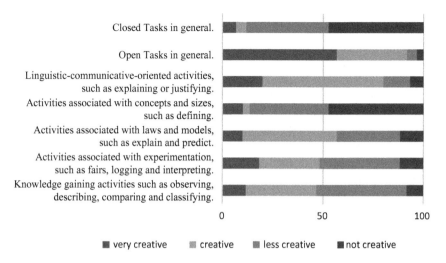

Fig. 4 Rating of the participants of different tasks in chemistry class. The tasks had to be rated on a four-point Likert scale from 1 = not creative up to 4 = very creative. The figure shows the descriptive results of the Likert-scaled items (N = 64)

Discussion and Implications

The own definitions of the teachers that took part in this study suggests a closed view on creativity. For them, creativity is closely related to a well-elaborated ability to solve problems showing versatile ideas. The underlying ability of divergent thinking is mentioned explicitly only by approximately 10% of the respondents. Following the questionnaire of the European Commission (Cachia and Ferrari 2010: 27), teachers think of creativity as the ability to produce something original. Here, the teachers could only rate three given definitions of creativity (creativity is the ability to produce something original; creativity is about finding connections between things that have not been connected before; creativity is the ability to produce something of value) and could not write down their own definition. Thus, we were able to get some more detailed information about the German teachers' definition of creativity. However, our study had only 64 participants; hence the results have to be interpreted carefully. In our example, teachers think more of solving problems than of a creative product, which is understandable in school context. Comparing the data of our study with the ones of Cachia and Ferrari (2010), the question about the views on creativity of chemistry teachers is not answered yet and has to be analysed in more detail. For this, a study using concept maps for getting information about the teachers' view on creativity and creativity in chemistry education amongst pre-service and in-service chemistry teachers is currently carried out (Semmler and Pietzner 2016).

The chemistry teachers that took part in this study do not think of themselves as a creative person; they tend to follow a certain structure in their teaching. This implies that if teachers use teaching methods that would enable the students to be creative, the opportunity is possibly left unused because they do not think of fostering creativity. In addition, a person who thinks of him/herself as less creative cannot foster the students' creativity properly. For the development of creativity in school, teachers need to believe that they are creative themselves and experiment with their teaching and show the students how to work creatively (Tanggaard 2011). At the same time, their attitude towards creativity in general and creativity in class is rather positive, as Cachia and Ferrari (2010) already could show and can (also) be seen in our study. They are aware that creativity enriches teaching, that it can be developed and that fostering creativity in class is an important superordinate goal; especially female teachers do so.

Looking at the ranking of character traits, the results of Westby and Dawson (1995) could be confirmed: teachers prefer character traits that belong to less-creative children like logical, good-natured or sincere, probably because these children tend to follow more the rules in class. On the other hand, they dislike typical character traits of creative students like impulsive, nonconformist or making their own rules. Therefore, creative children are in danger to be rejected by the teachers. As a consequence, creative children might not be able to develop their full potential. However, the teachers who participated in our study have a positive attitude towards creativity, but it seems that they do not know how to implement or to use it in class. Future research should investigate teachers' dealing with creativity in class in more detail. Because creativity in science is different from arts or music, subject-related studies should provide a better view on what is happening in class.

Looking at the differences between teachers regarding their assessment of their own creativity, interesting results can be found. Teachers who consider themselves as creative use significantly more methods that have the potential to foster creative working of students. In addition, they realised that important scientific methods like defining or interpreting data have creative aspects. However, like the other teachers we could interview in this study, they favour the same character traits as little or noncreative teachers. This result has to be interpreted carefully. Creative teachers might realise that scientific work in chemistry also includes creative phases. However, the creative methods were perhaps not used for fostering the students' creativity consciously but with the aim to let them work more independently. This would fit with the result that they favour character traits of creative children as the little and noncreative teachers do.

The study could show that teachers who participated in this study know and, to some extent, already use creative teaching methods like project work or explorative teaching. However, they do not seem to realise that these teaching methods are suitable to foster creativity, which was also shown by the study of Cachia and Ferrari (2010). Therefore, it is necessary to help teachers in this respect. As a consequence, professional development courses for in-service teachers should be developed that could strengthen both their own perceived competence in terms of creativity and ways to demonstrate how to create more open learning situations in class. Here, all different types of creativity in class should be taken into account: teaching creatively, creativity teaching, teaching for creativity and creative learning.

References

Barron, F., & Harrington, D. (1981). Creativity, intelligence, and personality. *Annual Review of Psychology, 32*, 439–476.

Cachia, R., & Ferrari, A. (2010). *Creativity in schools: A survey of teachers in Europe*. Luxembourg: Publications Office of the European Union.

Craft, A. (2005). *Creativity in schools - tensions and dilemmas*. London: Routledge.

Cropley, A. J. (1982). Kreativität: Entstehungsbedingungen und Einflußfaktoren [Creativity: Formation conditions and influencing factors]. In W. Wieczerkowski & H. zur Oeveste (Eds.), *Lehrbuch der Entwicklungspsychologie* [Textbook of developmental psychology] (Band 2, pp. 259–274). Düsseldorf: Schwann.

Cropley, A. J. (1991). *Unterricht ohne Schablone – Wege zur Kreativität* [Teaching without template – Ways to creativity]. München: Ehrenwirth.

Cropley, A. J. (2001). *Creativity in education & learning: A guide for teachers and educators*. Cornwall: Kogan Page.

Drevdahl, J. E. (1956). Factors of importance for creativity. *Journal of Clinical Psychology, 12*, 21–26.

European Council. (2009). Council conclusions of 12 May 2009 on a strategic framework for European cooperation in education and training ('ET 2020') (No. (2009/C 119/02)).

Flick, U., von Kardorff, E., & Steinke, I. (2004). *A companion to qualitative research*. London: SAGE Publications.

Gärtner, H. J. (1997). Kreativität im Chemieunterricht [Creativity in chemistry classes]. *Naturwissenschaften im Unterricht: Chemie, 42*, 12–20.

Guilford, J. P. (1987). Creativity research: Past, present, and future. In S. G. Isaksen (Ed.), *Frontiers of creativity research: Beyond the basics* (pp. 33–65). Buffalo: Bearly Limited.

Hallmann, R. J. (1970). Techniken des kreativen Lehrens [Techniques of creative teaching]. In G. Mühle & C. Schell (Eds.), *Kreativität und Schule* [Creativity and school] (pp. 175–180). München: Piper.

Hilgard, E. R. (1959). Creativity and problem-solving. In H. Anderson (Ed.), *Creativity and its cultivation* (pp. 162–180). New York: Harper & Row.

Jeffrey, B., & Craft, A. (2004). Teaching creatively and teaching for creativity: Distinctions and relationships. *Educational Studies, 30*(1), 77–87.

Kind, P. M., & Kind, V. (2007). Creativity in science education: Perspectives and challenges for developing school science. *Studies in Science Education, 43*, 1–37.

Mackinnon, D. W. (1962). The personality correlates of creativity: A study of American architects. In G. Nielson (Ed.), *Proceedings of the XIV International Congress of Applied Psychology. Vol. 2. Personality Research* (pp. 11–39). Oxford: Munksgaard.

Mackinnon, D. W. (1967). The study of creative persons: A method and some results. In J. Kagan (Ed.), *Creativity and learning* (Vol. 8, pp. 20–35). Boston: Houghton Mifflin.

Mayring, P. (2014). Qualitative content analysis: Theoretical foundation, basic procedures and software solution. Retrieved from http://nbn-resolving.de/urn:nbn:de:0168-ssoar-395173 (28.10.2016).

N.A.C.C.C.E. (1999). *All our futures: Creativity, culture and education*. London: DfEE.

Newton, D. P., & Newton, L. D. (2009). Some student teachers' conceptions of creativity in school science. *Research in Science Technological Education, 27*(1), 45–60.

Rhodes, M. (1961). An analysis of creativity. *The Phi Delta Kappan, 42*(7), 305–310.

Runco, M. A. (2004). Creativity. *Annual Review of Psychology, 55*, 657–687.

Sawyer, R. K. (2012). *Explaining creativity – The science of human innovation*. New York: Oxford University Press.

Semmler, L., & Pietzner, V. (2016). Untersuchung zu Auffassungen von Kreativität mit Concept Maps [Investigation of views on creativity with concept maps]. In C. Maurer (Ed.), *Authentizität und Lernen – das Fach in der Fachdidaktik* [Authenticity and learning – The subject in subject didactics]. (Gesellschaft für Didaktik der Chemie und Physik, Band 36, pp. 479–481). Kiel: IPN.

Simonton, D. K. (2012). Teaching creativity: Current findings, trends, and controversies in the psychology of creativity. *Teaching of Psychology, 39*(3), 217–222.

Tanggaard, L. (2011). Stories about creative teaching and productive learning. *European Journal of Teacher Education, 34*(2), 219–232.

Torrance, E. P. (1970). *Education and the creative potential.* Minneapolis: University of Minnesota Press.

Urban, K. K. (2004). *Kreativität: Herausforderung für Schule, Wissenschaft und Gesellschaft* [Creativity: A challenge for education, science and society]. Münster: LIT.

Weisberg, P. S., & Springer, K. J. (1961). Environmental factors in creative function: A study of gifted children. *Archives of General Psychiatry, 5*(6), 554–564.

Westby, E. L., & Dawson, V. L. (1995). Creativity: Asset or burden in the classroom? *Creativity Research Journal, 8*(1), 1–10.

Williams, G., & Wyburd, J. (2006). *Facilitating creativity in higher education – The views of National Teaching Fellows.* A research report by the Creativity Centre Ltd. Retrieved from http://www.academia.edu/284128/Facilitating_Creativity_In_Higher_Education (30.11.2016).

"Creating Creativity": Improving Pre-service Teachers' Conceptions About Creativity in Chemistry

Markus Bliersbach and Christiane S. Reiners

Creativity as an Important Component of Teaching Chemistry

Creativity is considered one of the key competencies of modern society (Kind and Kind 2007). Due to global economic restructuring, accompanied by fundamental and rapid changes, life will be characterised more and more by complexity and uncertainty (Hodson 2003). Creativity – as the ability to produce something new and relevant (Sternberg and Lubart 1999) – is a necessary condition to meet the requirements of an uncertain future. This is also a challenge for general education. Who if not the schools and universities should prepare students to handle these demands and become possible future innovators?

In chemistry education, the implementation of creativity is also important in terms of representing an adequate image of the discipline. According to research on nature of science (NOS), students "should appreciate that science is an activity that involves creativity and imagination as much as many other human activities" (Osborne et al. 2003: 702). This is even more important, as school students as well as pre- and in-service teachers reveal certain misconceptions about the role of creativity in chemistry (Lederman 2007). In contrast to its real nature, they characterise the discipline as solely logical and analytical (McComas 1998). The following is a typical student statement: "Chemists think logically but are not necessarily creative. Creative people are good in languages, and chemists are good logically" (Becker et al. 2014: 358; translation by the authors). As a consequence, students, who prefer open and inventive tasks, reject chemistry as a potential career aspiration (Tobias 1990). If chemistry education addresses creative elements, this would not only promote an adequate image of chemistry and prepare the students for future tasks in

M. Bliersbach (✉) • C.S. Reiners
University of Cologne, Cologne, Germany
e-mail: markus.bliersbach@uni-koeln.de

© Springer International Publishing AG 2017
K. Hahl et al. (eds.), *Cognitive and Affective Aspects in Science Education Research*, Contributions from Science Education Research 3,
DOI 10.1007/978-3-319-58685-4_5

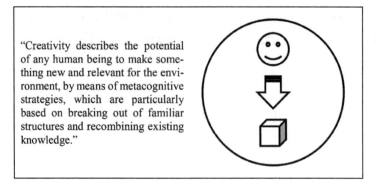

"Creativity describes the potential of any human being to make something new and relevant for the environment, by means of metacognitive strategies, which are particularly based on breaking out of familiar structures and recombining existing knowledge."

Fig. 1 A definition and the four components of creativity

this profession, but could also lead to a broader acceptance of the discipline and motivate young people to become chemists.

The Notion of Creativity

As one of the first steps in this research project, a definition of the term creativity was formulated. It is based on an analysis of general, educational and science-specific literature as well as on a survey among lecturers of a regional educational institution. The definition includes all four central components of creativity described by Rhodes (1961): the *creative person*, who in a *creative process* creates a *creative product* and, during the process, is embedded in a *creative environment*. For the definition and a visualisation of the relationship between the four components, see Fig. 1.

Creativity in Chemistry

Research on the epistemology of science confirms that creativity is a key component in the development of scientific knowledge. The next question then is: where exactly in chemistry are we depending on creativity? Based on NOS literature (e.g. Lederman 2007; Osborne et al. 2003), international documents on educational standards (National Research Council 2012) and on our own considerations, five moments can be emphasised: *generating research questions, planning experimental investigations, formulating hypotheses, generating theories and models* and *presenting research results*. Thus, creativity is required throughout the whole scientific process.

For two reasons, the generation of theories and models should have an exceptional position. Firstly, it is presumably the most creative act in scientific enterprises.

The history of the Nobel Prize in chemistry reveals that its winners usually come up with new, often revolutionary theories or models, which clarify long-existing problems and/or open new research fields. Hence, according to our definition, such theories and models represent creative products of very high *relevance for their environment*, which are often strongly characterised by *breaking out of familiar structures*. The second reason has to do with the representation of the four creative moments in current chemistry education practice. Generating research questions, formulating hypotheses and planning experimental investigations are essential parts of inquiry-based learning and similar approaches (Abd-El-Khalick et al. 2004). In contrast to their outstanding position in chemical research, model building processes are much harder to find in contemporary chemistry classes.

A Qualitative Study: Production Versus Reproduction of Models

The long-term aim of this research project is to support school students in appreciating creativity as a part of chemistry. Therefore, creativity has to be implemented into chemistry education practice. Based on the assumption that chemistry teachers' adequate conceptions represent a necessary condition on the way to implement NOS into chemistry education appropriately (Lederman 2007), we focus on pre-service chemistry teachers. The principle aim is to find out how prospective chemistry teachers can be supported in developing appropriate conceptions about creativity, about the role of creativity in scientific research processes and about possibilities to implement creativity into chemistry lessons. For this purpose, different approaches are evaluated in the project, concerning their impact on the development of the teacher students' views and competencies (for one of the earlier studies, cf. Bliersbach and Reiners 2015). The qualitative study presented here focusses on the use of historical case studies (Abd-El-Khalick and Lederman 2000) on the one hand and student-generated analogies (Haglund 2013) on the other. Both interventions spotlight the construction of models. The following investigation questions were formulated:

1. How does the *reproduction* of model building processes by historical case studies contribute to the development of prospective teachers' conceptions about creativity in chemistry?
2. How does the *production* of analogical models as an activity, in which prospective teachers actively go through model building processes by themselves, contribute to the development of their conceptions about creativity in chemistry?

Method

The study was carried out in two pre-service chemistry teacher courses in winter term 2014/2015. All participants were between 19 and 26 years old and in the second year of their bachelor studies (intended degree: chemistry education at higher secondary level). Overall, 38 teacher students participated (21 in course 1 and 17 in course 2; 24 male and 14 female).

Among other objectives (e.g. scientific literacy, experiments in chemistry education, teaching methods, etc.), each course contained two sessions of 90 min about the topic "models in chemistry". By different means, both courses aimed at promoting a deeper understanding of the role of creativity in the generation of chemical models. It must be emphasised that this aim was not transparent to the students. They only knew that they had to deal with concrete examples of model building processes in the two sessions. Neither the term creativity nor any synonyms or periphrases were used.

Course 1: Reproduction of a Model Building Process

Course 1 included a historical case study concerning Jacobus Henricus van 't Hoff (1852–1911), the first Nobel Prize winner in chemistry in 1901, and his models about the spatial arrangement of molecules. In 1874, van 't Hoff accounted for the phenomenon of optical activity by assuming that the chemical bonds between carbon atoms and their neighbours were directed towards the corners of a regular tetrahedron. He formulated a revolutionary theory (Cohen 1912: 82–92), which laid the foundation of stereochemistry and stereoisomerism. He shares the reputation for this with the French chemist Joseph Le Bel (1847–1930), who independently came up with the same idea. van 't Hoff can be considered as a pioneer in reflecting on the role of creativity in chemistry. This becomes obvious particularly in his lecture on "Imagination in Science", held in Amsterdam in 1878 (Cohen 1912: 150–165). In this lecture, he compares about 200 historical scientists and reflects on their use of imagination to come to scientific achievements.

To enable the students to appreciate van 't Hoff's achievements and to understand how new and revolutionary his models were, the intervention itself began with an introduction to the history of isomerism and stereochemistry before van 't Hoff and Le Bel. After that, the courses were structured by the *Jigsaw Teaching Technique* (Aronson 1978). The teacher students were separated into three expert groups which focused on different aspects concerning van 't Hoff and his model building process. In the expert groups, they had to read certain texts and try to answer associated guiding questions. The subsequent jigsaw groups consisted of three students from the different expert groups, who presented their respective results to the other two students. The members of Group A had to reflect on van 't Hoff's model itself. With the aid of his original publication, they had to elaborate on how he came up with the model and which arguments he used to justify it. Group B had to focus on biographical issues and the context of the model building process. The students had to reflect on van 't Hoff's personality, on his scientific career and on obstacles on the way to establish his revolutionary model. The members of Group C dealt with the aforementioned lecture "Imagination in Science". They had to concentrate on van 't Hoff's views on scientific processes and try to assess them from a contemporary perspective. After the students presented and discussed their results in the jigsaw groups, they had to reflect on some additional questions, for example, what they had learned about model building processes in general and whether this could have any consequences for chemistry education. The intervention ended with a plenary discussion about general characteristics of scientific model building processes.

Course 2: Production of Analogical Models

While the students in course 1 considered a creative model building process from the exterior, course 2 offered them the opportunity to create a model by themselves. As mentioned before, there are not many approaches focusing on autonomous model building processes in chemistry education. Black box activities (Lederman and Abd-El-Khalick 1998) are one of them. Though they are suitable to mediate some important characteristics of chemical models (e.g. that models are not a copy of reality and are based on individual interpretations and preferences), they are less suitable to convey the notion of creativity in model building processes. A possible way to implement autonomous model building processes into chemistry education, which stresses creativity on the one hand and is not too ambitious on the other, could be the generation of analogical models. Why analogical models? Firstly, analogies are a crucial compound of scientific models. Every appropriate model includes certain analogies to the real phenomenon that it represents (Gilbert 2004). Secondly, analogies often play a crucial role in creative model building processes. One common example is August Kekulé (1829–1896), who declared that he had discovered the ring shape of the benzene molecule after having a daydream of a snake seizing its own tail (Rocke 2010). Thirdly, empirical studies concerning the use of self-generated analogies in science education show positive results with regard to students' learning and enjoyment (Haglund 2013). Finally, analogical models can be used in school to introduce relevant scientific models (e.g. the electric circuit, which is compared with a model of a water circuit in physics education) (Olson 1958).

According to this, the participants of course 2 had to create analogical models of the valence shell electron pair repulsion (VSEPR) model (Gillespie 2004). In order to ensure that they knew what the theoretical model implies – normally every student in the course should be familiar with it – a short introduction to the model and its underlying assumptions was given. Then the teacher students were separated into groups of three. The task was to generate one or several analogical models that could be used to introduce or explain the VSEPR model to school students and to present it to the rest of the course in the next session. The groups were asked to

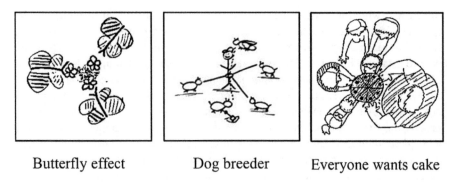

Butterfly effect   Dog breeder   Everyone wants cake

Fig. 2 Excerpts of some student-generated analogical models related to the VSEPR model

cover as much characteristics of the original model as possible but were free to use any analogy they wanted. They could relate to everyday life contexts, other chemical models or anything else they could think of. The students had 90 min to develop their models. The way they presented it in the next lesson was also up to them. The analogical models could be visualised via posters, animations or role plays (actually, every group presented a poster). After every group had presented, the models were compared and evaluated. The intervention ended with a plenary discussion about model building processes in general. Figure 2 documents excerpts of a few examples of the analogical models that were developed.

Data Collection

In our study, we tried to investigate if the described interventions can influence the teacher students' conceptions about the role of creativity in chemistry. To be able to evaluate a possible influence, a comparison between pre- and postconceptions of the students was necessary. In order to get individual and deep insights into the teacher students' conceptions, data were collected qualitatively, by open-ended questionnaires, semi-structured interviews and participant observation. The different methods were used to triangulate data by validating the obtained results through the combination of data collected from different sources (Mayring 2000). For an overview of the study, see Fig. 3.

During the data collection, the aim of the study was not transparent to the students. Neither the questionnaires nor the interviews included the term creativity (or any synonyms or periphrases). The assumption behind this decision was that most students, if they had known what was investigated, probably would have answered in the way they "should" and would have stated the importance of creativity in chemistry without really being convinced.

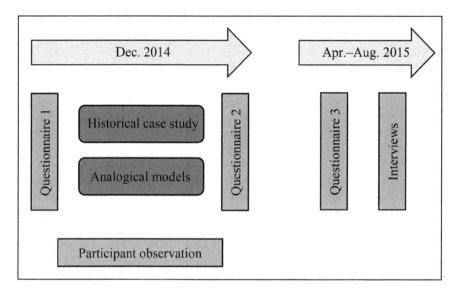

Fig. 3 Overview of the study: interventions and data collection

Open-Ended Questionnaires

In order to collect the teacher students' pre- and postconceptions, open-ended questionnaires were used immediately before (questionnaire 1) and after the respective interventions (questionnaire 2). To evaluate the sustainability of the students' perceptions, a sample of 21 teacher students were asked to complete a follow-up questionnaire (questionnaire 3) about half a year later (the extended period is related to the different availability of the students). Apart from the acquisition of some preliminary information (age, gender, other subjects, previous experiences in chemistry didactics), all three questionnaires contained the same two questions:

1. What are the similarities and what are the differences between chemistry and artistic disciplines (e.g. art, music and architecture)?
2. Which skills and/or abilities do chemists need to be successful researchers?

The students had about 15 min to answer them. They are based on the *Views of Nature of Science* questionnaires (Lederman et al. 2002). The questionnaire was previously tested with students as well as with further science education researchers. As indicated above, an open-ended format was chosen to allow free and autonomous answers and to avoid forcing a choice upon the students. For the data analysis, all questionnaires were transcribed.

Interviews

Half a year after the respective interventions, problem-centred interviews (Mayring 2002: 67–72) were conducted with a sample of six participants (three out of each course). To achieve a high diversity, both teacher students, whose conceptions changed according to the results of questionnaires 1 and 2 (students who did not refer to creativity in questionnaire 1 but in questionnaire 2), as well as students, whose conceptions did not change (students who did not refer to creativity in questionnaire 1 nor in questionnaire 2), were selected. In addition to the validation of the responses to the questionnaires, this nonstandardised, semi-structured interview format was used to facilitate the generation of in-depth profiles of some teacher students. Compared to the questionnaires, they had more time to describe their viewpoints and conceptions. Besides, if certain statements seemed to need further clarification, additional questions were asked by the researcher.

In addition to a more extensive discussion of the items of the questionnaire, the students were asked what they remembered from the respective interventions and had to reflect their learning processes during the courses (e.g. "In the course, did you learn something new about chemistry?", "What do you remember from the sessions where we dealt with model building processes in chemistry?", "Did the sessions influence your view on chemistry?", etc.). Besides, some additional aspects that are not directly connected with the investigation questions were addressed (e.g. "Did you like the sessions?", "Do you think this approach could be applied in school chemistry?", etc.). In order to structure the interviews and to ensure that all planned questions were really asked, a question manual was used. For the data analysis, all audio recordings were transcribed.

Participant Observation

As an additional means of triangulation, participant observation was used, which is a qualitative investigation method in which the observer is an active part of the social situation she/he is investigating (Mayring 2002: 80–84). In our study, it served to directly capture the students' communications – especially the discussions at the end of the respective interventions. Moreover, some other aspects not directly connected with the investigation questions could be observed (appropriateness of the respective tasks and activities, motivational aspects, etc.). Similar to the question manual for the interviews, an observation manual was formulated to determine the dimensions of interest and to stay focused during the observations. In order to record the observations, field notes were written.

Table 1 Overview category system: occurrence of terms related to creativity in chemistry

Category	Subcategory	Anchor example
Explicit confirmation of the role of creativity in chemistry		"Chemists need to be creative to be successful researchers"
Implicit confirmation 1: "synonyms" for creativity	Phantasy	"Chemists need to use their phantasy"
	Imagination	"Chemists need to use their imagination"
	Ingenuity	"Ingenuity is important in chemistry"
Implicit confirmation 2: components of creativity	Creating something new	"Chemists need to create something new if they want to be successful"
	Breaking out of familiar structures	"Sometimes it is important to break with existing knowledge"
	Recombining knowledge	"Chemists need to combine existing knowledge in new ways"
	Working on new problems	"Chemists need to explore unknown areas"
	Using interdisciplinary approaches	"Chemists need to use approaches that are used in other disciplines"
Implicit confirmation 3: preconditions of creativity	Courage	"Chemists need to be courageous"
	Curiosity	"Chemists need to be curious"
	Openness	"Chemists need an open mind"
No reference to creativity		(neither category is mentioned)
Explicit denial of the role of creativity in chemistry		"In artistic disciplines, you can be creative; in chemistry, you have to think logically and follow certain rules"

Data Analysis

In order to be able to answer the question of whether and how the different interventions affect teacher students' conceptions about creativity in chemistry, the collected data was analysed following the rules of qualitative content analysis, a convenient and frequently used approach to systematically structure and summarise qualitative, text-based material via categories (Mayring 2000). In our case, the content was analysed concerning the occurrence of terms related to the role of creativity in chemistry.

During the analysis of the various data sources, all corresponding student statements were collected and arranged in four groups: explicit confirmation of the role of creativity in chemistry, implicit confirmation, no reference to creativity and explicit denial of the role of creativity in chemistry. Implicit statements were further subdivided into synonyms, components and preconditions of creativity (see Table 1 for an overview of all categories, subcategories and anchor examples).

In order to assess the learning progress of every student, their answers to the questionnaires before and after the respective interventions were compared. In the case of the participants, who additionally completed the follow-up questionnaires or interviews, these data, as well as the results of the participant observation in both courses, were incorporated, too. Finally, in order to evaluate whether there were differences between both courses, the collected results of the respective courses were compared.

All data were analysed by the authors and a further science education researcher to ensure reliability in the coding. The interpretations were mostly consistent. Only a few differences emerged, and consensus was reached after discussions and further consultation of the data.

Results

The results of the three different forms of data collection are presented separately.

Questionnaires

In questionnaire 1, which was handed to the teacher students before the respective interventions, only few of the students mentioned creativity as a characteristic of chemistry, neither explicitly nor via synonyms, components or preconditions of creativity. Several students explicitly negated creativity as a part of chemistry, so the typical misconceptions could be confirmed. In answering question 1, student BR05 (we used anonymous codes to be able to compare the students' answers in the different questionnaires) wrote, for example, "In arts, the own creativity can be unfolded, open space is provided, where you can create for your own. In chemistry you have to act according to and argue with the scientific laws" (this and all following student statements were translated by the authors).

In questionnaire 2, which was handed to the students after the interventions, such formulations were barely used. The number of explicit and implicit mentions of creativity increased significantly in the student answers, although – comparing both courses – in different forms and in different extents. In course 1 (historical case study), the increase of explicit mentions as well as synonyms was very obvious. In course 2 (analogical models), the growth of explicit expressions was not as big. The participants mentioned creativity rather implicitly, with categories that were assigned to components and preconditions of creativity.

Questionnaire 3 was handed out about half a year after the respective interventions. The comparison of the students' answers in this questionnaire with their answers in questionnaire 2 reveals only a very light decrease of the categories mentioned, no matter which course they participated in.

Interviews

Both types of teacher students – those whose conceptions changed during the interventions as well as those students whose conceptions did not change according to the results of the questionnaires – were interviewed. The students, who changed their conceptions, still appreciated the role of creativity in chemistry explicitly. Upon request, some additional implicit categories, mainly components of creativity, were mentioned. When the students were asked if they learned something new about chemistry, all students referred to the respective interventions. Student UO67 (course 1) stated, for example, "Yes. The example with van 't Hoff. Before the course, I would not have connected creativity with chemistry".

What about the students who did not change their conceptions? When they were asked the two questions of the questionnaire, no difference could be determined. This impression changed, when further questions about the respective interventions were asked. Student MO11 (course 2) stated, for example, "Especially as we made the model for ourselves, it became obvious that it is not easy (…), you need creativity to find something that represents the model adequately".

Participant Observation

With regard to the investigation questions, the participant observation was mainly important during the discussions at the end of the respective interventions, where the students reflected on model building processes in chemistry. In both courses, creativity was mentioned during the reflections. In course 1, a discussion about the role of creativity in chemistry arose, in which the students compared van 't Hoff's case and his perceptions about imagination in science with contemporary scientific research. In course 2, creativity was mentioned two times. One student stated that his group needed a certain amount of creativity to find an analogy. He also assumed that it would be similar in scientific model building processes. Another student liked the fact that the course offered the opportunity to act in an open environment ("It was a nice variety to become creative in chemistry"). He also stated that the activity – for the same reason – could be implemented into chemistry education.

All in all, both interventions were well accepted by most teacher students. The participants of course 1 appreciated van 't Hoff's achievements and were fascinated by the fact that he was so successful at the age of only 22 years. The participants in course 2, as indicated above, were fond of the open environment.

Discussion and Conclusions

Because of the small number of participants, the study is far from being representative. It is one of several preliminary studies within the project, in which first indications concerning the suitability of different approaches were collected. The following conclusions should be treated with caution and have to prove themselves in further studies.

A comparison between the data obtained in the different courses indicates that the reflection on historical case studies as well as the production of analogical models can contribute to more informed views of the teacher students. Many of the students, who did not mention or who even denied creativity as a part of chemistry in questionnaire 1, changed their minds after the respective course elements. According to the follow-up questionnaires and interviews, these changed perceptions proved to be relatively robust. With the aid of concrete requests concerning the course content and its influence on the students' perceptions in the semi-structured interviews, it could also be assumed that these changed perceptions were based on the respective interventions.

The different results in the questionnaires of courses 1 and 2 are probably based on the methodological decision that during the courses and during the different forms of interrogation, it should not become apparent to the students what the different interventions were aimed at. Comparing the two approaches from this perspective, the historical case study – in providing a concrete example of a creative person in chemistry (van 't Hoff), who furthermore reflects on the role of imagination in science – more explicitly draws the students' attention to creativity. The production of analogical models rather implicitly mediates the role of creativity in chemistry. According to research results, conceptions of NOS are best learned through explicit, reflective instruction (Lederman 2007). In our case, that would have meant to tell the students that they had developed the models in order to appreciate the role of creativity in chemistry. For the aforementioned reason, this was not possible. Keeping this in mind, it is not surprising that the latter approach, even though it also contributed to more informed views of the participating students, was not quite as successful as the historical case study.

In contrast to questionnaires, the results of the participant observation and the interviews revealed the actual potential of the production of analogical models. In the reflection period after the model building process, creativity was mentioned as a necessary ability to generate a scientific model. Besides, the teacher students mentioned that the intervention could be used to implement creative processes into chemistry education. The interviews confirmed this impression. When the students were asked half a year after the intervention what they learned new about chemistry in the course, they mentioned the same both aspects.

Summing it up, the use of historical case studies – in providing explicit examples of creative persons and their creative products – is a great possibility to enable teacher students to recognise *that* creativity is important in chemistry. The production of analogical models, however, offers the opportunity to generate a creative environment. Accordingly, this approach, where students actively go through chemistry-specific model building processes for themselves, is appropriate to realise *how* creativity works in chemistry and how creative processes can be implemented in school chemistry. To exploit its full potential, the latter, rather implicit approach, should be combined with explicit instructions. In order to develop comprehensive conceptions about creativity in chemistry, a combined approach of both course elements seems to be the most suitable.

These conclusions were already put into practice in winter term 2015/2016, in which the final and main study of the research project was conducted. Supplemented with some other elements that proved themselves in previous studies, a combined course unit about creativity in chemistry was evaluated concerning its potential to support ongoing teacher students in developing appropriate conceptions about creativity, about the role of creativity in scientific research processes and about possibilities to implement creativity into chemistry lessons. In the main study, we could rely on a bigger sample of participants and a more extensive data collection, for example, in the form of more comprehensive questionnaires and additional interviews at the beginning of the course. As a concrete outcome of the research project, we want to implement this course unit into regular chemistry education studies at our university as well as into our training programmes for in-service chemistry teachers. We hope that this will contribute to an increased implementation of creativity into chemistry education practice.

References

Abd-El-Khalick, F., & Lederman, N. G. (2000). The influence of history of science courses on students' views of nature of science. *Journal of Research in Science Teaching, 37*, 1057–1095.

Abd-El-Khalick, F., BouJaoude, S., Duschl, R., Lederman, N. G., Mamlok-Naaman, R., Hofstein, A., et al. (2004). Inquiry in science education: International perspectives. *Science Education, 88*, 397–419.

Aronson, E. (1978). *The jigsaw classroom*. Beverly Hills: Sage.

Becker, H. J., Kühlmann, J. K., & Parchmann, I. (2014). Trendbericht Chemiedidaktik. Der Kompetenzbereich "Erkenntnisgewinnung" [Trend report chemistry didactics. The area of competence "epistemological and methodological knowledge"]. *Nachrichten aus der Chemie, 62*, 356–359.

Bliersbach, M., & Reiners, Ch. S. (2015). Implementierung von Kreativität in den Chemieunterricht?! [Implementation of creativity into chemistry education?!]. In S. Bernholt (Ed.), Heterogenität und Diversität – Vielfalt der Voraussetzungen im naturwissenschaftlichen Unterricht [Heterogeneity and diversity – Variety of requirements in science education] (pp. 193–195). Kiel: IPN.

Cohen, E. (1912). *Jacobus Henricus van 't Hoff. Sein Leben und Wirken* [Jacobus Henricus van 't Hoff. His life and his work]. Leipzig: Akademische Verlagsgesellschaft.

Gilbert, J. K. (2004). Models and modelling: Routes to more authentic science education. *International Journal of Science and Mathematics Education, 2*, 115–130.

Gillespie, R. J. (2004). Teaching molecular geometry with the VSEPR model. *Journal of Chemical Education, 81*, 298–304.

Haglund, J. (2013). Collaborative and self-generated analogies in science education. *Studies in Science Education, 49*, 35–68.

Hodson, D. (2003). Time for action: Science education for an alternative future. *International Journal of Science Education, 25*, 645–670.

Kind, P. M., & Kind, V. (2007). Creativity in science education: Perspectives and challenges for developing school science. *Studies in Science Education, 43*, 1–37.

Lederman, N. G. (2007). Nature of science: Past, present, and future. In S. K. Abell & N. G. Lederman (Eds.), *Handbook of research on science education* (pp. 831–879). New York: Lawrence Erlbaum Associates.

Lederman, N. G., & Abd-El-Khalick, F. (1998). Avoiding denatured science: Activities that promote understandings of the nature of science. In W. F. McComas (Ed.), *The nature of science in science education. Rationales and strategies* (pp. 83–126). Dordrecht: Kluwer.

Lederman, N. G., Abd-El-Khalick, F., Bell, R. L., & Schwartz, R. S. (2002). Views of nature of science questionnaire: Toward valid and meaningful assessment of learners' conceptions of nature of science. *Journal of Research in Science Teaching, 39*, 497–421.

Mayring, P. (2000). Qualitative content analysis. *Forum: Qualitative Social Research, 1*(2). http://www.qualitative-research.net/index.php/fqs/article/view/1089/2385. Accessed 25 Apr 2016.

Mayring, P. (2002). *Einführung in die qualitative Sozialforschung*. Weinheim: Beltz.

McComas, W. F. (1998). The principal elements of the nature of science: Dispelling the myths. In W. F. McComas (Ed.), *The nature of science in science education. Rationales and strategies* (pp. 53–72). Dordrecht: Kluwer.

National Research Council. (2012). *A framework for K-12 science education: Practices, crosscutting concepts, and core ideas*. Washington, DC: National Academy Press.

Olson, H. F. (1958). *Dynamical analogies*. Princeton: Van Nostrand.

Osborne, J., Ratcliffe, M., Collins, S., Millar, R., & Duschl, R. (2003). What 'ideas-about-science' should be taught in school science? A Delphi study of the expert community. *Journal of Research in Science Teaching, 40*, 692–720.

Rhodes, M. (1961). An analysis of creativity. *Phi Delta Kappan, 42*, 305–310.

Rocke, A. J. (2010). *Image and reality. Kekulé, Kopp, and the scientific imagination*. Chicago: The University of Chicago Press.

Sternberg, R. J., & Lubart, L. A. (1999). The concept of creativity: Prospects and paradigms. In R. J. Sternberg (Ed.), *Handbook of creativity* (pp. 3–15). Cambridge: Cambridge University Press.

Tobias, S. (1990). *They're not dumb, they're different: Stalking the second tier*. Tucson: The Research Corporation.

Is Topic-Specific PCK Unique to Teachers?

Marissa Rollnick, Bette Davidowitz, and Marietjie Potgieter

Background, Framework and Rationale for Study

The construct of pedagogical content knowledge, PCK, is well established in the education research community. Since the phrase was first coined by Shulman (1986), multiple understandings of the construct have emerged (Kind 2009). There is, however, general consensus around Shulman's conception that PCK is unique to teachers and different from but supported by content knowledge, CK. Content knowledge is thus a necessary but not the only requirement for development of PCK. In other words, the possession of good CK does not necessarily imply good PCK. For some time, the authors have been investigating teachers' specialised knowledge at topic level, using the construct of TSPCK articulated below. Instruments have been designed and validated for several topics such as chemical equilibrium (Mavhunga and Rollnick 2013), particulate nature of matter (Pitjeng 2014), electrochemistry (Rollnick and Mavhunga 2014) and organic chemistry (Davidowitz and Vokwana 2014). Each instrument consists of paper-and-pencil tests to assess CK and PCK in a particular topic. Given that PCK is conceptualised as knowledge unique to teachers, we decided to test our instruments on non-teachers to find out how they would perform. The basic premise of the current study was that subject specialists with no formal training and limited teaching experience are likely to show lower proficiency than practising teachers on the PCK test.

M. Rollnick (✉)
University of the Witwatersrand, Johannesburg, South Africa
e-mail: marissa.rollnick@wits.ac.za

B. Davidowitz
University of Cape Town, Cape Town, South Africa

M. Potgieter
University of Pretoria, Pretoria, South Africa

© Springer International Publishing AG 2017   69
K. Hahl et al. (eds.), *Cognitive and Affective Aspects in Science Education Research*, Contributions from Science Education Research 3,
DOI 10.1007/978-3-319-58685-4_6

The topic-specific nature of PCK is apparent from several studies. For example, Van Driel et al. (1998) explored the development of PCK of pre-service teachers in chemical equilibrium, while Park et al. (2011) investigated photosynthesis. Another study on two chemistry topics (electrochemical cells and redox reactions) conducted by Aydin et al. (2014) to identify PCK components revealed the topic-specific nature of PCK. Their findings showed that teacher competence varied between the two topics investigated. These findings provide empirical evidence for the idea that PCK is experienced at different levels, as suggested by the PCK taxonomies of Veal and Makinster (1999). It is therefore important to distinguish between PCK as a construct at a disciplinary or topic level.

While most researchers agree that it is possible to point to features that characterise quality PCK, there are debates about whether PCK is purely personal or whether its quality can be defined and thus measured. As argued above, we consider PCK as a construct at topic level and adopt the notion of topic-specific PCK, TSPCK, as a product of transformation of content knowledge through five components (Mavhunga 2014). These are:

(i) Students' prior knowledge: the knowledge of conceptions and preconceptions that students bring with them
(ii) Curricular saliency: the ability to identify the big ideas of a topic, to sequence them and realise their importance
(iii) What is difficult to teach: what makes the learning of specific topics easy or difficult
(iv) Representations: analogies, illustrations, examples, explanations and demonstrations
(v) Conceptual teaching strategies: implying knowledge of teaching strategies most likely to be fruitful

Organic chemistry has long been regarded as a difficult topic to teach and learn at both secondary and tertiary levels (Green and Rollnick 2006; Katz 1996). The topic currently constitutes more than 30% of the South African Grade 12 curriculum and 27% of the first year curriculum at the university where this study was conducted. It was thus thought that this would be an important topic for further investigation.

The organic chemistry instrument consists of two paper-and-pencil tests designed to assess Grade 12 teachers' CK and PCK in organic chemistry. The content of the questions in the CK test was informed by the curriculum for Grade 12 as well as by chemistry textbooks in South Africa while those in the PCK test consisted of questions corresponding to the five components of TSPCK (Mavhunga 2014). Davidowitz and Potgieter (2016) found this instrument to be valid in that it measured a single construct, had face validity and was reliable when administered to a sample of practising teachers which was diverse with respect to training and experience.

The subject matter specialists who participated in this study are graduate teaching assistants at a South African university. They are referred to as chemists in line with the policy of the South African Chemical Institute which regards a 4-year degree as an entry level requirement to be considered a chemist (http://www.saci.co.za/membership_requirements.html). Graduate teaching assistants can be considered as subject matter specialists with little training as teachers. They assist students to complete tasks set by the lecturers and provide guidance in practical sessions as well as grade students' written work. As chemists studying at an advanced level, they would be expected to perform well in the CK test but might find the TSPCK test challenging. In this study, the CK and TSPCK of chemistry subject matter specialists in organic chemistry were investigated in comparison to a group of school teachers. If the chemists were able to perform at the same level as teachers in the TSPCK test, it would call into question the consensus that TSPCK is unique to teachers. We thus pose the question "How do the CK and TSPCK of chemists compare to teachers of Grade 12 organic chemistry?" with the following sub-questions:

1. What is the level of the chemists' and teachers' CK?
2. What is the level of their TSPCK?
3. How does their CK relate to their TSPCK?

Methodology and Sample

Instrument Design

Two previously validated tests, one for CK and the other for TSPCK (Davidowitz and Potgieter 2016), were used to collect both quantitative and qualitative data for this study. The CK test consisted of 5 questions comprising 23 items and covered the main subtopics taught at school level, viz. drawing and naming of isomers, functional groups, types of reactions and the relationship between structure, intermolecular forces and boiling points of organic compounds. These topics are also included in the first year curriculum at university level. An example of a question probing knowledge of isomers is presented in Fig. 1.

2	Molecules with the same molecular formula can have different structures, this is known as isomerism. There are several kinds of isomerism such as structural and geometric. Answer the following questions about isomers.

a) Draw the following isomers for 2-butene.

Structural isomer of 2-butene	Geometric isomer of 2-butene

Fig. 1 Example of an item from the CK questionnaire

A1.   You hand out a worksheet to be used in a classroom activity and ask students to select compounds that are alcohols from the table below:

1	2	3	4
$CH_3\text{-}OH$	$CH_3\text{-}\overset{\overset{\displaystyle O}{\|}}{C}\text{-}OH$	$CH_3\text{-}\overset{\overset{\displaystyle O}{\|}}{C}\text{-}H$	$CH_3\text{-}CH_2\text{-}OH$

5	6	7	8
$CH_3\text{-}\overset{\overset{\displaystyle O}{\|}}{C}\text{-}CH_3$	$H\text{-}\overset{\overset{\displaystyle O}{\|}}{C}\text{-}O\text{-}CH_3$	$NaOH$	$CH_3\text{-}\overset{\overset{\displaystyle O}{\|}}{C}\text{-}O\text{-}CH_3$

Sipho selects compounds 1, 2, 4 and 7. You then realize that other students in the class have given the same response. How would you explain to the students in the class how to distinguish alcohols from other compounds?

Fig. 2 TSPCK item A1, probing the students' prior knowledge component

The TSPCK questionnaire consisted of ten items based on the five components of Mavhunga's (2014) TSPCK model and presented the respondent with tasks set in a teaching context. Figure 2 shows an example of an item testing the students' prior knowledge component of TSPCK.

The CK test was assessed against a memorandum, while the TSPCK test was scored using a rubric with criteria on a scale of 1–4 corresponding to limited, basic, developing and exemplary PCK. An extract of the rubric used to score the TSPCK item, A1, (Fig. 2) is shown in Fig. 3.

Scores for the TSPCK test were peer validated by independent raters. An interrater reliability analysis using Fleiss' kappa statistic was carried out using MedCalc version 15.11.0 (https://www.medcalc.org; 2015) to determine consistency among raters. The value of kappa is 0.82 ($p < 0.001$), 95% CI 0.779 to 0.856, which is considered to be very good agreement.

Rasch Analysis of Data

The Rasch measurement model (Bond and Fox 2007) can be used to generate linear item measures from ordinal data to reflect the relative difficulty of test items and linear person measures relating to the ability a person demonstrates on a particular construct, in this case proficiency in organic chemistry in the CK test and TSPCK in the PCK test. In addition to providing important validation data, Rasch analysis enables useful representations of the data. Person abilities and item difficulties can be aligned on the same linear scale in a Wright map using a unit called a logit (log-odds unit), with poor performance and easy items at the bottom of the map and good

TSPCK Component, A1	Students' Prior Knowledge including misconceptions (may be implied)
(1) Limited	• Content-based explanation only (correct answer) • No consideration of learner prior knowledge or misconceptions
(2) Basic	• Identifies misconception or prior knowledge • Provides standardized knowledge as definition
(3) Developing	• Identifies misconception or prior knowledge • Provides standardized knowledge as definition • Expands explanation
(4) Exemplary	• Identifies misconception or prior knowledge • Provides standardized knowledge as definition • Expands explanation • Explanation demonstrates conceptual understanding

Fig. 3 Extract from the rubric used for scoring the TSPCK item A1, probing students' prior knowledge

performance and difficult items at the top. The Wright map allows a qualitative evaluation of a test instrument in terms of the match between the range of item difficulties and the range of performance abilities of the respondents. In the case of the TSPCK test, the ordinal data derived by applying the rubric are converted to linear data. The Rasch analysis of the data for both CK and PCK was carried out using RUMM2030 software (Andrich et al. 2011). The CK data fitted the model well (total item chi-square 48.56, df 46, prob 0.37), and Cronbach's alpha was good, 0.82. However, three items showed differential item functioning (DIF), and two items showed strong response dependence (Q2bii and Q2cii). The PCK data also fitted the model well (total item chi-square 14.02, df 20, prob 0.83), and Cronbach's alpha was good, 0.86, but one item showed DIF, namely, item D1. These observations of DIF and response dependence are significant and will be discussed below. This means that we have succeeded in constructing meaningful interval scales for the measurement of CK and PCK in organic chemistry. A complete account of the Rasch analysis of the teacher data has been reported elsewhere (Davidowitz and Potgieter 2016).

Sample

The instrument was administered to 28 chemistry graduate teaching assistants at a South African university. All participants had at least a 4-year degree in chemistry which includes intensive training in organic chemistry and achieved their status as graduate students by being placed in the top 33% of the undergraduate cohort. At the time of the study, the chemists were engaged in chemistry research on a variety of topics. It is thus to be expected that they would possess good knowledge of chemistry content. They are required to spend 6 h a week tutoring and demonstrating to undergraduate students in laboratory sessions. They are exposed to limited training in the form of a workshop at the beginning of the academic year to introduce them to facilitation of small group work. Predesigned tasks together with comprehensive solutions are provided at weekly meetings held with the course lecturer, and chemists are expected to work through the tasks thereafter. The chemists do not participate in the design or sequencing of the worksheets but assist groups of undergraduate students to master the tasks for a particular course. They also grade written tasks of students according to a memorandum and are expected to provide verbal feedback to the students. The tests were given to the chemists to complete on their own, and the reported duration of the task was 1.5–2 h.

The teacher cohort for this study was very diverse, consisting of 89 teachers from all over South Africa with a wide variety of qualifications ranging from a 3-year teaching diploma to a science degree followed by a 1-year teaching certificate. Typically those with the 3-year diploma would have content knowledge equivalent to a first year university student, while teachers with science degrees would have studied their chemistry to at least senior undergraduate level. While the teachers had different qualifications and levels of experience, what they had in common is their practice, their close interaction with students, their involvement in curriculum planning and their daily enactment of their teaching strategies. The data was collected in three separate studies (Davidowitz and Vokwana 2014; Bam and Mavhunga 2015; Van der Merwe and Rollnick 2015). Evidence in support of treating the sample as one group despite their wide range in chemistry training was derived from the Rasch analysis. The fact that the CK data fitted the model well, but particularly the absence of the so-called misfitting persons and items, indicates that all respondents engaged with the test items in a systematic manner determined only by item difficulty and person ability.

Results

Analysis of the CK Test

A summary of the CK scores for the teachers and chemists is presented in Table 1.

The overall scores for teachers and chemists were relatively similar, the biggest difference being for question 2. Q2a required respondents to draw isomers, while items Q2b and Q2c probed the ability to generate and name two structural isomers for propanoic acid, each isomer having a different functional group. The latter task (Q2b and Q2c) seemed to cause difficulty for teachers. Chemists were far more adept at generating and naming isomers for a given compound. Table 2 presents the performance of the teachers versus the chemists for questions 2 and 4b in the CK test.

Table 1 Summary of CK scores (%) of the two groups

Task	Drawing structures, Q1	Drawing and naming isomers, Q2	Types of reactions, Q3	Functional groups, Q4	Intermol. forces and Bpt, Q5	Mean and standard deviation for the CK test
No. of items	2	6	6	6	3	
Chemists	96.3	91.9	79.2	92.3	66.4	78.0 ± 11.0
Teachers	88.2	50.2	85.2	92.3	66.0	71.0 ± 18.6

Table 2 Performance for chemists vs. teachers for Q2 and Q4b in the CK test

Question	Responses	% correct answers	
		Chemists	Teachers
Q2a	Geometric and structural isomers for 2-butene	82.1	59.5
Q2bi	First isomer for propanoic acid	100.0	87.5
Q2bii	Second isomer for propanoic acid	96.2	18.0
Q2ci	Correct name for isomer Q2bi	96.3	73.9
Q2cii	Correct name for isomer Q2bii	85.1	11.2
Q4b	Identification of aldehyde	75.0	94.4

As can be seen in Table 2, there was a big difference between teachers' and chemists' performance on Q2a which can be attributed to the fact that geometric isomers were not included in the school curriculum. Teachers also had particular difficulty in generating the second structural isomer of propanoic acid (Q2bii); many of them drew two esters instead of two compounds with different functional groups. Consequently the teachers' scores for Q2bii and Q2cii were low. The chemists on the other hand performed very well on item Q2bii with many of them drawing novel structures compared with those drawn by the teachers. They were also successful at naming the compounds they drew. Unexpectedly, the chemists struggled more with recognising aldehydes (Q4b) among the Kekulé representations in question 4 than the teachers, possibly because such representations are rarely used by chemists. These items displayed differential item functioning (DIF) in the Rasch analysis, which means that one group of respondents experienced each of these items as uncharacteristically easy or difficult as compared to their responses on the rest of the items in the test. The RUMM2030 software has a feature whereby the data for these items can be split for the two subsets of respondents, teachers and chemists, before estimating the respective item difficulties. The different locations of the split items in the Wright map give a visual representation of the difference in difficulty of each item as experienced by the two subgroups. For example, the Wright map presented in Fig. 4 shows that there is more than 4-logit difference in the difficulty of item Q2cii as experienced by the teachers (δ 4.7 logits) and the chemists (δ 0.4 logits), indicated by the arrows in Fig. 4. Similar comparisons can be made for the other two sets of split items.

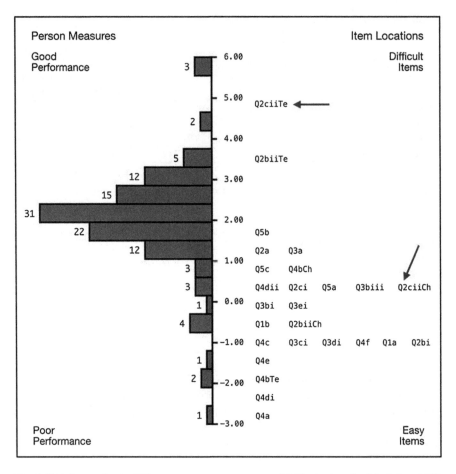

Fig. 4 Wright map for the CK test, mean person measure = 2.00 logits (standard deviation, 1.34). *Ch* chemists, *Te* teachers

Both chemists and teachers found item Q4f, identification of the isomer of ethanol from a list of structures, to be a reasonably easy item which shows that for teachers, there was a much greater conceptual challenge related to generating structural isomers, item Q2b, than recognising and identifying them, item Q4f.

A deeper qualitative examination of the responses to the CK questions provides an interesting insight into the differences between the two groups. Question 5, a moderately difficult item, required respondents to provide an explanation for their answers to a question on physical properties of alcohols. Both groups obtained very similar scores (see Table 1), but many teachers gave a full explanation, while many of the chemists gave compressed answers as shown by the two sample responses to Q5a presented in Figs. 5 and 6.

Teacher 80 provided the following answer as presented in Fig. 5 which scored 100%.

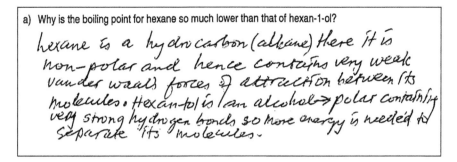

Fig. 5 Teacher 80's response to Q5a on physical properties

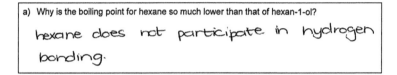

Fig. 6 Chemist 25's response to Q5a on physical properties

This response can be contrasted with the chemist 25's answer in Fig. 6 which scored 50%.

It is clear that chemist 25 understood the concept but failed to make explicit the link between the strength of intermolecular forces and the energy required to vaporise an organic compound which was required to achieve full marks.

Analysis of the TSPCK Test

As mentioned above, the TSPCK test was scored using a rubric assigning scores of 1–4 where 1 is limited, 2 is basic, 3 is developing and 4 is exemplary. A zero was assigned for no response. Although the teachers' average scores were marginally higher, both averages lay about midway between basic and developing. However, the averaging process conceals important differences between the groups. Figure 7 shows the distribution of the scores for selected items for the two groups, with component codes as follows:

A. Students' prior knowledge
B. Curricular saliency
D. Representations
E. Conceptual teaching strategies.

Figure 7 shows a striking difference in performance for item D1 which probed the use of representations in teaching where the performance of the chemists was superior to that of the teachers. Sixty-one percent of chemists performed at the

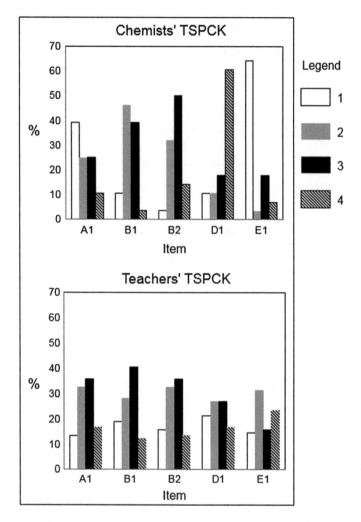

Fig. 7 Comparison of teachers' and chemists' TSPCK for specific items; *1* limited, *2* basic, *3* developing, *4* exemplary

exemplary level in item D1 which required them to identify appropriate situations in which different representations of molecules would be used. Only 17% of teachers were classified as exemplary for this item.

On the other hand, the teachers' scores were higher for item A1, student's prior knowledge, with more than 50% of teachers scoring 3 or 4 for this item, while the majority of chemists scored 1 (39%) or 2 (25%). Teachers' performance in item E1 was also better than the chemists; the majority of teachers scored 2, 3 or 4, while most chemists scored 1 for this item. A possible reason for these differences can be seen in Figs. 8 and 9 which exemplify limited and exemplary responses to item A1 (see Fig. 2) which required respondents to explain why not all compounds containing an OH group are alcohols. As shown in Fig. 8, the chemist provides a response

The carbon attached to the hydroxy group should be fully substituted & only bonded to H/C.

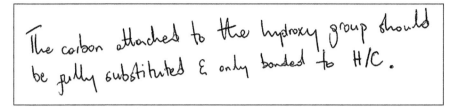

Fig. 8 Chemist 4's response to TSPCK item, A1; coded as limited

Learners should know that it is not enough to identify the alcohols only by their functional group (-OH). They also need to know the general formula. ($C_nH_{2n+1}OH$). This therefore means that an alcohol always has 1 Oxygen atom (except for the di-and triols, which are not necessary for the NCS).

It should be highlighted that an alcohol has only carbon, Hydrogen and Oxygen atom so as to avoid confusion with Hydroxides.

In alcohols, the O-atom is attached to a C- atom and an H-atom (—C-O-H) and no double bond is involved between O and C- atoms (the carbonyl group, -C=O).

Fig. 9 Teacher 1's response to TSPCK item, A1; coded as exemplary

in terms of a definition of the structure of an alcohol, while the response from the teacher (Fig. 9) demonstrates the awareness of students' misconceptions related to compounds containing a –OH group.

Chemists' performance on items probing curricular saliency, namely, the ability to choose prior concepts for teaching organic chemistry, item B1, and choosing big ideas in organic chemistry and sequencing them for teaching, item B2, was better than expected. The performance of the two cohorts on item B1 was similar with 44% of chemists and 54% of teachers being scored as developing or exemplary. The pattern of responses was somewhat different for item B2 where 64% of chemists and 49% of teachers were coded as developing or exemplary.

The Wright map for the TSPCK test is shown in Fig. 10. This map provides a more nuanced picture of the difference in performance in certain items than that derived from the raw data depicted in the charts in Fig. 7. The analysis of the raw data revealed a differential performance for D1, the item probing knowledge of the use of representations. Figure 10 reveals that in fact, this was the easiest item for the chemists, D1Ch, and one of the most difficult for the teachers, D1Te.

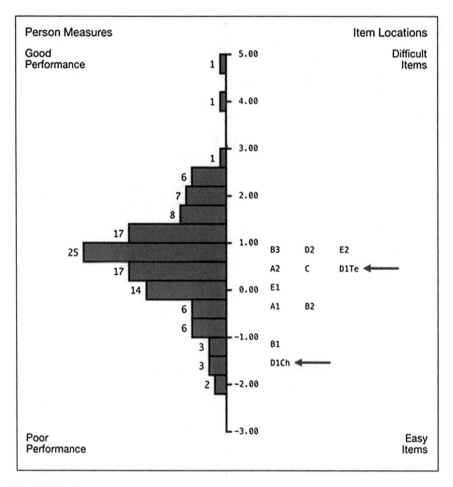

Fig. 10 Wright map for the TSPCK test, mean person measure = 0.70 logits (standard deviation, 1.10). Differential performance for item D1 is indicated by *arrows*

The Relationship Between CK and TSPCK

A scatterplot of CK person measures versus TSPCK measures was constructed to explore the relationship between the CK and TSPCK measures for the two groups of participants (see Fig. 11). CK measures are represented on the x-axis and TSPCK measures on the y-axis. The average person measure for the CK test was 2.00 logits, while the average measure for the TSPCK test was 0.70 logits. We obtained a moderately strong sample correlation of 0.61, r (p < 0.001), between teacher person measures for TSPCK and CK. For chemists the sample correlation was 0.54, r (p < 0.001). One-way analysis of variance (ANOVA) for both the CK and TSPCK measures confirmed that there was no significant difference between the means for the

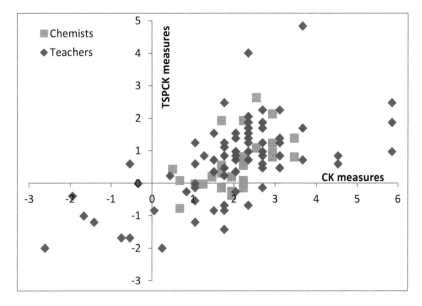

Fig. 11 Scatterplot showing relationship between CK and TSPCK for the two groups

teachers and the chemists, but as can be seen in Fig. 11, the spread for the teachers was larger on both axes.

Discussion

Level of the Chemists' and Teachers' CK of Grade 12 Organic Chemistry

In answering the first part of the research question on the level of content knowledge of the two groups of respondents, it can be seen that the average overall performance is relatively high. As expected the average score of the chemists is slightly higher. The spread of the teachers is far greater with some very low scores from teachers in the rural areas who are inadequately qualified and three teachers who achieved a perfect score. The mean person measure derived from the Rasch analysis, 2.00 (standard deviation, 1.34), reflects the fact that most teachers and chemists in the sample experienced the test as being easy (see Fig. 4). This is to be anticipated from both groups as both are expected to have mastered the content knowledge for use in different contexts.

The Wright map (Fig. 4) illustrates the difference in performance for the teachers and chemists for Q2bii and Q2cii which were easier for the chemists but challenging for the teachers. The chemists were more proficient at applying their knowledge of organic chemistry to produce the required structural isomers and name them. We

noted that the teachers found the generation of structural isomers (Q2b) more difficult than selecting a structural isomer from potential answers (Q4f). The chemists
however were less proficient than the teachers in the use of more basic Kekulé
structures for simple chemical structures. The vertical positions of the split items in
the Wright map (Fig. 4) demonstrate these differences in difficulties for the three
items in question.

The almost identical average score of 66% for the questions on physical properties (Table 1, Q5) conceals an interesting difference between the groups where the
chemists tend to provide more compressed answers than teachers as illustrated in
Figs. 5 and 6 where conceptual understanding can be inferred, but unlike the teacher,
the chemist does not make explicit the link between intermolecular forces and the
energy required for the vaporisation of a substance.

Chemists' and Teachers' Level of TSPCK

While there was no significant difference in the overall performance on the TSPCK
test, there was a difference in the response patterns for the chemists and teachers. The
chemists are involved in a limited amount of teaching where they assist undergraduate students with specific tasks which have been designed and selected by a lecturer.
Thus, it may be expected that they would gain some understanding of students' prior
knowledge in their interactions with students. Figure 7 shows, however, that the
chemists' performance in category A1, students' prior knowledge, is largely limited
or basic, while more than half the teachers are at the developing or exemplary level.
Figures 8 and 9 provide examples of their different level of responses to item A1
where the chemist's response was coded as limited, while the teacher's response was
coded as exemplary. The chemist simply states the facts, while the teacher provides
an explanation of the structure of alcohols which also demonstrates his or her awareness of learners' misconceptions for this topic. Similarly on conceptual teaching
strategies (item E1), the chemists are predominantly at the limited level. While the
teachers' performance is disappointing, there are appreciable numbers who achieve
at the higher levels for this item with almost a quarter being scored as exemplary.

Figure 7 shows that the majority of chemists perform at the exemplary level in
component D1, representations, an ability not shared by the teachers where only a
minority could be classified as exemplary. The exemplary performance on item D1
is not surprising as the chemists are well versed in using the various representations
of molecules for the situations where they would be applicable. Most unexpected
however was the chemists' ability to choose important big ideas in organic chemistry and sequence them for teaching, item B2, given that they have no formal training
for teaching. Big ideas are defined as science ideas that teachers consider essential
in underpinning understanding of a particular topic. While big ideas in teaching
may be the same as big ideas in science, the two are not necessarily equal as the
curriculum impacts how teachers conceptualise the big ideas for teaching the topic
(Loughran et al. 2006: 22). According to Green and Rollnick (2006), organic chem-

istry can be considered to be a linear discipline where "a single basic concept has the potential to hinder or enable access to the whole module" (Green and Rollnick 2006: 1379–1380). It is possible that the chemists benefit from a well-developed oversight of the discipline when responding to items probing curricular saliency and are able to use their knowledge to select the big ideas for teaching organic chemistry and to suggest an appropriate sequence in which they should be taught. While the small sample size limits any claims about transferability of these findings, Linacre (1994) noted that useful exploratory work using Rasch analysis is possible with a small sample.

The Relationship Between CK and TSPCK for Chemists and Teachers

Figure 10 shows that CK and TSPCK are positively correlated in the population of teachers and chemists in this study; the correlation for teachers was 0.61, while that for the chemists was 0.54. The lack of a significant difference in the average measures for teachers and chemists on the TSPCK test leads us to reject our initial hypothesis that chemists who have little in the way of teaching experience are likely to perform at a lower level than teachers on the TSPCK test, though deeper analysis referred to above suggests that the chemists' performance is uneven. We found that for the chemists, in general, an above-average performance in CK was associated with above-average performance in TSPCK and similarly below-average performance in CK with below average in TSPCK. We did not see the expected association between high CK and low TSPCK. The latter identifies the subject matter experts and is where one would have expected non-teachers to lie. With the teachers, the numbers in the low CK/low TSPCK region are a cause for concern counteracted by a pleasing number in the top right-hand region of the graph. Teachers lacking adequate CK cannot be expected to develop TSPCK, but it is concerning to note that there are teachers who, despite their experience, have not acquired the expected expertise in CK. While the purpose of this study was to determine whether our instrument could be used to answer the question posed in the title, the findings reveal that there is a need for pre-service and in-service training for teachers to improve their CK in the topics which they are required to teach. Implementation of such training is evident in other work by our group (e.g. Mavhunga and Rollnick 2013).

Conclusions

The study has yielded some expected and some unanticipated results in attempting to answer the question posed in the title. Shulman's (1986) fundamental claim about PCK was that it is knowledge unique to teachers. Hence the expectation would be that chemists would possess good content knowledge but would be limited in their ability to answer questions assessing PCK. The chemists did obtain high CK scores, higher than the teachers but with less spread. However they showed exemplary ability to identify and describe the use of different representations in the TSPCK test. This could be ascribed to their familiarity with multiple representations of organic molecules, an integral part of their practice. They also performed well in the curricular saliency component, particularly in their ability to select big ideas and sequence them for instruction. Nothing in their exposure to teaching gives them experience with this kind of task; thus, it raises the question whether this ability is purely teaching knowledge or whether their performance can be ascribed to possessing a good conceptual understanding of a topic. They have little or no teacher training; thus their weaker performance on the learner prior knowledge and conceptual teaching strategy components could have been anticipated.

This study of a group of organic chemists acting as graduate teaching assistants has revealed that, contrary to expectations, certain components of teacher knowledge such as curricular saliency and representations appear to develop without formal teacher training. Further research is required to determine whether these findings are unique to the topic of organic chemistry where possession of a good conceptual understanding of the discipline may have provided chemists with the tools for this important knowledge base for teaching.

Notes: This work was supported by a grant from the National Research Foundation (NRF) of South Africa. The findings are those of the author(s) and not of the NRF.

References

Andrich, D., Sheridan, B., & Luo, G. (2011). *RUMM2030 software and manuals*. Perth: University of Western Australia.

Aydin, S., Friedrichsen, P. M., Boz, Y., & Hanuscin, D. L. (2014). Examination of the topic-specific nature of pedagogical content knowledge in teaching electrochemical cells and nuclear reactions. *Chemistry Education Research and Practice, 15*, 658–674.

Bam, N., & Mavhunga, E. (2015). *Measuring the quality of TSPCK of Grade 12 educators in organic chemistry*. Paper presented at the 22nd Annual Conference of the Southern African Association for Research in Mathematics, Science and Technology Education, Maputo.

Bond, T. G., & Fox, C. M. (2007). *Applying the Rasch model: Fundamental measurement in the human sciences* (2nd ed.). Mahwah: Lawrence Erlbaum Associates.

Davidowitz, B., & Potgieter, M. (2016). Use of the Rasch measurement model to explore the relationship between content knowledge and topic-specific pedagogical content knowledge for organic chemistry. *International Journal of Science Education, 38*(9), 1483–1503.

Davidowitz, B., & Vokwana, N. (2014). Developing an instrument to assess Grade 12 teachers' TSPCK in organic chemistry. In H. Venkat, M. Rollnick, M. Askew, & J. Loughran (Eds.), *Exploring mathematics and science teachers' knowledge: Windows into teacher thinking* (pp. 178–194). Oxford: Routledge.

Green, G., & Rollnick, M. (2006). The role of structure of the discipline in improving student understanding: The case of organic chemistry. *Journal of Chemical Education, 83*(9), 1376–1381.

Katz, M. (1996). Teaching organic chemistry via student-directed learning: A technique that promotes independence and responsibility in the student. *Journal of Chemical Education, 73*(5), 440–445.

Kind, V. (2009). Pedagogical content knowledge in science education: Perspectives and potential for progress. *Studies in Science Education, 45*(2), 169–204.

Linacre, J. M. (1994). Sample size and item calibration [or person measure] stability. *Rasch Measurement Transactions*. Available at http://mmm1406.sanjose14-verio.com/rascho/rmt/rmt74m.htm [8 July 2016].

Loughran, J., Mulhall, P., & Berry, A. (2006). *Understanding and developing science teachers' pedagogical content knowledge*. Rotterdam: Sense Publishers.

Mavhunga, E. (2014). Improving PCK and CK in pre-service chemistry teachers. In H. Venkat, M. Rollnick, M. Askew, & J. Loughran (Eds.), *Exploring Mathematics and science teachers' knowledge: Windows into teacher thinking* (pp. 31–48). Oxford: Routledge.

Mavhunga, E., & Rollnick, M. (2013). Improving PCK of chemical equilibrium in pre-service teachers. *African Journal of Research in Mathematics, Science and Technology Education, 17*(1–2), 113–125.

Park, S., Jang, J.-Y., Chen, Y.-C., & Jung, J. (2011). Is pedagogical content knowledge (PCK) necessary for reformed science teaching?: Evidence from an empirical study. *Research in Science Education, 41*, 245–260.

Pitjeng, P. (2014). Novice unqualified graduate science teachers' topic specific pedagogical content knowledge and their beliefs about teaching. In H. Venkat, M. Rollnick, M. Askew, & J. Loughran (Eds.), *Exploring mathematics and science teachers' knowledge: Windows into teacher thinking* (pp. 65–83). Oxford: Routledge.

Rollnick, M., & Mavhunga, E. (2014). PCK of teaching electrochemistry in chemistry teachers: A case in Johannesburg, Gauteng Province, South Africa. *Educacíon Química, 25*(3), 354–362.

Shulman, L. S. (1986). Those who understand: Knowledge growth in teaching. *Educational Researcher, 15*(2), 4–14.

Van Driel, J. H., Verloop, N., & de Vos, W. (1998). Developing science teachers' pedagogical content knowledge. *Journal of Research in Science Teaching, 35*, 673–695.

Van der Merwe, D., & Rollnick, M. (2015). *Measuring teacher pedagogical content knowledge in organic chemistry*. Paper presented at the 22nd Annual Conference of the Southern African Association for Research in Mathematics, Science and Technology Education, Maputo.

Veal W. R., & MaKinster J. G. (1999). Pedagogical content knowledge taxonomies. *Electronic Journal of Science Education, 3*. Available at: http://wolfweb.unr.edu/homepage/crowther/ejse/vealmak.html

Pedagogical Content Knowledge and Constructivist Teaching: A Hot Potato's Last Straw in Swiss Science Classrooms

Alexander F. Koch

Background

Generally speaking, teaching development projects and educational research projects aim to support teachers in competence development, improve instruction, and support student learning. In order to engage with teacher competences, many researchers build up on Shulman's (1987) notion of essential teacher knowledge dimensions, e.g., content knowledge, pedagogical knowledge, and pedagogical content knowledge. Based on this triad, research into teacher competences, in particular with regard to pedagogical content knowledge (PCK) and content knowledge (CK) and their effect on teaching practice, is plentiful in educational science.

In theory, one could assume that content-related knowledge, such as CK and PCK, has a direct effect on teaching. For example, substantial research in German-speaking Europe has shown that content knowledge can be a predictor for meaningful adaptation of and activating teaching practice (e.g., Lipowsky 2006). Kunter et al. (2013) are in line with this notion and further specify that PCK in particular can predict cognitively activating and supportive teaching practice. From these perspectives, knowledge seems to play an important role when it comes to student-oriented teaching.

Additionally, other research draws upon the importance of attitudes and beliefs in relation to teaching practice, from which one can conclude that teacher beliefs depict additional relevant variables in terms of teaching practice (Blömeke et al. 2009; Dubberke et al. 2008; Hartinger et al. 2006; Stipek et al. 2001). From a wider perspective, one can then start to envisage how these two praxis-relevant issues (PCK and teacher beliefs) relate with each other. Van Merrienboer and Paas (2003: 8) assert that "effective performance relies on an integration of skills, knowledge

A.F. Koch (✉)
University of Applied Sciences and Arts Northwestern Switzerland, Basel, Switzerland
e-mail: alexander.koch@fhnw.ch

© Springer International Publishing AG 2017   87
K. Hahl et al. (eds.), *Cognitive and Affective Aspects in Science Education Research*, Contributions from Science Education Research 3,
DOI 10.1007/978-3-319-58685-4_7

and attitudes […]." In summary, teaching practice does seem to depend on not only teachers' knowledge but also their attitude toward implementing a particular way of teaching. But how strong is the relation between knowledge and teaching practice? In this study, we relate this question to constructivist teaching and analyze data with reference to a moderate constructivist teaching approach.

Research Questions and Theory

In this study, we want to address the question of the predictive power of PCK on constructivist teaching habits in science classrooms. With this we hope to add to the notion of the "PCK effect" in science teaching. Currently, researchers focus their studies on the relevance of pedagogical content knowledge in teacher professional development (TPD) and teaching practice or teaching behavior (e.g., Kunter et al. 2013; Van Driel and Berry 2012). In this field, they debate on whether there is a direct or indirect link between teacher knowledge and teaching practice.

Kunter et al. (2013) draw upon these issues in their analyses of the Professional Competence of Teachers, Cognitively Activating Instruction, and the Development of Students' Mathematical Literacy (COACTIV) project data and have found that PCK can be predictive for cognitive activating instruction and learning support in mathematics teachers. Yet, teacher instruction was measured and based upon a student-based assessment which means the students were asked about their cognitive activation and on other variables that affected them individually. On critical reflection, one can no doubt understand how students could be utilized to obtain reliable data on teaching quality issues (e.g., Fauth et al. 2014; Wagner et al. 2013). However in terms of teacher's pedagogical intentions, one would imagine that teachers themselves are able to provide more meaningful data in regard to their intentions and strategies in class. So why not ask the teacher about their intended instruction?

Blömeke et al. (2009) and Fishbein and Ajzen (2010) support the notion that variables such as attitudes, beliefs, and skills can be used to predict behavior. Fishbein and Ajzen (2010) also propose to evaluate general practice with self-assessment measures. The authors claim that self-description – although defective – should be a good proxy to estimate general and habitual behavior, especially when huge data would need to be collected. This is particularly true for the case of long-term teaching observations. If one wanted to describe general teaching, one would have to assess every lesson of every teacher during a school year. Thus, it may be more efficient and cost-effective to rely upon self-description. Yet, if a self-description is to be a reliable measure of a general teaching approach, what can then be the predictor of teaching practice?

With respect to science teaching strategies, Van Driel et al. (1998) state that – when it comes to teaching practice – teachers, albeit being on similar PCK levels, differ in their instructional strategies. One could argue that these differences in the translation of knowledge into practice are rather related to content representations in individual teachers. This led Van Driel et al. (1998) to the conclusion that PCK should be investigated on a more general level, namely, "on the conceptions about the teaching and learning of science" (Van Driel et al. 1998: 681).

Therefore one can conclude that teacher attitudes are relevant when it comes to teaching and also that PCK has an impact on teaching practice. In order to accurately study PCK in relation to classroom practice, we approach PCK on a general level and synchronize its measurement with constructivist teaching processes. This means that PCK items will be paralleled with constructivist teaching items. Our hypothesis is that what we now call constructivist pedagogical content knowledge (c-PCK) is a predictor of constructivist teaching or one of its subdimensions. Our underlying conception of constructivism follows the idea of a methodological theory for instruction that is based on adaptive and situation-specific student-oriented teaching. The teacher's main goal is then to foster knowledge construction in students. In this conception, we follow the ideas of Piaget (1970) and Vygotsky (1978) and build our definition of constructivism on the basis of cognitive apprenticeship (see also Davis and Samura 2002). Our operationalization of cognitive apprenticeship relies upon the categories defined by Muijs and Reynolds (2011), e.g., modeling, scaffolding, articulation, exploration, etc. Here, for example, for a person that says he/she generally teaches according to the individual student's needs and adapts tasks to the individual's level of expertise, we define that individual as a constructivism-oriented teacher. In a preliminary study, we used ten items in a questionnaire to assess teachers' constructivist orientation (see Koch 2013) and found three subdimensions of constructivist teaching (see section Preliminary Work in this chapter).

Method

The Swiss Science Education project (SWiSE) is focused on competence development in in-service teacher education and wants to foster activating, inquiry-based teaching practice. One hundred fifty-nine experienced science teachers from kindergarten, primary school, and lower secondary school were evaluated with multiple measures regularly for 3 years. In this context, the nature and relevance of pedagogical content knowledge (PCK) in everyday teaching practice was investigated. The evaluation began at the beginning of the school year 2012/2013 and ended at the end of the school year 2014/2015. Teacher knowledge was measured pre and post, i.e., in November 2012 and May 2015. Due to time-intensive nature of coding of this knowledge test, this study here will only address the pre-measures in the SWiSE project evaluation.

Measures

Heinzer and Oser (2013) suggest that action-relevant knowledge can be assessed indirectly via a person's asserted interpretation of a situation that another person finds himself or herself in. In other words, an observer's comment on a situation is a projection of that person's own behavior if they were in a similar situation. The authors call this principle "advocatory," i.e., whenever someone comments on somebody else's behavior, these comments can be interpreted as an indication of the commenting person's alternative behavior. Based on this assumption, c-PCK was assessed in an online-based vignette test that was developed and validated by colleagues at the University of Teacher Education in Lucerne, Switzerland (Bölsterli et al. 2011; Brovelli et al. 2013, 2014).

The vignettes describe extracted but representative classroom situations where teachers and students interact. As is described in detail in Bölsterli et al. (2011), the vignettes were derived from video-based material and utilized expert decisions on the quality and content of teaching with reference to constructivist teaching. Each vignette contained three to eight "hidden" problems related to constructivist teaching practice. Each hidden test item or "problem" could be interpreted as being deficient or lacking in constructivist-oriented teaching. Teachers were asked to comment on these vignettes/situations in an open question format. A detailed example of a vignette can be found in Koch and Labudde (2014).

The prompt to answer a vignette was chosen to be as open as possible and designed so as to not trigger any bias, i.e., to allude that a "problem" had in fact occurred. Therefore the prompt just read: "Please, comment on this situation." Teachers then commented on all vignettes one after the other, and their answers were rated by two trained and independent coders. The coding was guided by a manual and had an intermediate inferential rating, using the codes 0 = problem not recognized, 1 = problem recognized, and 2 = problem solved. These codes were based on clear-cut examples from the preliminary validation study, but "twilight zones" could not be eliminated with the manual. For example, when the teacher in the vignette started the lesson asking for the students' preconceptions of a phenomenon but did not work with these further, the manual considers this a "problem" and assumes a constructivist-oriented and competent teacher would in fact use these preconceptions by trying to incorporate them meaningfully into the lesson. If a tested teacher comments in a way such as "Just asking is perfectly fine," the score derived would be zero; if the comment was "The teacher should have picked up on the preconception again," the score would be 1. A score of 2 would be given if the above comment was given but also included a qualitatively useful suggestion, such as "Later in the situation the teacher could have picked up the preconception, because...." About 20% of all items in the test were double-coded by both coding persons. Inter-rater reliabilities (Krippendorff's alpha) were acceptable at $K-\alpha = [0.65, 0.72]$. The final ratings were aggregated to individual sum scores and used for further analyses as can be seen in the results section below.

With regard to constructivist teaching, we relied upon Fishbein and Ajzen (2010) and operationalized general constructivist teaching behavior in self-assessment as a proxy for habitual actions in class. For example, a teacher might usually ask for students' preconceptions and build up on them, but he or she would presumably not do that for every topic, on every day and in every lesson.

Preliminary Work

With reference to the knowledge test used in this study, recent analyses by Wilhelm et al. (2015) have shown that PCK/c-PCK can be divided into three subdimensions of science teaching competences: (a) methodological teaching competence (mc), (b) diagnostic competence (dc), and (c) adaptive teaching competence (ac). This allows us to analyze their individual contribution to any outcome variable as, for example, constructivist teaching. In this paper, we will draw upon the scale of constructivist teaching that was first introduced by Koch (2013).

The criterion variable in this study is named *constructivist teaching*. In the preliminary analysis, Koch (2013) used exploratory, varimax-rotated principal component analyses. Koch (2013) showed that habitual constructivist teaching consists of three student-oriented dimensions: (a) initiation of learning processes (CILP), (b) orientation toward student understanding (COSU), and (c) orientation toward student knowledge increase (CSKI). In general, these dimensions are consistent and in line with several developed criteria for cognitive apprenticeship (see, e.g., Muijs and Reynolds 2011 or Davis and Samura 2002). Koch (2013) also showed that these subdimensions were related to a teacher's attitude toward implementing inquiry-based learning (IBL) and the preceding methods used by the teacher to access new topics or learning fields (methods to access new topics, MNT), for example, "I try to recruit adequate contacts to find new places for learning activities."

Table 1 lists all variables and their abbreviations used in this work. The variables are categorized as either being criterion or predictor variables in nature.

Sample

In the pre-evaluation of the SWiSE project, 159 teachers (57% male) in primary and lower secondary schools in German-speaking Switzerland were assessed online with respect to c-PCK through a vignette test and filled in a questionnaire. Inter-rater reliability coefficients for the test were acceptable (Krippendorff's alpha = [0.59, 0.72]) (Krippendorff 2004); Cronbach's alphas of the self-assessment scales on teaching habits were acceptable: CLIP $\alpha = 0.70$, COSU $\alpha = 0.73$, CSKI $\alpha = 0.64$, IBL $\alpha = 0.70$, MNT $\alpha = 0.71$ (see Koch 2013).

Table 1 List of all variables used in this text

	Variables	Abbreviation
Criterion variables	Constructivist teaching	CT
	CT subdimension initiation of learning processes	CLIP
	CT subdimension orientation toward student understanding	COSU
	CT subdimension orientation toward student knowledge increase	CSKI
Predictor variables	Constructivist pedagogical content knowledge	c-PCK
	c-PCK subdimension methodological teaching competence	mc
	c-PCK subdimension diagnostic competence	dc
	c-PCK subdimension adaptive teaching competence	ac
	Attitude toward implementing inquiry-based learning	IBL
	Methods to access new topics	MNT

Procedure of Analyses

MPlus 7.11 software was used for the statistical analyses. All measures are parametric, i.e., all correlations are Pearson correlations. Preliminary analyses had shown that c-PCK as well as constructivist teaching (CT) could be seen as general constructs or as constructs with subdimensions. Therefore, we coded structural equation models in a stepwise procedure from firstly a general model to then more elaborated models. In the elaborated models, we used the subdimensions of the general constructs.

In Table 2 all predictor and criterion variables used in this study are shown. Due to the assumption that the predictors are positively correlated with the criterion variables, one-tailed Pearson correlations are computed. In Table 2 the general constructs c-PCK and CT are significantly correlated. With reference to their subdimensions, one can also see that this correlation can be attributed to correlations between the variables *diagnostic/adaptive teaching competence* and *orientation toward student knowledge increase*.

Also, the variables *attitude toward implementing inquiry-based learning* and *methods to access new topics* significantly correlate with teaching activities. Therefore, these variables were added as predictors in each model, in order to control for their effects.

In the first "general" models, we had overall *c-PCK* regress on overall *constructivist teaching (CT)* and on the subdimensions of constructivist teaching (model 1a and 1b). Then, we split *c-PCK* into its subdimensions and ran a second model (model 2). Last, we also divided constructivist teaching into its three subdimensions as described above (model 3). *IBL* and *MNT* were predictors in all three models simultaneously. In the final model, a linear regression model was specified in order to further investigate the individual variance as explained by the predictor variable

Table 2 Predictor/criterion variables and one-tailed Pearson intercorrelations between predictor and criterion variables

		Criterion variables			
		CT	CLIP	COSU	CSKI
Predictors	c-PCK	0.21**	0.11	0.15*	0.28*
	mc	0.13	0.08	0.10	0.16*
	dc	0.11	0.05	0.02	0.22**
	ac	0.20*	0.10	0.16*	0.25**
	IBL	0.57**	0.44**	0.54**	0.39**
	MNT	0.52**	0.45**	0.40**	0.37**

Notes: Please see Table 1 for clarification of abbreviations; * = $p < 0.05$; ** = $p < 0.01$

diagnostic competence. In each model, we first specified all paths and correlations and stepwise eliminated nonsignificant correlations.

All models were evaluated according to their fit statistics (chi-square, CFI, RMSEA). A good model fit is interpreted as so when the chi-square statistics were measured to be insignificant ($p > 0.05$), CFI > 0.95, and RMSEA < 0.05. We chose to use single-level models, as we had not assumed a shared/clustered way of teaching in schools, nor had we clustered the data for knowledge/competence. In all models, we have shown intercorrelations among predictor variables and intercorrelations among criterion variables as two-way arrows, between the relevant predictor and criterion variables with one-way arrows.

Results

Model 1a

The results from the first linear regression model showed that general *c-PCK* could not predict *constructivist teaching* (model 1a, Fig. 1). Only *IBL* and *MNT* had a significant influence on *CT*. Also IBL and MNT were highly and significantly intercorrelated. One should also mention that *c-PCK* was not significantly related to the other predictors, but the correlation between *c-PCK* and *IBL* was $r = 0.16$, $p = 0.063$.

Model 1b

As model 1a did not reveal a significant relation between constructivist pedagogical content knowledge and constructivist teaching, we split the criterion variable into its three subdimensions (CSKI, CLIP, and COSU). In model 1b (Fig. 2), c-PCK was shown to be now significantly correlated with IBL ($r = 0.16$, $p < 0.05$). IBL and MNT still significantly predicted each *subdimension of constructivist teaching*. Yet,

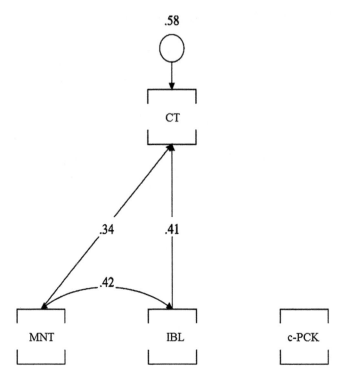

Fig. 1 Model 1a – c-PCK does not predict constructivist teaching (only significant correlations and paths shown, p < 0.05)

in this model, we found a significant path from *c-PCK* to the teaching subdimension *orientation toward student knowledge increase.*

Model 2

We then further subdivided c-PCK in model 2 (Fig. 3), which fitted the data very well ($X^2(17) = 19.534$, p = 0.299, RMSEA = 0.03, CFI = 0.99). It revealed that the c-PCK subdimension *diagnostic competence (dc)* predicted *orientation toward student knowledge increase (CSKI)*, but not any of the other criterion variables. Yet, *dc* seemed to be as good a predictor for teaching as *IBL* and *MNT*, because the estimates did not differ largely (*dc* = 0.22, *IBL* = 0.29, *MNT* = 0.24). These three variables explained 25% of the variance of *CSKI*.

Yet, *dc* was also significantly correlated with *ac*, and *ac* was significantly correlated with *mc*. *IBL* and *MNT* were as before also correlated (r = 0.41) and in this case significant predictors for *CSKI* but also for *CLIP* and *COSU*.

Surprisingly, adaptive and methodological teaching competences were not directly related with general teaching practice but seemed to inform each other, as

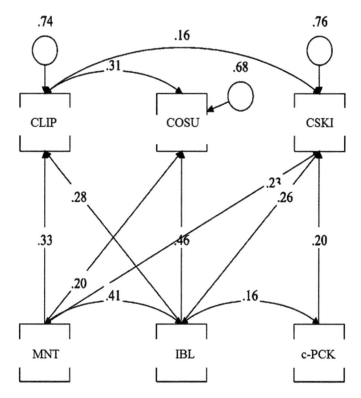

Fig. 2 Model 1b – c-PCK predicts subdimensions of constructivist teaching (only significant correlations and paths shown, p < 0.05)

well as the diagnostic competence, which then translated into teaching practices intended to increase students' knowledge.

Model 3

In order to further investigate the relevance of *diagnostic competence*, we specified an isolated model for *diagnostic competence* as a predictor of *teaching for students' knowledge increase*. In this model 3 (Fig. 4), one can see that all estimators were rather stable and that the three predictors could explain 25% of the outcome variable. The model fit was $X^2(2) = 0.067$, $p = 0.967$, RMSEA = 0.00, CFI = 1.00.

Finally we wanted to investigate the variance explained by *dc* on its own. Therefore, we specified a simple linear regression path between *diagnostic competence (dc)* and teaching for *students' knowledge increase (CSKI)* and found that *dc* could predict 5% of unique variance in *CSKI*. That meant that a detailed analysis on these variables showed that diagnosing a learning need does result in teacher activities that are related to *students' knowledge increase*.

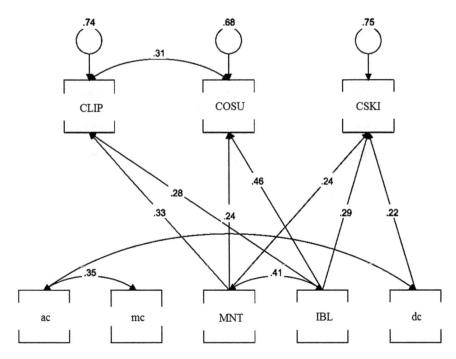

Fig. 3 Model 2 – Subdimensions of c-PCK predict subdimensions of constructivist teaching (only significant correlations and paths shown, p < 0.05). Model fit: X2(17) = 19.534, p = 0.299, RMSEA = 0.03, CFI = 0.99

Discussion

In general, the analyses conducted have shown that constructivist pedagogical content knowledge as a general factor is not related to constructivist classroom instruction. Yet, when investigating in more detail by separating both instruction and pedagogical content knowledge into subdimensions, one can find that teachers' knowledge is transformed to behavior that in fact intends to increase students' knowledge (Fig. 4). Although *students' knowledge increase* differs from other goals like *initiating learning* or *understanding,* our results suggest that teaching (with respect to competences) aims to increase teachers' attempts to increase students' learning behavior, their understanding, and their knowledge. Thus, teachers' knowledge is crucial for the intention to foster learning and understanding in students. Similar results, where teachers' PCK influenced the cognitive activation in students, were found in physics (Ergönec et al. 2014), in mathematics (Kunter et al. 2013), and in technology education (Jones and Moreland 2003) classrooms. Yet, most of these studies relied upon a more general notion of PCK.

In this study, our results also indicate the advantage of a more differentiated assessment of PCK, especially if one is interested in evaluating the relevance of knowledge on teaching processes. This study was able to replicate what COACTIV

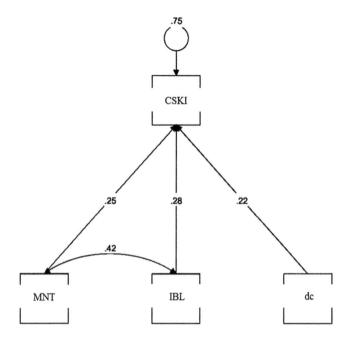

Fig. 4 Model 3 – Diagnostic competence as a predictor of teaching to increase students' knowledge (only significant correlations and paths shown, p < 0.05). Model fit: X2(2) = 0.067, p = 0.967, RMSEA = 0.00, CFI = 1.00

(Dubberke et al. 2008) found in mathematics classes and furthermore could transfer the understanding of the knowledge-action relationship to Swiss science classrooms. As the scale *orientation toward student knowledge increase* mostly assesses teaching procedures that are in line with cognitive activating teaching, one could suggest that a teacher first has to diagnose a student's learning activation before he/she initiates any teaching processes. As the intention of teachers is influenced by a learning goal, these results are in line with the idea that teachers usually want to teach something and follow an educational aim (see Eun et al. 2008). Olszewski (2010) found that physics teachers use their PCK "for designing adequate tasks during a lesson and are thereby able to enhance students' learning" (Olszewski 2010: 98). This, in our view, would make diagnostic competence a mediator between "basic" competences (here methodology and adaptivity) and *student-oriented teaching*. Also, the dependency of *recruiting new learning possibilities (MNT)* and the *attitude toward the implementation of inquiry learning (IBL)* seems plausible; the greater the repertoire of learning possibilities, the more one sees opportunities to do inquiry-based learning activities and vice versa. As *MNT* and *IBL* predict all constructivist teaching activities – the initiation of learning processes, the orientation toward student understanding, and the orientation toward student knowledge increase – we believe that these results and others found within the SWiSE project can be used to meet the aim at fostering activating and inquiry-based teaching and can have a positive influence on teachers' ways of teaching.

With reference to what we summarized in the introduction, this study shows that teacher knowledge is only partly relevant for student-oriented, constructivist teaching. Due to economic reasons, we did not assess in this study the highly relevant content knowledge. We believe that this type of knowledge is particularly relevant in higher educational levels. But in lower-level schools, what may become more important is in fact pedagogy. One could say that the older the students, the more important the content becomes. This is what Lipowsky (2006) finds. However in lower-level schools, one could also assume that the drive for learning does not initiate from content, but rather from motivation, which is the main driving force for further learning (Deci and Ryan 1990).

Unfortunately, a limitation of this study was the lack of German-speaking or even Swiss-based studies to compare our results. Even though one could argue that Europe shares some general attributes with respect to teacher practice, studies have shown significant differences in teaching and teachers' attitudes toward teaching. Leuchter et al. (2008) found contradicting results. In their study, constructivist teaching was not related (zero correlation) to student understanding in Germany. In contrast, in Switzerland a statistically significant and negative association between student understanding and constructivist instruction was revealed in the same study. That suggests that the more "constructivist" the teachers were, the less they cared about student understanding.

Our study presented here supports the notion that a teacher who can diagnose the needs of an individual student can also meaningfully utilize material to help this student develop his/her knowledge. This interpretation is in line with international research on the relevance of diagnostic teacher competence. Klug et al. (2013) assert that diagnostic competence is crucial, especially in student-centered or individualized teaching. Yet, in these interactions, teachers are not only supposed to diagnose student achievement, but they also have to assess the student learning practices and progress in order to adapt the instruction. Edelenbos and Kubanek-German (2004) found that goal- and task-oriented assessment of students accounts for 24% of a teacher's diagnostic activity, while another 35% relates to asking content-related questions. This makes diagnostic behavior a part of pedagogical content knowledge and a relevant competence in teacher professional development (see Busch et al. 2015). From this perspective, the results presented in this study positively support the view of the significance of supporting the development of diagnostic abilities during teacher education or in further education programs, in order to facilitate individualized instruction and enhancing student learning outcomes.

References

Blömeke, S., Felbrich, A., & Müller, C. (2009). Future teachers' beliefs on the nature of mathematics. In F. Achtenhagen, F. K. Oser, & U. Renold (Eds.), *Teachers' professional development: Aims, modules, evaluation* (pp. 25–46). Rotterdam: Sense Publishers.

Bölsterli, K., Brovelli, D., Rehm, M., & Wilhelm, M. (2011). Vignettentest zur Erhebung professioneller Kompetenz [A vignette test for professional competence assessment]. In D. Höttecke (Ed.), *Naturwissenschaftliche Bildung als Beitrag zur Gestaltung partizipativer Demokratie. Gesellschaft für Didaktik der Chemie und Physik. Jahrestagung in Potsdam 2010* (pp. 285–287). LIT Verlag: Berlin.

Brovelli, D., Bölsterli, K., Rehm, M., & Wilhelm, M. (2013). Erfassen professioneller Kompetenzen für den naturwissenschaftlichen Unterricht – ein Vignettentest mit authentisch komplexen Unterrichtssituationen und offenem Antwortformat [Assessing professional competencies for science teaching: A vignette test using authentically complex teaching contexts and an open-ended answer format]. *Unterrichtswissenschaft, 41*(4), 306–329. doi:09201304306.

Brovelli, D., Bölsterli, K., Rehm, M., & Wilhelm, M. (2014). Using vignette testing to measure student science teachers' professional competencies. *American Journal of Educational Research, 2*(7), 555–558. doi:10.12691/education-2-7-20.

Busch, J., Barzel, B., & Leuders, T. (2015). Promoting secondary teachers' diagnostic competence with respect to functions: Development of a scalable unit in Continuous Professional Development. *ZDM Mathematics Education, 47*, 53–64. doi:10.1007/s11858-014-0647-2.

Davis, B., & Samura, D. (2002). Constructivist discourses and the field of education: Problems and possibilities. *Educational Theory, 52*, 409–428. doi:10.1111/j.1741-5446.2002.00409.x.

Deci, E. L., & Ryan, R. M. (1990). *Intrinsic motivation and self-determination in human behavior* (3rd printing ed.). New York: Plenum Press.

Dubberke, T., Kunter, M., McElvany, N., Brunner, M., & Baumert, J. (2008). Lerntheoretische Überzeugungen von Mathematiklehrkräften. Einflüsse auf die Unterrichtsgestaltung und den Lernerfolg von Schülerinnen und Schülern [Mathematics teachers' beliefs and their impact on instructional quality and student achievement]. *Zeitschrift für Pädagogische Psychologie, 22*(2–3), 193–206. doi:10.1024/1010-0652.22.34.193.

Edelenbos, P., & Kubanek-German, A. (2004). Teacher assessment: The concept of 'diagnostic competence'. *Language Testing, 21*(3), 259–283.

Ergönec, J., Neumann, K., & Fischer, H. E. (2014). The impact of pedagogical content knowledge on cognitive activation and student learning. In H. E. Fischer, P. Labudde, K. Neumann, & J. Viiri (Hrsg.), Quality of instruction in physics: Comparing Finland, Germany and Switzerland (pp. 145–159). Münster: Waxmann.

Eun, B., Knotek, S. E., & Heining-Boynton, A. L. (2008). Reconceptualizing the zone of proximal development: The importance of the third voice. *Educational Psychology Review, 20*(2), 133–147.

Fauth, B., Decristan, J., Rieser, S., Klieme, E., & Büttner, G. (2014). Student ratings of teaching quality in primary school: Dimensions and prediction of student outcomes. *Learning and Instruction, 29*, 1–9. doi:10.1016/j.learninstruc.2013.07.001.

Fishbein, M., & Ajzen, I. (2010). *Predicting and changing behavior: The reasoned action approach*. New York: Psychology Press.

Hartinger, A., Kleickmann, T., & Hawelka, B. (2006). Der Einfluss von Lehrervorstellungen zum Lernen und Lehren auf die Gestaltung des Unterrichts und auf motivationale Schülervariablen [The influence of teachers' conceptions of teaching and learning on the design of lessons and on motivational pupil variables]. *Zeitschrift für Erziehungswissenschaft, 9*(1), 110–126. doi:10.1007/s11618-006-0008-1.

Heinzer, S., & Oser, F. (2013). Das Advokatorische Messverfahren: Die stellvertretende Art Kompetenzen zu messen [The advocatory method for assessment]. In F. Oser, T. Bauder, P. Salzmann, & S. Heinzer (Hrsg.), *Ohne Kompetenz jeine Qualität. Entwickeln und Einschätzen von Kompetenzprofilen bei Lehrpersonen und Berufsbildungsverantwortlichen.* (pp. 139–168). Bad Heilbrunn: Klinkhardt.

Jones, A., & Moreland, J. (2003). Considering pedagogical content knowledge in the context of research on teaching: An example from technology. *Waikato Journal of Education, 9*, 77–89.

Klug, J., Bruder, S., Kelava, A., Spiel, C., & Schmitz, B. (2013). Diagnostic competence of teachers: A process model that accounts for diagnosing learning behavior tested by means of a case scenario. *Teaching and Teacher Education, 30*, 38–46. doi:10.1016/j.tate.2012.09.004.

Koch, A. F. (2013). *Wie Lehrpersonen Konstruktivismus interpretieren* [How teachers interpret constructivism]. Paper presented at the Forschungstag der Pädagogischen Hochschule Fachhochschule Nordwestschweiz, Basel.

Koch, A. F., & Labudde, P. (2014). Strike the iron when it's hot – Teacher professionalization in Swiss science education. In C. P. Constantinou, N. Papadouris, & A. Hadjigeorgiou (Eds.), *E-Book Proceedings of the ESERA 2013 Conference: Science Education Research For Evidence-Based Teaching and Coherence in Learning* (11 p.). Nicosia: European Science Education Research Association.

Krippendorff, K. (2004). Reliability in content analysis: Some common misconceptions and recommendations. *Human Communication Research, 30*(3), 411–433. doi:10.1111/j.1468-2958.2004. tb00738.x.

Kunter, M., Klusmann, U., Baumert, J., Richter, D., Voss, T., & Hachfeld, A. (2013). Professional competence of teachers: Effects on instructional quality and student development. *Journal of Educational Psychology, 105*(3), 805–820. doi:10.1037/a0032583.

Leuchter, M., Reusser, K., Pauli, C., & Klieme, E. (2008). Zusammenhänge zwischen unterrichtsbezogenen Kognitionen und Handlungen von Lehrpersonen [Correlations between teaching beliefs and action in teachers]. In M. Gläser-Zikuda & J. Seifried (Hrsg.), *Lehrerexpertise: Analyse und Bedeutung unterrichtlichen Handelns* (pp. 165–185). Münster: Waxmann.

Lipowsky, F. (2006). Auf den Lehrer kommt es an: Empirische Evidenzen für Zusammenhänge zwischen Lehrerkompetenzen, Lehrhandeln und dem Lernen der Schüler [It depends on the teacher: Empirical evidence for correlations between teacher competences, teaching, and student learning]. In C. Allemann-Ghionda & E. Terhart (Eds.), *Kompetenzen und Kompetenzentwicklung von Lehrerinnen und Lehrern* (pp. 47–70). Weinheim: Beltz.

Muijs, D., & Reynolds, D. (2011). *Effective teaching: Evidence and practice* (2nd ed.). London: Sage.

Olszewski, J. (2010). *The impact of physics Teachers' pedagogical content knowledge on teacher actions and student outcomes*. Berlin: Logos.

Piaget, J. (1970). *Science of education and the psychology of the child*. New York: Orion Press.

Shulman, L. S. (1987). Knowledge and teaching: Foundations of the new reform. *Harvard Educational Review*, 1–21. doi:10.17763/haer.57.1.j463w79r56455411.

Stipek, D., Givvin, K., Salmon, J., & MacGyvers, V. (2001). Teachers' beliefs and practices related to mathematics instruction. *Teacher and Teacher Education, 17*, 213–226. doi:10.1016/ S0742-051X(00)00052-4.

Van Driel, J., & Berry, A. (2012). Teacher professional development focusing on pedagogical content knowledge. *Educational Researcher, 41*(1), 26–38. doi:10.3102/0013189X11431010.

Van Driel, J., Verloop, N., & de Vos, W. (1998). Developing science teachers' pedagogical content knowledge. *Journal of Research in Science Teaching, 35*(6), 673–695. doi:10.1002/ (SICI)1098-2736(199808)35:6<673::AID-TEA5>3.0.CO;2-J.

Van Merrienboer, J. J. G., & Paas, F. (2003). Powerful learning and the many faces of instructional design: Toward a framework for the design of powerful learning environments. In E. De Corte, L. Verschaffel, N. Entwistle, & J. J. G. Van Merriënboer (Eds.), *Powerful learning environments: Unravelling basic components and dimensions* (pp. 3–20). Amsterdam: Erli-Pergamon.

Vygotsky, L. (1978). *Mind in society: The development of higher psychological processes*. Cambridge, MA: Harvard University Press.

Wagner, W., Göllner, R., Helmke, A., Trautwein, U., & Lüdtke, O. (2013). Construct validity of student perceptions of instructional quality is high, but not perfect: Dimensionality and generalizability of domain-independent assessments. *Learning and Instruction, 28*, 1–11. doi:10.1016/j.learninstruc.2013.03.003.

Wilhelm, M., Brovelli, D., & Bölsterli, K. (2015). *Naturwissenschaften unterrichten können – empirische Validierung eines Modells professioneller Kompetenzen für den naturwissenschaftlichen Unterricht* [Being able to teach natural sciences – Empirical validation of model of professional competences in science education]. Paper presented at the Tagung Fachdidaktiken 2015, Bern.

Part II
Student Engagement

Quality of SSI Scenarios Designed by Science Teachers

Kari Sormunen, Anu Hartikainen-Ahia, and Ilpo Jäppinen

Introduction

Promoting students' interest and motivation towards science learning has been a focus of various recent studies (e.g. Abrahams 2009; Krapp and Prenzel 2011; cf. European Commission 2015). Students seem to lack the interest and motivation towards academic science studies, and hence the challenge is to improve cognitive and affective aspects of science instruction. A review about science education research (Potvin and Hasni 2014) concluded that real-life issues are beneficial for triggering students' interest and motivation. Thus one way to enhance students' interest and motivation towards science is to design teaching modules that have a connection to real life. The socio-scientific issue (SSI) approach emphasises societally significant science issues in school lessons (Aikenhead 2005; Sadler 2011; Potvin and Hasni 2014). Socio-scientific issues are controversial social issues with conceptual and/or procedural links to science; they are open-ended problems, which tend to have multiple plausible solutions (Sadler 2011).

Zeidler et al. (2005: 360) describe the focus of the SSI movement being "on empowering students to consider how science-based issues reflect, in part, moral principles and elements of virtue that encompass their own lives, as well as the physical and social world around them". The authors identify four central aspects in the teaching of SSI: nature of science issues, classroom discourse issues, cultural issues, and case-based issues. These issues act as entry points in the science curriculum, which can contribute to a student's personal intellectual development and in turn help to inform pedagogy in science education to promote so-called functional scientific literacy (Zeidler et al. 2005).

K. Sormunen (✉) • A. Hartikainen-Ahia • I. Jäppinen
University of Eastern Finland, Joensuu, Finland
e-mail: kari.sormunen@uef.fi

© Springer International Publishing AG 2017   103
K. Hahl et al. (eds.), *Cognitive and Affective Aspects in Science Education Research*, Contributions from Science Education Research 3,
DOI 10.1007/978-3-319-58685-4_8

Earlier studies suggest that the SSI approach enhances students' scientific literacy, reasoning, and decision-making (Hodson 2003; Zohar and Nemet 2002; Lewis and Leach 2006) in the context of real-life issues. Connecting science learning to everyday life from a societal viewpoint to a local or global problem triggers students' interest (Sadler 2004). Allowing students to generate problems and questions, as in the SSI approach, has a positive impact on motivation (Swarat et al. 2012). In addition, students could gain motivation in a SSI context when role-play, group work, discussion, or other social interactions are connected to the inquiry (Osborne et al. 2003; Toplis 2012). Recognising the inherent complexity of SSI, examining issues from multiple perspectives, appreciating that SSI are subject to ongoing inquiry, and exhibiting scepticism if there is potentially biased information are important practices for decision-making in the context of SSI (Sadler et al. 2007). Furthermore, motivation and fundamental conceptual understanding increase as students present their arguments to others (Benware and Deci 1984).

Our study focuses on the use of so-called scenarios in connecting socio-scientific issues to science learning. Scenarios, sometimes referred to as problems, cases, or vignettes relating to real-life situations (Abrandt Dahlgren and Öberg 2001), have, for instance, been used as starting points for problem-based learning (Akınoğlu and Tandoğan 2007) or engaging students in socio-scientific inquiry (cf. Sadler et al. 2007). Scenarios have been considered to provide a meaningful context for science concepts and principles (Abrandt Dahlgren and Öberg 2001). Thus, one way to enhance students' interest and motivation towards science is to design scenarios that have connection to socio-scientific contexts. With a scenario as a starting point, learning becomes more than the gaining of factual knowledge as students participate in the process of posing questions about the problems emerging from the scenario.

Much of the science education research addressing SSI has focused primarily on students and how they make decisions on such issues (Saunders and Rennie 2013); less attention has been paid to teachers' pedagogical practices with SSI. It has been argued that there is a challenge for teachers, who play a critical role in shaping how curricula are implemented in a classroom and experienced by students, to move beyond traditional modes of science teaching (Sadler 2009). However, there are some studies related to science teachers' perceptions of the SSI approach in science teaching. Ekborg et al. (2013) studied seven Swedish science teachers who conducted SSI teaching modules and found that the teachers appreciated the idea of SSI: the teachers interpreted the modules as a way to increase students' interest in school science. They all included elements of SSI but mostly to introduce the regular science content. Ekborg et al. (2013) interpreted that the teachers had the driving force to do something different but they did not exploit their freedom to make changes in the content.

Similar findings were found among 86 Korean science teachers whose perceptions revealed a disparity between participants' beliefs about the need to address SSI and their actual commitment and teaching of these issues in their classrooms (Lee and Abd-El-Khalick 2006). Furthermore, the Korean teachers expressed low confidence in their abilities to develop materials for teaching about SSI. Lee and Witz (2009) conducted in-depth interviews of four science teachers in Illinois, and they found that the teachers' initial motivations for teaching SSI are basically disconnected from the SSI reform efforts. In addition, the teachers developed their own approaches to SSI according to their own values and ideals during the study.

Our study is related to the PROFILES project (www.profiles-project.eu), which supports science teachers' continuous professional development in regard to implementing SSI in science classrooms as its core idea (see Bolte et al. 2011). Our aim is to find out how Finnish science teachers have succeeded in creating SSI scenarios that could trigger students' situational interest and intrinsic motivation.

Scenario-Based Science Teaching

Triggering students' situational interest (Krapp and Prenzel 2011) through positive, affective, and cognitive experiences could be seen as an essential goal for a scenario. On the other hand, activation of intrinsic motivation (Ryan and Deci 2009) is needed for maintaining readiness to acquire new information. From an educational perspective, motivation can be interpreted as any process that activates and maintains learning (Palmer 2005). Intrinsic motivation can be maintained by presenting possibilities in a task within the range of students' skills (Ryan and Deci 2009). Situational interest is triggered by the environment and also considered to be motivating, although teachers often struggle with how to influence students' interest (Hidi and Renninger 2006). Another challenge for situational interest is that it is unlikely to endure beyond a particular lesson (Abrahams 2009). Hence a high-quality scenario should be designed to stimulate cognitive and affective features of learning that will encourage the students to investigate the problem in depth. The instructional innovation of the PROFILES project is the so-called three-stage model (TSM) which aims to arouse students' intrinsic motivation undertaken in a familiar, socio-scientific context (scenario), to offer a meaningful inquiry-based learning environment (inquiry), and to use the science learning in solving socio-scientific problems (decision-making) (cf. Bolte et al. 2011).

Three-Stage Model

The three-stage model (Fig. 1) is designed to promote students' interest and motivation in learning science content and to undertake inquiry learning and, in particular, to meet the aims of "education through science" (cf. Valdman et al. 2012). The following description is based on the descriptions of the TSM by Bolte et al. (2011) and Sormunen, Keinonen, and Holbrook (2014).

The intention of Stage 1 (scenario) is to involve students in undertaking activities that lead to better understanding of the issue – an issue seen by students as relevant to their lives, not simply relevant to the curriculum – and worthy of greater appreciation. The motivation is intended to be activated by a scenario including an appealing title and a purposely chosen phenomenon related to nature, everyday life, or socio-scientific issues. In facilitating the move to the second stage, the initial

Fig. 1 The central role of
a scenario in the TSM
approach

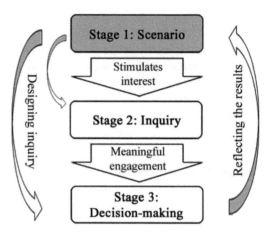

motivation forms a key launch platform for the intended science learning. It seeks
to draw the students' attention to think about deficiencies in their prior knowledge
and to undertake a meaningful discussion related to the scenario. This facilitates the
posing of the scientific question or questions intended for investigation.

Stage 2 (inquiry) is expected to maintain the motivational learning from Stage 1
and to meet science learning outcomes that relate to cognitive processes, operationalize scientific process skills through the intended inquiry-based learning, develop
personal attributes (e.g. creativity, showing initiative, perseverance, and safe working), and also promote students' social development through collaborative teamwork. These processes, together with the learning outcomes from inquiry, facilitate
the move to consolidation that can be enacted through, for instance, interpretation
of the outcomes, presentation of the findings, and discussion of the relevance and
reliability of the outcomes.

Stage 3 (decision-making) is the consolidation phase for the science learning, in
which the acquired science ideas are given relevance by including them back into
the socio-scientific scenario, which provided the initial student motivation. This
enables the students to reflect on the issues, while placing the newly learned science
knowledge alongside other attributes important for participating in argumentation
and reasoning to reach consensus, first within a small group and then for the class
as a whole. This can take place in a range of formats, e.g. argumentation debates,
role-playing, or discussion, so as to derive a justified, society-relevant decision or a
consideration seen as reasonable by the class.

Towards an Ideal Scenario

We have chosen the scenario stage for our focus in this study because of its central
role in creating a socio-scientific context for science learning. We have found that
quite a remarkable part of the Finnish teachers' concerns related to the TSM have
been interconnected with the scenario stage (Sormunen et al. 2014). In the

Fig. 2 The components of
an ideal scenario

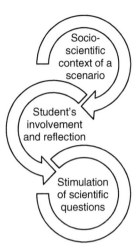

following, we describe the components of an ideal scenario (Fig. 2) that comprise our focus points for analysing the scenarios created by teachers.

The scenario stage of the TSM is meant to trigger students' situational interest and intrinsic motivation in a familiar daily life or socio-scientific context. The *context of a scenario* should be a real-life problem in an open-ended form in order to arouse a sense of curiosity (Akınoğlu and Tandoğan 2007). Scenarios should be complex enough but not overloaded or too structured (Abrandt Dahlgren and Öberg 2001). Furthermore, the scenarios should lead to multiple plausible solutions (Sadler 2011) in the decision-making stage. The intention is to *involve* students in undertaking activities that relate to better understanding of the issue; scenarios should help students to *reflect on* their prior knowledge and share their conceptions and views with peers (cf. Abrandt Dahlgren and Öberg 2001; Akınoğlu and Tandoğan 2007). This all facilitates the students to *pose scientific questions* intended for investigation (Bolte et al. 2011); the initial interest and motivation in the scenario stage form a key launch platform for the intended science learning (cf. Fig. 1).

Altogether, when the context of a scenario is carefully chosen and the scientific ideas are embedded in it, the actual science learning can begin after the ideas are decontextualised from the initial context and an inquiry-based approach is then applied (Bolte et al. 2011). Based on the importance of SSI in science teaching, teachers' role in implementing interventions such as TSM instruction, and the components of scenarios, our research question is what kinds of scenarios did the Finnish PROFILES teachers design?

Method

The data for this qualitative case study was gathered from 30 Finnish teachers who participated in the PROFILES project. We went through all the teaching modules that the teachers planned and implemented, concentrating only on those 24 teaching modules that the teachers created by themselves; the rest of them (6) were based on ready-made materials. The curricular content of the scenarios in our sample was mainly related to physics (18), chemistry (3), biology (2), and earth science (1); 2 of them were implemented at the primary level, 12 at the lower secondary, and 10 at the upper secondary level. The themes of the scenarios concerned energy production (7) and consumption (6), environmental issues (5), water (2) and other natural resources (1), motion of an object (1), and a practical everyday problem (1). The scenarios were presented in the forms of a realistic (11) or fictional story (9), an authentic news article (3), or an attitude questionnaire (1); three of the realistic stories were illustrated with photos.

Deductive content analysis, including the preparation, organising, and typification phases (cf. Elo and Kyngäs 2008), was used to compare the features of an ideal scenario to the scenarios designed by the teachers. In the preparation phase, the units of analysis were selected from the teaching modules. Besides the scenario stage, we also had to pay attention to the inquiry and decision-making stages in the modules in order to analyse the nature and use of the scenario as a whole, because the three stages are interconnected. Next in the analysis process, data was read through several times in order to be thoroughly acquainted with it.

Next, the categorisation matrix was developed in the organising phase. The leading focus points of our analysis were constructed on the basis of the components of an ideal scenario (cf. Fig. 2) with categories derived from Bolte et al. (2011), the focus points being relevance of the scenario (categories: title, interdisciplinary content, meaningful socio-scientific context, and interesting introductory materials), students' involvement and reflection (interesting involvement activities, collective thinking), and facilitating scientific questions (enabling students' open-ended questions and enhancing several solutions) (see Table 1). The data from the scenarios was gathered according to the categories. Next, the features of the scenarios related to the categories were inductively analysed according to their quality. As three researchers, we have utilised investigator triangulation (cf. Gibbs 2007); the features describing the quality of the scenarios in each category were checked regularly by all the authors.

After deductive content analysis, we typified the scenarios: by using scenario-by-scenario comparisons according to how extensive the quality categories emerged (cf. Table 1), we constructed three typologies (cf. Gibbs 2007): high-quality scenarios scored seven or more quality categories, mediocre scenarios four to six, and low-quality scenarios three or less. The aim of the typologies is to describe how the focus points were considered in a scenario in order to trigger students' situational interest and activate intrinsic motivation.

Table 1 Scenario types

Scenario	Relevance of the scenario			Meaningful socio-scientific context				Involvement and reflection		Facilitating scientific questions		Scenario type
	Appealing title		Inter-disciplinary content	Everyday issue	Local issue	Global issue	Interesting introductory materials	Interesting activities	Collective thinking and reflection	Leading to 1–2 open-ended question(s)	Enhancing several solutions	
	Relatedness to the students' life	In the form of an appropriate question										
S20	×	×	×	×		×	×	×	×	×	×	High-quality (n = 8)
S2	×		×	×	×		×	×	×	×	×	
S17	×	×	×		×			×	×	×	×	
S3	×	×		×			×	×	×	×		
S1	×		×	×	×				×	×	×	
S10	×		×	×	×				×	×	×	
S14	×	×	×	×					×	×	×	
S19	×	×	×	×	×				×		×	

(continued)

Table 1 (continued)

Scenario	Relevance of the scenario							Involvement and reflection		Facilitating scientific questions		Scenario type
	Appealing title		Inter-disciplinary content	Meaningful socio-scientific context			Interesting introductory materials					
	Relatedness to the students' life	In the form of an appropriate question		Everyday issue	Local issue	Global issue		Interesting activities	Collective thinking and reflection	Leading to 1–2 open-ended question(s)	Enhancing several solutions	
S7	x		x		x		x	x	x		x	Mediocre (n = 14)
S15	x	x	x						x	x	x	
S18			x		x		x		x	x	x	
S23	x		x	x					x	x	x	
S22			x				x	x	x	x	x	
S9			x			x		x	x	x	x	
S5	x		x	x				x	x	x	x	
S11				x			x	x	x	x	x	
S16				x				x	x	x	x	
S24						x		x	x	x	x	
S6			x						x	x	x	
S12					x		x	x		x		
S21						x	x		x	x	x	
S8				x			x	x		x		
S4				x						x	x	Low-quality (n = 2)
S13					x						x	
	11	6	14	13	9	4	9	12	20	21	20	

Results

In the following, we describe our findings according to the three focus points of our analysis, which in turn are related to the components of an ideal scenario, i.e. relevance of the scenario, students' involvement and reflection, and stimulation of scientific questions.

Firstly, regarding our first focus point (relevance of the scenario; see Table 1), the teachers were interpreted as using appealing or relevant scenario titles related to the students' life (11), such as "Window and wall element options in Alice's wonderland" (the scenario S1, cf. Table 1) or "Competition: conserve electricity!" (S2), and/or they included an appropriate question (6), e.g. "How to protect your hearing while using an mp3-player?" (S5) and "Does an electricity invoice make my family happy?" (S19). Many of the scenario themes were interdisciplinary (14), and almost all of them were related to sustainable development: the students were supposed to ponder how power production and/or consumption is related to sustainability. The socio-scientific context in the scenarios was related to everyday (13) and/or local issues (9), e.g. writing an exemplary article about the water cycle for a summer job application in a popular science journal or solving a local environmental issue related to a pond, which is polluted and eutrophic. Four scenarios included a global problem, e.g. "You only consumed one cubic meter of water for food preparation before noon: are you to blame for the global water crisis now and in the future?" (S20) The everyday contexts were usually based on popular events or phenomena in youth culture, e.g. a scenario titled as "Why do the Dudesons fall?" (S3). The scenarios included also interesting introductory materials (9), e.g. a piece of news or an article.

Secondly, relating to students' involvement and reflection, interesting activities were planned to be based on reading and discussion, role-play, fieldwork, or the use of pictures; in many cases (12), there was a combination of these activities. Collective thinking and reflection on prior knowledge were also included in most scenarios (20). In the scenario stage, the actual scenarios were followed by small and/or whole group discussions, interesting activities, collective thinking, and reflection, e.g. in an electricity conserving competition in which the students individually responded to an attitude questionnaire and then pondered their views together and formed a consensus standpoint for conserving electricity (S2).

Thirdly, in regard to the stimulation of scientific questions, the majority of the scenarios (15) included one problem, which could stimulate students to pose research questions. Some scenarios (6) included a problem with ready-made research questions, although the students had a choice to form some research questions of their own. The rest of the scenarios (3) included several predetermined questions, which restricted the students from framing their own questions. In sum, most of the scenarios (21) were interpreted to enable the students to form open-ended research questions for the inquiry stage. In regard to the possible solutions, four of the scenarios were closed in their nature, leading to only one "correct" solution. The themes of the majority of the scenarios (20) were complex enough to enable several solutions at the decision-making stage.

After the content analysis, the scenarios were typified. Eight of them were of *high quality*, i.e. they took variously into account most of the features in each focus point (cf. Table 1). We consider that the higher the quality of a scenario is, the better the scenario triggers students' situational interest and activates their intrinsic motivation. These kinds of scenarios are related to the students' everyday life, are interdisciplinary, use interesting activities, and involve them in collective thinking and reflection on their prior knowledge, as well as stimulate them to ponder open-ended scientific questions with multiple solutions. The majority of the scenarios (14) were *mediocre* scenarios, which weakly included the features related to the relevance of a scenario. The mediocre scenarios did not include all the necessary features that describe a relevant scenario, i.e. an appealing title, interdisciplinary content, meaningful socio-scientific context, or interesting introductory materials (see Table 1). The *low-quality* scenarios (2) lacked an interdisciplinary approach, appealing titles, and interesting introductory materials; the scenarios lacked interesting involvement activities and they did not encourage students to think collectively nor to reflect on their prior knowledge.

Discussion

The current problem related to science education is that students seem to lack interest and motivation towards science studies both in secondary and higher education (cf., European Commission 2015; Potvin and Hasni 2014). The SSI approach is related to the promotion of scientific literacy and the improvement of science learning experiences (Sadler 2011), which in turn could trigger students' interest and activate their motivation.

The goal of the presented TSM-based instruction follows a SSI approach, and it has been developed to arouse students' situational interest and intrinsic motivation (cf. Valdman et al. 2012). The TSM approach is likely to maintain students' interest over several lessons in contrast to a short-term effect of situational interest (cf., Abrahams 2009). Instead of handing out ready-made teaching materials for teachers (cf., e.g. Ekborg et al. 2013), the novelty in our research setting was that the teachers designed SSI modules themselves during their participation in the PROFILES project. In this study, we focused on the quality of the scenarios as they should evoke students' affective involvement and adjust their cognitive evaluation of the science content to be more meaningful (Bolte et al. 2011).

The theme and the form of a scenario are important in developing students' interest and motivation. The themes of a total of 24 scenarios created by science teachers were mostly about energy production and consumption, environmental issues, or natural resources, which are in line with crucial areas of global concern (cf. Hodson 2003; Hogan 2002). Most of the scenarios (20/24) were in a form of a realistic or fictional written story; some of them were illustrated with photos (cf. Sadler et al. 2007); news articles were also used in scenarios (cf. Abrandt Dahlgren and Öberg 2001).

Relevance of the Scenario for Students Most of the teachers (22/24) succeeded in creating meaningful socio-scientific contexts in their scenarios by connecting curricular content to everyday, local, or global issues. The fact that the scenarios are connected to students' daily lives enables students to understand how science classes are interrelated with real life (Zeidler et al. 2005; Akınoğlu and Tandoğan 2007). However, it seems that some of the teachers did not include an interdisciplinary approach in their scenarios; that might be due to the Finnish subject-divided science teaching in secondary level (Lampiselkä et al. 2007). It is also challenging for teachers to use materials and formulate scenario titles in ways that attract students' attention and interest (cf. Sormunen et al. 2014); one way might be to use scenarios that are provocative or evoke emotional involvement (Abrandt Dahlgren and Öberg 2001).

Students' Involvement and Reflection The teachers used strategies such as reading and discussion, role-play, and use of pictures in their scenario activities, i.e. strategies to engage learners in SSI (Sadler 2011). Twenty of the scenarios were based on instructions that encouraged students to think collectively and reflect on the topics in light of their prior knowledge (cf. Abrandt Dahlgren and Öberg 2001) fulfilling the aim of the SSI approach to engage students in dialogue, discussion, and debate (Zeidler and Nichols 2009). This provides a venue where students simultaneously develop their critical thinking and moral reasoning skills while learning curricular content (Zeidler et al. 2005).

Stimulating Scientific Questions Almost all of the analysed scenarios (21/24) were interpreted to enable students to formulate scientific questions of their own for the inquiry stage in the TSM approach; if a scenario is relevant enough for the students, they treat it as their own and thus they are willing to solve the questions that emerge (Akınoğlu and Tandoğan 2007). Furthermore, when designing a scenario, the level of students' cognitive skills should be considered so that they are able to formulate questions related to the problem of a scenario. Intrinsic motivation should be maintained with a suitable challenge (Ryan and Deci 2009). The majority of the scenario themes (20/24) were complex enough to enable several solutions at the decision-making stage in the TSM approach; complexity is an important feature of scenarios that makes students problematize them in depth (Abrandt Dahlgren and Öberg 2001). This result highly supports the idea of maintaining intrinsic motivation with possibilities to make choices (Ryan and Deci 2009).

Possible Reasons for Modest Quality of Scenarios One third of the scenarios were typified as high quality; they paid attention to almost all of the features of an ideal scenario. Yet, the majority (14/24) of the teachers' scenarios were mediocre in regard to their quality as they did not fully take into account the relevance of an ideal scenario, i.e. a title related to students' life in the form of an appropriate question, interdisciplinary nature of the content, meaningful scientific context, or interesting introductory materials (see Table 1). However, some of them were interpreted to include activities for student involvement, and almost all of them encouraged students in collective thinking and reflection. Two of the scenarios were of low quality as almost all of the features of an ideal scenario were missing. The teachers tend to lack confidence in addressing scenario-based teaching in their science classrooms (Sormunen et al. 2014); this might be a consequence of constraints such as a lack of awareness of pedagogical choices and guidance of how to apply them (cf. Saunders and Rennie 2013). Findings in earlier studies also indicate that science teachers appreciate the SSI approach and teaching modules for that purpose (Ekborg et al. 2013), but the actual use of an SSI approach is dependent on their values (Lee and Witz 2009) or teachers' pedagogical knowledge and skills (Lee and Abd-El-Khalick 2006).

Conclusion

The quality of SSI scenario in our study seemed to depend on its relevance to students (see Table 1). The majority of the scenarios acknowledged the cognitive aspects such as collective thinking and reflection, and practically all of the scenarios facilitated students' scientific questions. In contrast, quite a few scenarios disregarded the affective features such as the interesting and appealing form and context of the problem. We consider this result to be significant from both research and theoretical perspectives; it extends our understanding of challenges related to SSI scenario design. Our findings indicate that it seems to be challenging but possible for science teachers to develop an SSI scenario that is interesting and motivational from the students' point of view. Although the present study does not concern itself with how students benefit from the scenarios, our preliminary findings (Jäppinen et al. 2015) support the conclusion that the scenarios trigger students' interest and motivation.

It can be noted that affective features in scenario-based science education need to be particularly emphasised. This cannot be achieved without teachers' awareness of the importance of affective aspects of science education. Therefore, there is a need for systematic and longitudinal support for science teachers to create SSI scenarios (cf. Saunders and Rennie 2013).

References

Abrahams, I. (2009). Does practical work really motivate? A study of the affective value of practical work in secondary school science. *International Journal of Science Education, 31*, 2335–2353.

Abrandt Dahlgren, M., & Öberg, G. (2001). Questioning to learn and learning to question: Structure and function of problem-based learning scenarios in environmental education. *Higher Education, 41*, 263–282.

Aikenhead, G. (2005). Research into STS science education. *Education Quimica, 16*, 384–397.

Akınoğlu, O., & Tandoğan, R. Ö. (2007). The effects of problem-based active learning in science education on students' academic achievement, attitude and concept learning. *Eurasia Journal of Mathematics, Science & Technology Education, 3*, 71–81.

Benware, C. A., & Deci, E. L. (1984). Quality of learning with an active versus passive motivational set. *American Educational Research Journal, 21*, 755–765.

Bolte, C., Steller, S., Holbrook, J., Rannikmäe, M., Hofstein, A., Mamlok-Naaman, R., & Rauch, F. (2011). Introduction into the PROFILES project and its philosophy. In C. Bolte, J. Holbrook, & F. Rauch (Eds.), *Inquiry-based science education in Europe; reflections from the PROFILES project* (pp. 31–42). Berlin: Freie Universität Berlin.

Ekborg, M., Ottander, C., Silfver, E., & Simon, S. (2013). Teachers' experience of working with socio-scientific issues: A large scale and in depth study. *Research in Science Education, 43*, 599–617.

Elo, S., & Kyngäs, H. (2008). The qualitative content analysis process. *Journal of Advanced Nursing, 62*, 107–115.

European Commission. (2015). *Science education for responsible citizenship*. Brussels: European Commission.

Gibbs, G. R. (2007). *Analyzing qualitative data*. London: Sage.

Hidi, S., & Renninger, K. A. (2006). The four-phase model of interest development. *Educational Psychologist, 41*, 111–127.

Hodson, D. (2003). Time for action: Science education for an alternative future. *International Journal of Science Education, 25*, 645–670.

Hogan, K. (2002). Small groups' ecological reasoning while making an environmental management decision. *Journal of Research in Science Teaching, 39*, 341–368.

Jäppinen, I., Keinonen, T., Sormunen, K., & Bolte, C. (2015, September). *The Finnish secondary school students' gains in context-based science education*. Paper presented at the ESERA European Science Education Research Association Conference, Helsinki.

Krapp, A., & Prenzel, M. (2011). Research on interest in science: Theories, methods, and findings. *International Journal of Science Education, 33*, 27–50.

Lampiselkä, J., Ahtee, M., Pehkonen, E., Meri, M., & Eloranta, V. (2007). Mathematics and science in Finnish comprehensive school. In E. Pehkonen, M. Ahteen, & J. Lavonen (Eds.), *How Finns learn mathematics and science* (pp. 35–47). Rotterdam: Sense Publishers.

Lee, H., & Abd-El-Khalick, F. (2006). Korean science teachers' perceptions of the introduction of socio-scientific issues into the science curriculum. *Canadian Journal of Science, Mathematics and Technology Education, 6*, 97–117.

Lee, H., & Witz, K. G. (2009). Science teachers' inspiration for teaching socio-scientific issues: disconnection with reform efforts. *International Journal of Science Education, 31*, 931–960.

Lewis, J., & Leach, J. (2006). Discussion of socio-scientific issues: The role of science knowledge. *International Journal of Science Education, 28*, 1267–1287.

Osborne, J., Simon, S., & Collins, S. (2003). Attitudes towards science: A review of the literature and its implication. *International Journal of Science Education, 25*, 1049–1079.

Palmer, D. (2005). A motivational view of constructivist informed classrooms. *International Journal of Science Education, 27*, 1853–1881.

Potvin, P., & Hasni, A. (2014). Interest, motivation and attitude towards science and technology at K-12 levels: A systematic review of 12 years of educational research. *Studies in Science Education, 50*, 85–129.

Ryan, R. M., & Deci, E. L. (2009). Promoting self-determined school engagement: Motivation, learning, and well-being. In K. R. Wentzel & A. Wigfield (Eds.), *Handbook of motivation at school* (pp. 171–195). London: Routledge.

Sadler, T. D. (2004). Informal reasoning regarding socioscientific issues: A critical review of research. *Journal of Research in Science Teaching, 41*, 513–536.

Sadler, T. D. (2009). Situated learning in science education: Socio-scientific issues as contexts for practice. *Studies in Science Education, 45*, 1–42.

Sadler, T. D. (2011). Situating socio-scientific issues in classrooms as a means of achieving goals of science education. In T. D. Sadler (Ed.), *Socio-scientific issues in the classroom: Teaching, learning and research* (pp. 1–9). Dordrecht: Springer.

Sadler, T. D., Barab, S. A., & Scott, B. (2007). What do students gain by engaging in socioscientific inquiry? *Research in Science Teaching, 37*, 371–391.

Saunders, K. J., & Rennie, L. J. (2013). A pedagogical model for ethical inquiry into socioscientific issues in science. *Research in Science Education, 43,* 253–274.

Sormunen, K., Keinonen, T., & Holbrook, J. (2014). Finnish science teachers' views on the three stage model. *Science Education International, 25,* 172–185.

Swarat, S., Ortony, A., & Revelle, W. (2012). Activity matters: Understanding student interest in school science. *Journal of Research in Science Teaching, 49,* 515–537.

Toplis, R. (2012). Students' views about secondary school science lessons: The role of practical work. *Research in Science Education, 42,* 531–549.

Valdman, A., Holbrook, J., & Rannikmäe, M. (2012). Evaluating the teaching impact of a prior, context-based, professional development programme. *Science Education International, 23,* 166–185.

Zeidler, D. L., & Nichols, B. R. (2009). Socioscientific issues: Theory and practice. *Journal of Elementary Science Education, 21,* 49–58.

Zeidler, D. L., Sadler, T. D., Simmons, M. L., & Howes, E. V. (2005). Beyond STS: A research-based framework for socioscientific issues education. *Science Education, 89,* 357–377.

Zohar, A., & Nemet, F. (2002). Fostering students' knowledge and argumentation skills through dilemmas in human genetics. *Journal of Research in Science Teaching, 39,* 35–62.

The Use of Drama in Socio-Scientific Inquiry-Based Learning

Roald P. Verhoeff

Introduction

Socio-scientific Inquiry-Based Learning

The educational framework socio-scientific inquiry-based learning (SSIBL) is a pedagogy, which connects the study of socio-scientific issues with inquiry-based learning and citizenship education (Levinson and The PARRISE Consortium 2014). Socially and ethically sensitive inquiry is at the core of this approach. This inquiry-based aspect of SSIBL means that it is question-driven and open-ended. It requires scaffolding and the generation of questions and/or issues, preferably by students themselves, which are authentic, that is, they emerge from pressing interests of the students. Issues should thus relate to real-world problems, engage the interest of young people and draw on scientific knowledge. As such, SSIBL is a way to bring the EU framework of Responsible Research and Innovation (RRI) to classroom practice. RRI stands for a comprehensive approach of research and innovation in ways that allow the engagement of stakeholders in the processes of research and innovation at an early stage. To address this approach in education presupposes the acknowledgement that non-scientists are, like scientists, concerned by changes influenced by technology in academic, professional and/or everyday life settings. Collaborative learning and reflecting on these socio-scientific issues related to emerging technologies is key. As outlined by Levinson et al. (2014), SSIBL can be assessed through four dimensions:

- Knowledge about an issue (both scientific and transdisciplinary)
- Skills in organizing a socio-scientific-based inquiry

R.P. Verhoeff (✉)
Utrecht University, Utrecht, The Netherlands
e-mail: r.p.verhoeff@uu.nl

© Springer International Publishing AG 2017   117
K. Hahl et al. (eds.), *Cognitive and Affective Aspects in Science Education Research*, Contributions from Science Education Research 3,
DOI 10.1007/978-3-319-58685-4_9

- Values which reflect issues of social justice and well-being
- Dispositions in terms of recognition of inclusivity and democratic deliberation

These are scaffolded in such a way that criticality of students progressively increases via inquiry-based learning in which students contextualise knowledge to help answering their own questions.

In this study we focused on (future) issues around neuroscience, as it reflects an innovative field that has become a key topic for ethical deliberation (Savulescu et al. 2011). Over the last two decades or so, neuroscience has developed new treatments and technologies for therapeutic purposes but may in principle also be applied to healthy individuals to optimize cognitive functions, for example, in professions such as pilots, top athletes or the army. Enhancement technologies include genetic engineering, nootropic drugs (e.g. Modafinil), brain-computer interaction (BCI), neural implants and so on. These technologies may boost human performance in the near future and already raise various ethical dilemmas concerning autonomy (do these technologies empower 'us', or will we be forced to use them in an era of increased cognitive competitiveness?) and justice (will these technologies foster social mobility or rather enlarge the socio-economic division between those who will and those who will not have access to them?). On a more fundamental level, it raises the question whether and how it will affect human nature and human identity (Zwart 2014).

Drama in Science Education

To encourage young people to think about socio-scientific issues, various techniques can be used. For example, Knippels et al. (2009) show that storylines with a human theme, in this case using a clip from the movie Gattaca, are extremely effective at prompting opinion forming among young people. It has likewise been argued that well-considered use of drama fosters learning of cognitive, procedural and affective knowledge in an integrated way (Ødegaard 2003; Dorion 2009). Moreover, drama may allow students to engage in 'simulation' exercises, where the societal impact of science can be explored and enacted in various ways, providing a test-bed for probing alternative (perhaps conflicting) perspectives, inviting students to explicitly reflect on the tensions and differences that are made visible and tangible this way (Colucci-Gray et al. 2006).

Emerging technologies often involve uncertainties when it comes to their potential (medical, environmental or economic) benefits and risks, and drama seems especially apt to capture and articulate the ambivalence this entails. McSharry and Jones (2000) argue that, driven by the teacher, role play in science education can utilize learners' lifetime 'play practice' to both express themselves in a scientific context and develop an understanding of difficult concepts. They argue that engaging learners in creation and performance of science drama provides a physical and creative experience that may be more appropriate for personal learning styles, offering them a sense of ownership of their education. They also underline its potential for effective learning about moral and ethical issues such as genetic modification in food production (McSharry and Jones 2000).

So far, only a few empirical studies have been published concerning the effectiveness of drama in science education (Shepherd-Barr 2006; Wieringa et al. 2011). Moreover, in most papers describing the use of drama (or other 'genres of the imagination', such as novels or cinema), students are typically involved as spectators and do not actively perform themselves. In our case, we followed the tradition of research practitioners in 'drama in education' such as Heathcote, Bolton, O'Neill and others by involving students not only as an audience but also as authors and actors. Drama in education supports a collaborative learning process in which students explore ideas and feelings and take different perspectives within a fictional context (Bolton 1984, 1985; Heathcote and Bolton 1995). Reflection and analysis of the drama is key to extend and deepen students understanding of social problems and their (enacted) solutions (O'Neill 1995). As O'Neill (1985) has argued, linking the (enacted) world of fiction with the real world is key to the success of learning via drama. This asks of teachers to allow 'space for student reflection on the extent to which their enacted roles, movements or talk are realistic presentations of the science represented' (Braund 2015: 115). Teachers are required to cross pedagogical borders from the pedagogy of drama to the pedagogy of science and vice versa (Braund 2015; Fels and Meyer 1997). This is not self-evident for most science teachers, as they may perceive a loss of control when their students are improvising in an experiential setting. For teachers and learners more used to traditional educational activities or rational science teaching, for example, university (science) students, McSharry and Jones (2000) suggest the use of structured games, simulations like organized debates or court cases or plays scripted by the teacher or students in advance.

Following the suggestion of McSharry and Jones, we decided that the performance should be based on a script that had to be adhered to, albeit that the script was written by the students themselves (see also Toonders et al. 2016). We invited them to explore future societal impacts of emerging neuro-technologies with the help of drama, scripted and performed by the students themselves. This way, our approach was envisioned as a 'dramatic' form of SSIBL, with the performance being a kind of experiment, starting from an initial situation (the 'control condition') in which a novelty or unexpected element is introduced (the experimental condition), which then unleashes a series of consequences, building up the dramatic plot (outcome). The emphasis was on doing, acting and reflecting on science with and for society, in which plays were used to explore 'what if… ' scenarios as a means of socio-scientific inquiry. The teachers' role was to safeguard the realness of the play, i.e. the links between their imagined world and reality (cf. O'Neill 1985). Based on this drama experiment, we address in this paper the question: What aspects of SSIs related to neuroscience do students include in their plays?

Our Approach

The Drama Experiment

Neuro-enhancement is a rapidly growing research domain and a key issue in the ethical and moral debate on emerging technologies in the European Union. Within the context of an elective graduate course on human enhancement (3 ECTS), 22 students from various science disciplines were involved in a drama experiment, performing multiple roles as audience, reviewers, authors and actors (see also Toonders et al. 2016). The course started with two introductory lectures on new technologies to enhance neural processes and the ethical issues as described above in the introduction. The drama experiment itself featured three collaborative learning activities with different student roles. Students were invited to fulfil the role of an author writing a play, an actor performing it to their peers and an audience watching and assessing the performance of others. Following O'Neill (1995), in these different roles, we expected students to actively reframe and adapt their perceptions on science in society in general and neuro-enhancement in particular. Student groups (n = 5/6) were instructed to design an 8-min one-act play and to write out the whole screenplay, including short descriptions of the main characters and the words spoken by them, as well as nonverbal expressions of emotions. With respect to content, students were asked to develop a storyline that would provide insight in various options and dilemma's connected to an available, experimental or hypothetical neuro-enhancer, employable in a particular context sometime in the near future. The storyline should consist of at least three scenic elements: (1) an initial situation, (2) an occurring event and (3) an ending and focus on a particular neuro-enhancer, e.g. a technical device, pill software program and chip enhancing cognitive functions. In addition, students were asked to think about the presentational aspects of their play, e.g. how could the audience be involved? We expected that like a scientific investigation, drama would allow students to try out and enact possible 'what if…' scenarios and dilemmas in a relatively safe (intrusion-free) environment and explore the consequences of a certain innovation or technological novelty.

The teacher was asked to safeguard the realness of the play, to prevent symbolic overtones of meaning and to stimulate a reflective attitude, i.e. taking distance to negotiate the different enacted perspectives and views and their own relationship to them. This role implies the use of the 'reality principle' (O'Neill 1985), i.e. assessing the plausibility and authenticity of the plays.

Data Collection and Analysis

Qualitative methods were used to collect data from participating students as we wanted to gain insight into the kind of ideas they had about SSIs in a particular personal or social (theatrical) setting. Data collection and analysis are built on the analytical framework developed by Dorion (2009). We distinguished three themes: (1) prior knowledge, drama experience and motivation to enrol in the course; (2) learning activities and performances; and (3) students' self-perceived learning. Inspired by the SSIBL framework (Levinson and The PARRISE Consortium (2014), the analytical focus of this paper is on students' reflexivity, i.e. linking scientific knowledge with personal assessments and finding out which of the following attributes of SSIs they included in their plays:

- Openness (i.e. no preset answer)
- Authenticity (reality principle)
- Comprise different and conflicting perspectives
- Links between personal and social relevance
- Epistemologically appropriate (i.e. it should draw on science knowledge which students have acquired or can be taught)

We handed out three questionnaires to all participants before, during and after the course. The first questionnaire mainly focused on students' awareness of socio-scientific issues and prior knowledge on neuro-enhancement. The second questionnaire asked students to reflect on the plays that had been performed. The final questionnaire included questions about their self-perceived learning, their attitude towards the enacted SSIs and their appreciation of drama experiment including their own performance. Initial findings (student evaluations) were cross-checked via interviews with the teacher after each session in which he was asked about his opinion on the attitudes, skills and insights acquired by the students. In addition, all class discussions and performances were videotaped, while audio recordings were made of group discussions.

Results

Most of the students enrolled in the course indicated that they took this course because they were interested in neuro-technologies and wanted to gain more insight in the ethical aspects around this emerging technology. Only two out of 22 students had prior experience with performing a play before an audience, and consequently, the prospect of a live performance raised some general concern among the students initially. On the other hand, when asked what students expected of this module in a questionnaire at the start of the module, they valued it as an innovative learning strategy that would allow them to express their views and opinions in a creative manner (n = 7). Other students expected that it would allow for more creativity in developing their own perspective on future innovations (n = 2) or they were mainly looking forward to seeing and hearing the views of their fellow students (n = 3).

In the scripting phase, students were prompted to answer the question 'What would happen if…?' with a social inquiry around a specific neuro-enhancer with the three scenic elements: initial situation, event and (open) ending.

Table 1 Overview of the plays enacted by the five student groups

Neuro-enhancer	Short description of the play/SSI
Memory chip	An enhanced (flawless, arrogant) and a non-enhanced (experienced, friendly) surgeon apply for the same job. The enhanced doctor gets it. Should chip implantation be seen as a moral professional obligation?
Enhancement pills	Two families are portrayed in their everyday lives to illustrate different views on life quality: a competitive family using smart pills to excel at many areas and a cooperative family cherishing their talents.
Robotics	In a café a robot bartender symbolizes technological advancement and loss of autonomy. Without empathy, it monitors the physical state of the visitors and knows exactly when they have had enough to drink.
Brain-computer intervention	A talk show is enacted on memory alteration: What would you do when mistakes could be erased and your mental health improved? No 100% safety guaranteed!
Smart pills	A heated parental discussion is staged about raising their child with or without smart pills. There is mutual distrust among the parents about the child's excellent piano play. Does she still take her performance drug?

Concrete examples of neuro-enhancement were discussed in all the student groups including possible social implications (see Table 1), such as the idea of a 'memory chip'. Would it lead to an enhanced elite? What would be the impact on human nature if important capacities such as memory no longer need to be trained? Would it be desirable to remember everything, or does forgetfulness serve a purpose? Also, issues of autonomy were addressed: would it be objectionable to implant memory chips in children? Who is to decide? Comparisons with already available types of implants were made such as cochlear implants.

In designing their play, all students decided to present concrete applications in particular settings. All plays featured contrasting views, albeit in various ways, e.g. by staging an accurate and fast-acting physician with a memory chip implant and a traditional (non-enhanced) physician struggling to keep up with the new generation of medical doctors (group 1); by comparing an enhanced family 'with individuals aiming to achieve as many goals as possible' with a non-enhanced family, supporting each other to utilize their talents (group 2); by presenting the story of a traditional bartender, outcompeted by a robot bartender (group 3); by staging a discussion between experts who supported and experts who criticized a new technology on 'memory alteration' (group 4); or, finally, with a discussion between a mother and father of a juvenile candidate for enhancement therapy because he has lost his love for playing the piano (group 5). As one of the participants phrased it: 'presenting multiple viewpoints on stage allowed us to shed more light on the topic from different angles'.

In staging the controversies on neuro-enhancement, various contexts were chosen, varying from professional settings up to private and educational environments. For instance, two groups (1 and 3) demonstrated how untreated humans may be outcompeted by robots or enhanced humans at work. Other students showed how enhancement could make life easier for humans (groups 2 and 5), by making people more creative, but less social (group 2), or enhancing their effectiveness in doing their profession (groups 1, 2, 3 and 5), although the question was also raised whether enhanced professionals could really be trusted in complex situations (group 1) and whether unforeseen collateral damage might be involved (group 4). These proved to be questions for which a 'dramatic experimental' laboratory seemed especially suitable.

When reporting on their performances in classroom setting, students remarked that they considered it important for their performance to reflect an authentic 'realistic' controversy. They discovered that 'theatrical' exaggerations and the absence of ambivalence and doubt in the characters' viewpoints could easily distract the audience as well as the actors themselves from the socio-scientific inquiry that was to be staged and discussed afterwards. Two groups tried to 'personalize' the arguments by placing themselves in the characters' positions, after having thoroughly discussed the controversy and the discussion they wanted to raise among the audience. Subsequently, they started to improvise to see what worked best (groups 3 and 5). One of the lessons learned during this 'trying out' was that 'the devil is in the details'. In the case of group 5, for instance, this involved the exact age of the child, the relation between the parents and the qualitative aspects of the enhancement pill with respect to its effectiveness and potential side effects. Also, students experienced that a logical and coherent sequence of events helped them in communicating their message in a clear and convincing way.

Although 'effectiveness' and 'side effects' refer to the science behind neuro-enhancement, little reference was made in the plays to science knowledge which students had acquired during the course or previous education. The exact 'workings' of the neuro-enhancers remained a black box in each play and didn't seem to be considered relevant for the short storylines to be more 'realistic'.

After the course, we asked students to fill in a questionnaire on their self-reported learning (n = 17 students). A majority of students (12 students) indicated that our module helped them to gain insight in complex ethical issues related to neuro-enhancement. Fifteen (15) students answered positively to the question whether it had improved their understanding of other people's opinions and arguments, while only half of the students indicated they had further explored people's interests or stakes. Almost all students (16) reported that the module had given them more insight in how emerging technologies could influence daily life. Moreover, it helped them to develop their own opinion about the dilemmas at hand (12), although the drama experiment had not per se stimulated them to formulate and substantiate their own opinion (9).

Three students explicitly evaluated the drama module in negative terms. They stated that they had experienced the drama module to be rather time-consuming and that they were not comfortable in performing before an audience.

Discussion and Conclusions

SSIBL proposes a model of concepts and practices central to inquiry, which supports teachers in integrating citizenship education and the EU framework of Responsible Research and Innovation in science classes. This paper underlines the potential for using drama as an educational tool to stimulate students to take a reflexive position on the socio-scientific particularities of science and technology. In our study drama engaged students in a socio-scientific investigation in which the various techniques of enacting and expressing emotions and dilemmas are the equivalent elements compared to the scientific equipment (Kottler 1994: 273). By allowing students themselves to play a more active role, processes of imagination, exploration and identification enabled them to 'try out' and experience their envisioned scenarios in a relatively 'safe' environment, before the new technology enters the real world.

Contextualising a controversy and placing oneself in a real-life situation, e.g. by enactment in a play, seems essential in linking scientific knowledge with personal assessments and views on societal issues. Our drama experiment seemed to activate students in opening up a future lifeworld which they could relate to. It allowed them to acquire additional insights in social and ethical implications of neuro-technologies and created awareness of different viewpoints that people can have, primarily in everyday life settings. Moreover, students tended to take a reflexive position on the personal and social implications of neuro-enhancement so as to present a 'realistic' dilemma to the audience. In doing so, as also illustrated by the plays (see Table 1), students tended to frame SSIs more as personal and ethical (values and norms) as opposed to technical and economical (risks and stakes). The latter implications were hardly considered.

In addition, the focus on everyday life contexts did not prompt students to consider the science aspects of SSIs or draw on science knowledge they had acquired earlier in their academic science education. In terms of our SSIBL attributes, the SSIs portrayed in the plays could not be considered epistemologically appropriate. This illustrates the difficulty for both teachers and students to cross pedagogical or educational borders from 'drama education' to 'science education' as Fels and Meyer (1997) and Braund (2015) already noted. In this respect, the 'reality principle' (O'Neill 1985) should not only address the enacted scenes but also the science represented, based on science knowledge that students have acquired or can be taught.

While students in their role as script writers and actors actively used their imagination, as audience (and to a lesser degree also as actors looking back at their performance) they represented the 'reality principle' by assessing and discussing the plausibility and credibility of the enacted scenes (cf. O'Neill 1985). As others have argued, drama enables the exploration of different perspectives and conflicts in socio-scientific issues, including students' own relationships with the conflict (Colucci-Gray et al. 2006; Wieringa et al. 2011).

It must be noted here that a significant amount of students reported that they had not gained insight into how to formulate and substantiate their own opinion, which is considered an important aspect of SSIBL. This could be explained by the fact that students are used to a 'rational science context' in which inquiry is based on planned observation (cf. Yoon 2006) and deserves further attention: How could this type of active involvement of science students contribute effectively to their learning on the social implications of their domain? This also refers to the challenge that in our module, learners perceived the design and performance of a play to be time-consuming and thus less effective as opposed to more traditional ways of learning.

We are aware that many variables affected students' views on the social and scientific issues surrounding neuro-enhancement, due to the complex nature of teaching and learning. Nonetheless, the results of our study underline the potential benefits of using drama as a tool for socio-scientific inquiry-based learning. When asked whether students would recommend the course to fellow students, an overwhelming majority gave a positive answer. Socio-scientific drama clearly stimulated our students to consider in depth the impacts of new technologies in everyday life and to develop arguments that would be relevant in an authentic personal or democratic deliberation.

References

Bolton, G. M. (1984). *Drama as education: An argument for placing drama at the centre of the curriculum.* London: Addison-Wesley Longman Ltd.

Bolton, G. M. (1985). Changes in thinking about drama in education. *Theory into Practice, 24*(3), 151–157.

Braund, M. (2015). Drama and learning science: An empty space? *British Educational Research Journal, 41*(1), 102–121.

Colucci-Gray, L., Camino, E., Barbiero, G., & Gray, D. (2006). From scientific literacy to sustainability literacy: An ecological framework for education. *Science Education, 90*(2), 227–252.

Dorion, K. R. (2009). Science through drama: A multiple case exploration of the characteristics of drama activities used in secondary science lessons. *International Journal of Science Education, 31*(16), 2247–2270.

Fels, L., & Meyer, K. (1997). On the edge of chaos: Co-evolving worlds of drama and science. *Teaching Education, 9*(1), 75–81.

Heathcote, D., & Bolton, G. (1995). *Drama for learning. Dorothy Heathcote's mantle of the expert approach to education.* Portsmouth: Heinemann.

Kottler, J. A. (1994). *Advanced group leadership.* Pacific Grove: Brooks/Cole.

Knippels, M. C. P., Severiens, S. E., & Klop, T. (2009). Education through fiction: Acquiring opinion-forming skills in the context of genomics. *International Journal of Science Education, 31*(15), 2057–2083.

Levinson, R., and The PARRISE Consortium (2014). *The SSIBL framework,* D1.2 PARRISE, co-funded by the European Commission under the 7th Framework Programme, Utrecht/Institute of Education, London.

McSharry, G., & Jones, S. (2000). Role-play in science teaching and learning. *School Science Review, 82*(298), 73–82.

Ødegaard, M. (2003). Dramatic science. A critical review of drama in science education. *Studies in Science Education, 39*, 75–102.

O'Neill, C. (1985). Imagined worlds in theatre and drama. *Theory into Practice, 24*(3), 158–165.

O'Neill, C. (1995). *Drama worlds: A framework for process drama.* Portsmouth: Heinemann Drama.

Savulescu, J., ter Meulen, R., & Kahane, G. (Eds.). (2011). *Enhancing human capacities.* Oxford: Wiley.

Shepherd-Barr, K. (2006). *Science on stage: From Doctor Faustus to Copenhagen.* Princeton: Princeton University Press.

Toonders, W., Verhoeff, R. P., & Zwart, H. (2016). Performing the future. On the use of drama in philosophy courses for science students. *Science & Education.* doi:10.1007/s11191-016-9853-3.

Wieringa, N. F., Swart, J. A., Maples, T., Witmondt, L., Tobi, H., & van der Windt, H. J. (2011). Science Theatre at School: Providing a context to learn about socio-scientific issues. *International Journal of Science Education, Part B, 1*(1), 71–96.

Zwart, H. (2014). Limitless as a neuro-pharmaceutical experiment and as a Daseinsanalyse: On the use of fiction in preparatory debates on cognitive enhancement. *Medicine, Health Care & Philosophy: A European Journal, 17*(1), 29–38.

Useful Plants as Potential Flagship Species to Counteract Plant Blindness

Peter Pany and Christine Heidinger

Introduction: Plant Blindness

"We are all more interested in animals [than in plants]" (Flannery 1991). This statement captures best the atmosphere a biology teacher is confronted with in any classroom when he/she starts a lesson on botanical content (Wandersee 1986). The low interest in plant science has been lamented for decades by biology educaters on every level – from primary school up to university level (Tunnicliffe and Ueckert 2007). Additionally, major studies on students' interests (e.g. ROSE; Sjøberg and Schreiner 2010) confirm that students do not consider plants to be interesting. Wandersee and Schussler (2001) have coined the term "plant blindness" for this phenomenon, describing how plants are overlooked in many peoples' everyday life. The fragmentary perception of herbal life has serious consequences because students, for example, do not perceive plants as creatures but consider them merely as a kind of "background image" for animal life (Flannery 2002; Kinchin 1999; Sanders et al. 2015).

Plant blindness is a serious problem in botany education, and special efforts must be made to make botanical content more attractive for students (Greenfield 1955; Hershey 1992). Numerous programmes emphasise the importance of plants in school (e.g. "PlantingScience" www.plantingscience.org or "Biological Sciences Curriculum Study – BSCS" www.bcsc.org). These programmes range from plant development observation programmes (Hershey 1992, 2002, 2005), the investigation

P. Pany (✉)
University of Vienna, Vienna, Austria

Botanical Garden of the University of Vienna, Vienna, Austria
e-mail: peter.pany@univie.ac.at

C. Heidinger
University of Vienna, Vienna, Austria

© Springer International Publishing AG 2017
K. Hahl et al. (eds.), *Cognitive and Affective Aspects in Science Education Research*, Contributions from Science Education Research 3,
DOI 10.1007/978-3-319-58685-4_10

of the diversity of plants through field trips (Dillon et al. 2006; Drissner et al. 2010; Fančovičová and Prokop 2010; Pany 2010; Vaughan et al. 2003) to activities of sorting plants (Frisch et al. 2010; Lindemann-Matthies 2005; Randler 2008).

Our approach to counteract plant blindness presented in this chapter is different. Instead of developing learning environments to enhance students' interest in and awareness of plants, we start one step earlier – by exploring the pre-existing interest of students in plants. Planning botany learning environments based on students' pre-existing interest in plants would have many advantages according to psychological theories on learning and interest: considering an object "interesting" is an important condition to deal with the object more intensively (Deci and Ryan 1993), and pre-existing interests are important keys for building new knowledge and developing long-lasting interests (Hidi and Baird 1986; Hidi 1990; Krapp 1999). Nonetheless, detailed studies about which plants students perceive as interesting are scarce. Hence, exploring which plants are interesting for students is a prerequisite for effectively counteracting plant blindness.

Students' Interest in Plants

A review of recent scientific literature on students' interest in plants is not encouraging for botanists. As noted above, the largest international study on students' interests in science and technology topics ("Relevance of Science Education" – ROSE; Schreiner 2006; Schreiner and Sjøberg 2004; Sjøberg and Schreiner 2010) demonstrates that botanical content is uninteresting. Zoology or human biology is much more interesting than plant science, a result also shown by earlier interest studies (Baram-Tsabari et al. 2010; Baram-Tsabari and Yarden 2005; Kinchin 1999; Wandersee 1986). Moreover, interest in biological content decreases with age (Baram-Tsabari and Yarden 2007, 2009; Kattmann 2000; Löwe 1987, 1992; Potvin and Hasni 2014).

Hence, some scientists in the field of biology education recommend teaching central biological concepts such as evolution only in the context of those organisms that students consider interesting (e.g. animals) (Baram-Tsabari and Yarden 2009). This, however, would lead to many biology lessons not addressing plants at all. Given that knowledge about plants is necessary to become scientifically literate and to understand the major global challenges our society is currently facing, this cannot be an acceptable solution for biology education. From a biological point of view, we need knowledge about plant anatomy and morphology, plant reproduction and flowering ecology in order to understand the role of plants in the world food supply or biofuel production. Furthermore, botany basics such as plant physiology (e.g. photosynthesis) are necessary for developing a deeper understanding of the carbon cycle and climate change. Consequently, students' lack of knowledge about plants hinders them from seeing the full extent of such important problems as global warming. We must therefore face the challenge to make presumably uninteresting but biologically important content interesting for students.

Analysing the recent literature reveals that plants are always treated in a holistic way as a homogeneous group (e.g. Blankenburg et al. 2015; Dawson 2000; Sjøberg 2000). In the ROSE-study, for example, students were asked very general questions about their interest in botanical topics, e.g. "How plants grow and reproduce" or "Plants in my locality" (Schreiner and Sjøberg 2004). Importantly, the context in which a specific content is presented is even more important for developing interest than the content itself (Elster 2007; Häussler and Hoffmann 1998; Sjøberg 2000). This calls for differentiating and identifying distinct plant groups and contexts that may be interesting for students rather than doing research on plants on a very general level.

The first hints that useful plants could be worth further examination came from Mayer and Horn (1993), who show that students prefer living organisms that are of value for human use. In addition, Krüger and Burmester (2005) determined that besides the "beauty of plants" (Kinchin 1999; Tunnicliffe and Reiss 2000), the "usefulness of plants" is the most prominent category students use to classify plants. The study by Lindemann-Matthies (2005) tends to support these findings: organisms which are useful for humans have a higher value for most people. Hammann (2011) also supports the hypothesis that useful plants are interesting for students by showing that students are highly interested in medicinal plants.

On this basis, we designed the present study. We chose the group of useful plants as a starting point in our exploration of students' pre-existing interest in plants. First, we studied whether useful plants are interesting for students and whether students differentiate between specific subgroups of useful plants. Then we investigated whether there are differences between different age groups and genders regarding the interest in useful plants. Based on these findings, we tackle the overall question which plant groups can be recommended as promising "flagship species" to teach and learn botanical content.

Method: The FEIN Questionnaire

In order to investigate students' interest in useful plants, we developed the FEIN questionnaire (Fragebogen zur **E**rhebung des **I**nteresses an **N**utzpflanzen; Pany 2014) since the research of Urhahne et al. (2004) suggests that questionnaires are appropriate tools to explore students' interest. The item development was based on the biological classification of useful plants (Lieberei et al. 2007) in which useful plants are defined as all plant species used by humans and in which various subgroups are differentiated according to their specific purpose (spice plants, edible plants, medicinal plants, etc.).

On this basis, we developed three items for each of the five scales of medicinal plants, stimulant herbal drugs, edible plants, spice plants and ornamental plants. The wording of the items is based on the ROSE questionnaire (Schreiner and Sjøberg 2004). They are formulated as headlines, and students indicate their interest

Table 1 Questionnaire items to investigate interest in useful plants (FEIN, Pany 2014); this translation gives an impression of the items used in the original German questionnaire: the English items are not linguistically validated

English translation of the FEIN questionnaire	German version (original language)
How interested are you in learning about the following?	*Wie interessiert bist Du an folgenden Bereichen?*
1. In which countries do vegetables (e.g. tomatoes) grow naturally (EP)	1. In welchen Ländern verschiedene Gemüsepflanzen (z.B. Tomate) in der freien Natur wachsen
2. Plants used to produce narcotics (SHD)	2. Pflanzen, aus denen Rauschmittel erzeugt werden können
3. Plants used to cure inflammations (e.g. a sore throat) (MP)	3. Pflanzen, die gegen Entzündungen (z.B. Halsschmerzen) helfen
4. Parts of plants used to produce oregano, chilli or caraway (SP)	4. Pflanzenteile zur Herstellung von z.B. Oregano, Chili oder Kümmel
5. Plants for decorating my room (OP)	5. Pflanzen zur Verschönerung meines Zimmers
6. Organic agriculture (EP)	6. Biologische Landwirtschaft
7. Plants which can cause hallucinations (SHD)	7. Pflanzen, die Halluzinationen erzeugen können
8. Plants which enhance the healing process of wounds (MP)	8. Pflanzen, welche die Heilung von Wunden unterstützen
9. Spice plants (SP)	9. Gewürzpflanzen
10. Taking care of house plants (OP)	10. Die Pflege von Zimmerpflanzen
11. Horticulture without pesticides (EP)	11. Gartenbau ohne Spritzmittel
12. Producing opium and heroin from opium poppy (SHD)	12. Die Gewinnung von Opium und Heroin aus dem Schlafmohn
13. Plants which can be used to produce a soothing infusion (e.g. against coughs) (MP)	13. Pflanzen, aus denen man einen heilenden Tee (z.B. gegen Husten) machen kann
14. Substances that make spices taste hot (SP)	14. Inhaltsstoffe, die Gewürze scharf schmecken lassen
15. Balcony flowers (OP)	15. Blumen an Fensterbänken

The assignment to the respective scale is given at the end of each item (*MP* medicinal plants, *SHD* stimulant herbal drugs, *SP* spice plants, *EP* edible plants, *OP* ornamental plants)

by choosing from a four-stage Likert scale (very interested, 4; rather interested, 3; rather not interested, 2; not interested, 1) (Table 1).

A principal component analysis (PCA) revealed a five-factor structure (Pany 2014), confirming the differentiation of subgroups of useful plants according to Lieberei et al. (2007). This shows that students' interest in useful plants is not homogeneous across all subgroups but has to be examined in a more differentiated way.

Reliability analyses (Cronbach's alpha) were calculated for each scale of the FEIN questionnaire. Cronbach's alpha shows values between 0.66 for spice plants and 0.76 for ornamental plants, which is appropriate for scales consisting of only three items each.

Table 2 Descriptive data of the investigated sample ($n = 1299$)

Age group	<13 years	13–14 years	15–16 years	>16 years	Total
Male students	245	197	159	62	*663*
Female students	236	193	137	70	*636*
Total	**481 (37%)**	**390 (30%)**	**296 (23%)**	**132 (10%)**	*1299*
Mean age (year)	**11.2**	**13.5**	**15.5**	**17.6**	*14.4*

The FEIN questionnaire was administered to 1417 students in and around Vienna, Austria. During spring 2010, 15 secondary schools participated in our study, providing a representative cross section of Viennese schools. Finally, 1299 questionnaires were filled in by 51% male and 49% female students aged between 10 and 19 years (Table 2).

Data Analysis

We analysed the questionnaire data on two levels. In the first step, we sought to identify significant differences between relevant groups in our sample (age and gender) regarding their interest in the five groups of useful plants. We therefore compared the means of interest per scale of the FEIN questionnaire of four age groups (<13, 13–14, 15–16 and >16 years) and the two gender groups using t-test, MANOVA, ANOVA and post hoc tests (Scheffé).

As mean values do not allow conclusions on an individual level (Valsiner 1986), we calculated in the second step an interest profile for every student in the sample. When planning stimulating and interesting learning environments in school, knowing what an "average student" is interested in is not very helpful. It is indispensable to know the interest structure of individual students in a particular class. Therefore, we developed a method to calculate interest profiles on an individual level. For this purpose, we first developed a method to reduce the complexity of the data per participant. This achieved a reduction level which also considers the variation of the individual interest structure of each student and enables clustering students to larger units showing identical patterns of interest in terms of "interest types" (= groups of students with similar interest structure). This process of complexity reduction is described in detail in Pany and Heidinger (2015).

The resulting interest profiles are based on each student's interest in three subgroups of useful plants: medicinal plants, stimulant herbal drugs and ornamental plants.[1] Per subgroup of useful plants, three interest levels are generated, ranging from "low interest – level 1", "medium interest – level 2" to "high interest – level 3". A student's interest profile of "321", for example, means this student has high

---

[1] We take into account only those subgroups of useful plants, which best enable differentiating between different interest types because they show a clear deviation from an equal distribution in the whole sample (Pany and Heidinger 2015).

interest in medicinal plants, medium interest in stimulant herbal drugs and only low interest in ornamental plants (the amount of interest is always given in the same order: medicinal plants, stimulant herbal drugs, ornamental plants). Subsequently, we calculated whether there are characteristic interest profiles in the whole sample and for different age groups using frequency analysis.

Results

Mean Values

Data analyses using ANOVA show that medicinal plants, stimulant herbal drugs, spice plants, edible plants and ornamental plants raise different degrees of interest ($F_{4, 6490} = 202.5$, $P < 0.001$). Mean values show that medicinal plants are the most interesting group, followed by stimulant herbal drugs (Table 3).

Additionally, MANOVA results show that there are differences in how interested students of different age groups are in the five plant groups (*Wilks's* $\Lambda = 0.922$, $F_{15, 3564} = 7.074$, $P < 0.001$). Subsequent univariate analysis (ANOVA) with post hoc Scheffé tests revealed significant differences regarding the interest in subgroups of useful plants between different age groups (Table 4). The interest in medicinal

Table 3 Means (*M*) and standard deviations (SD) of interest in different plant groups measured with the FEIN questionnaire; means above 2.5 indicate above-average interest; all means are significantly different from each other ($P < 0.001$)

Plant group	M	SD
Medicinal plants	3.09	0.75
Stimulant herbal drugs	2.90	0.88
Spice plants	2.56	0.78
Edible plants	2.43	0.78
Ornamental plants	2.32	0.89

$F_{4, 6490} = 202.5$, $P < 0.001$

Table 4 Means (*M*), standard deviations (SD) and univariate F-statistics of interest in the subscales of the FEIN questionnaire for different age groups

Subscale	<13 years M	SD	13–14 years M	SD	15–16 years M	SD	>16 years M	SD	$F_{3, 1295}$	
Medicinal plants	3.19	0.73	2.96	0.76	3.01	0.78	3.26	0.63	14.268	**
Stimulant herbal drugs	2.87	0.88	2.89	0.90	2.90	0.86	3.01	0.86	0.873	
Spice plants	2.67	0.80	2.50	0.80	2.50	0.71	2.52	0.76	10.631	*
Edible plants	2.61	0.79	2.32	0.80	2.30	0.70	2.39	0.74	4.623	**
Ornamental plants	2.54	0.90	2.31	0.90	2.05	0.80	2.16	0.82	20.906	**

*$P < 0.01$, **$P < 0.001$

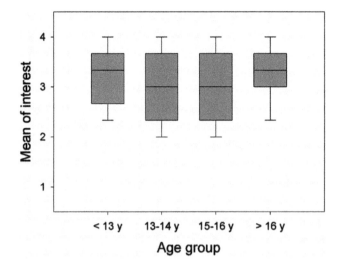

Fig. 1 Means of interest of the subscale "medicinal plants" for all age groups

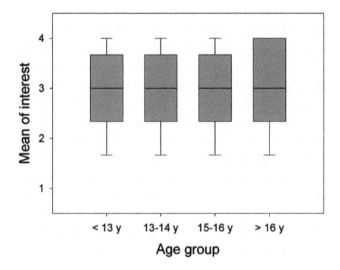

Fig. 2 Means of interest of the subscale "stimulant herbal drugs" for all age groups

plants is higher in younger students (<13 years) and older students (>16 years) but lower in the other age groups (Fig. 1), whereas the interest in stimulant herbal drugs shows no significant differences between the age groups (Fig. 2). Furthermore, only ornamental plants show significant gender differences. They are more interesting for girls than for boys ($t = -11.72$, df = 1298, $P < 0.001$) (Fig. 3).

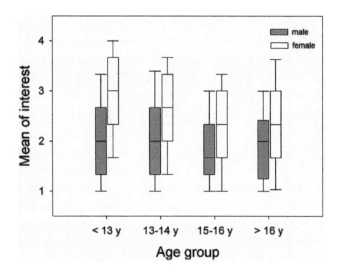

Fig. 3 Means of interest of the subscale "ornamental plants" for all age groups and both genders shown separately

Table 5 Characteristics of the ten most frequent interest types in the whole sample; marked interest types (*) are not evenly distributed among the age groups

	Interest in				
	Medicinal plants	Stimulant herbal drugs	Ornamental plants	Frequency percent	
331	*High*	*High*	*Low*	18.9	*
333	*High*	*High*	*High*	11.6	*
332	High	High	Medium	8.9	
313	*High*	*Low*	*High*	6.7	*
231	Medium	High	Low	6.2	
323	*High*	*Medium*	*High*	5.5	*
311	High	Low	Low	4.8	
131	*Low*	*High*	*Low*	4.3	*
321	High	Medium	Low	4.2	
232	Medium	High	Medium	3.8	

Interest Profiles

The ten most frequent interest profiles were chosen to give an overall impression of the sample. Table 5 shows that most of the students have low interest in at least one of the subgroups of useful plants.

In contrast, we found that most of the students are interested in at least one plant group more than in the others. Moreover, five of the interest types (331, 333, 313, 323 and 131) are not evenly distributed among the age groups (see Fig. 4 and Table 6).

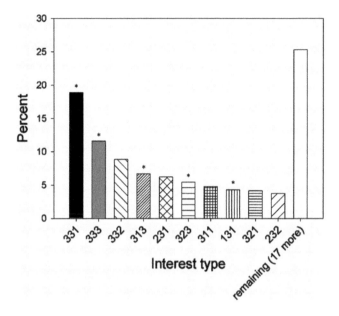

Fig. 4 Frequencies of the ten most frequent interest types within the whole sample; marked interest types (*) are not evenly distributed among the age groups. The amount of interest is always given in the same order: medicinal plants, stimulant herbal drugs, ornamental plants; 3 – high interest, 2 – medium interest, 1 – low interest

The interest profiles 313 and 323 are typical for lower age groups, which means that many younger students are most interested in medicinal plants and ornamental plants but show low interest in stimulant herbal drugs. In contrast, the interest profiles 311 and 321, which indicate a narrow interest restricted solely to medicinal plants, are frequent only in higher age groups. For students between 13 and 16 years old, we found another interest profile occurring only here within the first five ranks: 131 (and also 231), indicating high interest only in stimulant herbal drugs. In summary, there are typical interest profiles for each age group (Fig. 5).

Discussion

Our results show that students do not perceive plants as one homogeneous group of (uninteresting) organisms, as they have often been treated in past studies investigating students' interest in biology (e.g. Blankenburg et al. 2015; Schreiner and Sjøberg 2004). Accordingly, past recommendations for planning biology lessons derived from such a rough scale should be treated with caution: they may yield the misleading conclusion that botanical content and plant science are not interesting for students at all. Quite the contrary, the group of useful plants contains many objects that are suitable to develop interesting learning contexts for botanical contents referring to students' interests.

Table 6 Chi-square values for the distribution of the ten most frequent interest types within the whole sample; marked interest types (*) are not evenly distributed among the age groups

	331	333	332	313	231	323	311	131	321	232
Chi-square	12.008	15.385	4.443	22.003	7.465	8.156	3.169	10.198	4.813	2.215
df	3	3	3	3	3	3	3	3	3	3
Significance	0.007*	0.002*	0.217	<0.001*	0.058	0.043*	0.366	0.017*	0.186	0.529

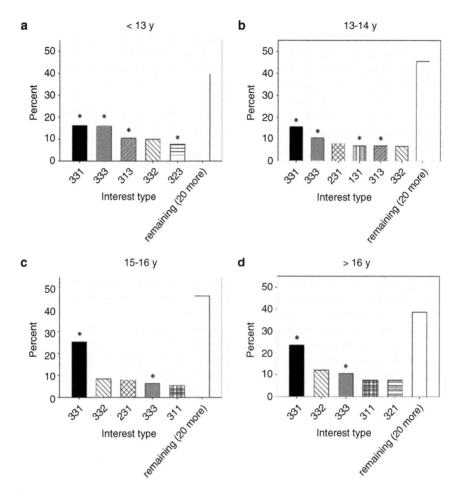

Fig. 5 Frequencies of all interest profiles representing more than 5% of an age group; marked interest types (*) are not evenly distributed among the age groups. The *bars* of identical interest types in the diagram are marked by the same patterns

The mean values seem to point to a clear strategy for botany lessons – medicinal plants and stimulant herbal drugs should be used as study objects in botany lessons. These two plant groups do not show the typical decrease of interest in higher age groups (Elster 2007; Kattmann 2000; Löwe 1987) but instead retain their high interest. However, the analysis of students' interest profiles shows a somewhat different picture. Nonetheless, the most frequent interest types still show high interest in medicinal plants, although stimulant herbal drugs seem to strongly polarise students. Especially in lower age groups, some students show no interest at all in stimulant herbal drugs (e.g. interest profile 313), whereas another group of students shows high interest only in simulant herbal drugs (e.g. interest profile 131). Furthermore, ornamental plants – raising only low interest when examining mean values – are highly interesting for a smaller group of students (e.g. interest profile 313).

Facing the difficulties that plant blindness presents to biology teachers, useful plants may open a wide field of reasonable gateways into botany. We identified interest profiles that are typical only for single age groups, which means that student interest can be addressed very specifically. Students younger than 13 years old may be addressed using medicinal plants (such as sage *Salvia officinalis*, hawthorn *Crataegus* spp. or marigold *Calendula officinalis*) and also ornamental plants (e.g. primroses *Primula* spp., tulips *Tulipa* sp.) but less via stimulant herbal drugs. A subgroup of students between 13 and 16 years can be specifically targeted by using study objects from the only plant subgroup interesting for them: stimulant herbal drugs (e.g. belladonna *Atropa belladonna*). The group of medicinal plants is very interesting for a large number of students across all age groups but especially for older students (above 16 years).

Some stumbling blocks still remain for botany lessons. Biology schoolbooks that cover botany topics listed in the biology curriculum (e.g. the structure of plants and flowers or photosynthesis) (Cholewa et al. 2010; Rogl and Bergmann 2003; Schirl and Möslinger 2011) and even biology textbooks at the university level (Campbell and Reece 2011) introduce such botanical content mostly based on ornamental plants. Our findings demonstrate that this choice complies with the interest of only a small part of learners.

Therefore, incorporating the results of the present study not only in learning environments but also in biology textbooks may help to create interesting contexts in botany lessons, supporting students to find access to botanical contents. One open field for prospective studies remains: experimental designs that enable testing the hypothesis whether study objects which take into account students' interest in plants indeed raise long-lasting interest and lead to higher learning outcome regarding botanical topics. At any rate, the present results offer a promising approach to counteract the unsatisfactory situation in which students neglect the vast majority of life on Earth.

References

Baram-Tsabari, A., & Yarden, A. (2005). Characterizing children's spontaneous interests in science and technology. *International Journal of Science Education, 27*(7), 803–826.

Baram-Tsabari, A., & Yarden, A. (2007). Interest in biology: A developmental shift characterized using self-generated questions. *The American Biology Teacher, 69*(9), 532–540.

Baram-Tsabari, A., & Yarden, A. (2009). Identifying meta-clusters of students' interest in science and their change with age. *Journal of Research in Science Teaching, 46*(9), 999–1022.

Baram-Tsabari, A., Sethi, R. J., Bry, L., & Yarden, A. (2010). Identifying students' interests in biology using a decade of self-generated questions. *Eurasia Journal of Mathematics, Science & Technology Education, 6*(1), 63–75.

Blankenburg, J. S., Höffler, T. N., & Parchmann, I. (2015). Fostering today what is needed tomorrow: Investigating students' interest in science. *Science Education, 100*(2), 364–391.

Campbell, N. A., & Reece, J. B. (2011). *Biology*. San Francisco: Pearson.

Cholewa, G., Driza, M., Einhorn, S., & Felling, J. (2010). Vom Leben [*About life*] (Vol. 1). Wien: Ed. Hölzel.

Dawson, C. (2000). Upper primary boys' and girls' interests in science: Have they changed since 1980? *International Journal of Science Education, 22*(6), 557–570.

Deci, E. L., & Ryan, R. M. (1993). Die Selbstbestimmungstheorie der Motivation und ihre Bedeutung für die Pädagogik [The self-determination-theory of motivation and its relevance for pedagogy]. *Zeitschrift Für Pädagogik, 39*(2), 223–238.

Dillon, J., Rickinson, M., Teamey, K., Morris, M., Choi, M. Y., Sanders, D., & Benefield, P. (2006). The value of outdoor learning: Evidence from research in the UK and elsewhere. *School Science Review, 87*(320), 107.

Drissner, J., Haase, H.-M., & Hille, K. (2010). Short-term environmental education – Does it work? – An evaluation of the 'Green Classroom'. *Journal of Biological Education, 44*(4), 149–155.

Elster, D. (2007). Student interests – The German and Austrian ROSE survey. *Journal of Biological Education, 42*(1), 5–10.

Fančovičová, J., & Prokop, P. (2010). Development and initial psychometric assessment of the plant attitude questionnaire. *Journal of Science Education and Technology, 19*(5), 415–421.

Flannery, M. C. (1991). Considering plants. *The American Biology Teacher, 53*(5), 306–309.

Flannery, M. C. (2002). Do plants have to be intelligent? *The American Biology Teacher, 64*(8), 628–633.

Frisch, J. K., Unwin, M. M., & Saunders, G. W. (2010). *Name that plant! Overcoming plant blindness and developing a sense of place using science and environmental education.* Dordrecht: Springer.

Greenfield, S. S. (1955). The challenge to botanists. *Challenge, 1*(1). Retrieved from https://secure.botany.org/plantsciencebulletin/psb-1955-01-1.php

Hammann, M. (2011). Wie groß ist das Interesse von Schülern an Heilpflanzen? [How interested are students in medicinal plants?]. *Zeitschrift Für Phytotherapie, 32*(01), 15–19.

Häussler, P., & Hoffmann, L. (1998). Chancengleichheit für Mädchen im Physikunterricht. Ergebnisse eines erweiterten BLK-Modellversuchs. [Equal opportunities for girls in physics education. Results from the extended BLK-pilot project]. *Zeitschrift Für Didaktik Der Naturwissenschaften, 4*(1), 51–67.

Hershey, D. R. (1992). Making plant biology curricula relevant. *Bioscience, 42*(3), 188–191.

Hershey, D. R. (2002). Plant blindness: 'We have met the enemy and he is us'. *Plant Science Bulletin, 48*(3), 78–84.

Hershey, D. R. (2005). *Plant content in the national science education standards.* Retrieved from http://www.Actionbioscience.Org/education/hershey2.Html. 20 Dec 2016.

Hidi, S. (1990). Interest and its contribution as a mental resource for learning. *Review of Educational Research, 60*(4), 549–571.

Hidi, S., & Baird, W. (1986). Interestingness-A neglected variable in discourse processing. *Cognitive Science, 10*(2), 179–194.

Kattmann, U. (2000). Lernmotivation und Interesse im Biologieunterricht [Motivation and interest in biology education]. *Lehren Und Lernen Im Biologieunterricht,* 13–31.

Kinchin, I. M. (1999). Educational research-investigating secondary-school girls' preferences for animals or plants: A simple' head-to-head' comparison using two unfamiliar organisms-A direct comparison of two. *Journal of Biological Education, 33*(2), 95–99.

Krapp, A. (1999). Interest, motivation and learning: An educational-psychological perspective. *European Journal of Psychology of Education, 14*(1), 23–40.

Krüger, D., & Burmester, A. (2005). Wie Schüler Pflanzen ordnen [How do students classify plants?]. *Zeitschrift Für Didaktik Der Naturwissenschaften, 11,* 85–102.

Lieberei, R., Reisdorff, C., & Franke, W. (2007). Nutzpflanzenkunde: Nutzbare Gewächse der gemässigten breiten, Subtropen und Tropen [*Useful plants: Useful plants of temperate regions, subtropics and tropics*]. Stuttgart: Georg Thieme Verlag.

Lindemann-Matthies, P. (2005). 'Loveable' mammals and 'lifeless' plants: How children's interest in common local organisms can be enhanced through observation of nature. *International Journal of Science Education, 27*(6), 655–677.

Löwe, B. (1987). Interessenverfall im Biologieunterricht [Decrease of interest in biology education]. *Unterricht Biologie, 124*, 62–65.

Löwe, B. (1992). *Biologieunterricht und Schülerinteressen an Biologie [Students' interest in biology]*. Weinheim: Dt. Studien-Verl.

Mayer, J., & Horn, F. (1993). Formenkenntnis–wozu [Knowledge about plants forms and taxonomy – What for?]. *Unterricht Biologie, 189*(17), 4–13.

Pany, P. (2010). Ausgedörrt und abgetreten. Über das widrige (?) Leben von Pflanzen in Pflasterritzen [Dried and trampled down. About the hard life of plants in paving cracks]. *Umwelt & Bildung, 1*, 19–21.

Pany, P. (2014). Students' interest in useful plants: A potential key to counteract plant blindness. *Plant Science Bulletin, 60*(1), 18–27.

Pany, P., & Heidinger, C. (2015). Uncovering patterns of interest in useful plants – Frequency analysis of individual students' interest types as a tool for planning botany teaching units. *Multidisciplinary Journal for Education, Social and Technological Sciences, 1*(1), 15–39.

Potvin, P., & Hasni, A. (2014). Analysis of the decline in interest towards school science and technology from grades 5 through 11. *Journal of Science Education and Technology, 23*(6), 784–802.

Randler, C. (2008). Teaching species identification—A prerequisite for learning biodiversity and understanding ecology. *Eurasia Journal of Mathematics, Science & Technology Education, 4*(3), 223–231.

Rogl, H., & Bergmann, L. (2003). Biologie aktiv [*Active biology*] (Vol. 1). Graz: Leykam.

Sanders, D., Nyberg, E., Eriksen, B., & Snæbjørnsdóttir, B. (2015). 'Plant blindness': Time to find a cure. *The Biologist, 62*(3), 9.

Schirl, K., & Möslinger, E. (2011). Expedition Biologie [*Biological expedition*] (Vol. 1). Wien: Dorner.

Schreiner, C. (2006). *Exploring a ROSE-Garden*. Oslo: Department of Teacher Education and School Development Faculty of Education, University of Oslo.

Schreiner, C., & Sjøberg, S. (2004). *Sowing the seeds of ROSE. Background, rationale, questionnaire development and data collection for ROSE (The relevance of science education)–A comparative study of students' views of science and science education (pdf)*, Acta Didactica 4/2004. Oslo: Department of Teacher Education and School Development, University of Oslo.

Sjøberg, S. (2000). Science and scientists: The SAS study. *Acta Didactica, 1*, 1–73.

Sjøberg, S., & Schreiner, C. (2010). *The ROSE project: An overview and key findings*. Oslo: University of Oslo.

Tunnicliffe, S. D., & Reiss, M. J. (2000). Building a model of the environment: How do children see plants? *Journal of Biological Education, 34*(4), 172–177.

Tunnicliffe, S. D., & Ueckert, C. (2007). Teaching biology—The great dilemma. *Journal of Biological Education, 41*(2), 51–52.

Urhahne, D., Jeschke, J., Krombaß, A., & Harms, U. (2004). Die Validierung von Fragebogenerhebungen zum Interesse an Tieren und Pflanzen durch computergestützte Messdaten [Using computer-based data to validate a questionnaire measuring interest in animals and plants]. *Zeitschrift Für Pädagogische Psychologie, 18*(3), 213–219.

Valsiner, J. (1986). Between groups and individuals. *The Individual Subject and Scientific Psychology*, 113–151.

Vaughan, C., Gack, J., Solorazano, H., & Ray, R. (2003). The effect of environmental education on schoolchildren, their parents, and community members: A study of intergenerational and intercommunity learning. *The Journal of Environmental Education, 34*(3), 12–21.

Wandersee, J. H. (1986). Plants or animals – Which do junior high school students prefer to study? *Journal of Research in Science Teaching, 23*(5), 415–426.

Wandersee, J. H., & Schussler, E. (2001). Toward a theory of plant blindness. *Plant Science Bulletin, 47*(1), 2–9.

Digital Videos of Experiments Produced by Students: Learning Possibilities

Wilmo Ernesto Francisco Junior

Introduction

The relevance of practical activities in science education seems to be undeniable. At the same time, a simple inclusion of experiments has been insufficient to ensure learning science, and thus the role of lab work is not self-evident (Hofstein and Lunetta 2004). Hodson (1993) describes that the emphasis in procedural aspects has interfered in work lab quality in terms of learning. In this manner, reflection, which has been frequently stated as more significant than interaction with materials, can become secondary. Additionally, the practical work needs to go beyond experiments that merely require following a "cookbook" recipe and move toward critical thinking about the results obtained during lab work, including the unexpected results.

Hence, it would be important to establish connections between practical work and discussion, analysis, interpretation, and social interactions as well as social validation and communication of the results. These elements serve to forge pathways for the production of science, including its processes and products. Moreover, the role of language as a way to learn science should be considered (Mortimer and Scott 2003).

In this context, digital technologies arise as additional tools for the inclusion of practical activities in the learning process. Some studies have pointed out positive aspects of audiovisual language such as pleasure, creativity, and social engagement (Pereira et al. 2012). Digital videos enable detailed observations of experiments or daily life events, which can make science more relevant to students. Besides, videos can also be used to improve experimental skills, encouraging students' engagement in different activities, both instrumental (i.e., hands on) and cognitive (i.e., minds on) (Rodrigues et al. 2001; Erdmann and March 2014).

W.E. Francisco Junior (✉)
Universidade Federal de Alagoas, Arapiraca, Brazil
e-mail: wilmojr@bol.com.br

© Springer International Publishing AG 2017
K. Hahl et al. (eds.), *Cognitive and Affective Aspects in Science Education Research*, Contributions from Science Education Research 3,
DOI 10.1007/978-3-319-58685-4_11

Movies have been used in many ways in chemistry education. For instance, back in 1973, Rouda recorded some experiments (e.g., the vacuum technique, determination of vapor pressure, and the kinetics of a hydrolysis reaction) performed by students themselves. Students who actively participated in these videos became familiarized with the experiments in different aspects including the apparatus, theories, calculations, and communication skills. Similar findings were described by Lichter (2012), who investigated a general chemistry course in which students created and uploaded (to the video-sharing website YouTube) videos about solubility. According to Lichter, the students who produced videos achieved significantly better learning than the rest of their classmates. However, most of these studies have not focused on the process of video production in order to evaluate the students' engagement.

The concept of school or student engagement has attracted growing interest in different realms. Some studies have investigated how the social contexts and school climate can interfere with students' learning (Vedder-Weiss and Fortus 2013; Sha et al. 2016) or cause dropping out of school (Connell et al. 1995). The relationship between instruction, teacher performance, and intellectual engagement has been also studied (Polman and Hope 2014). On account of this wide set of factors, student engagement is a complex construct that can be identified from different perspectives. In spite of these differences, there are some characteristics that allow classifying student engagement in three major groups: behavioral, emotional, and cognitive engagement (Fredricks et al. 2004).

Behavioral engagement is usually associated with commitment to learning and academic tasks, unveiling attitudes such as effort, persistence, concentration, and attention (Birch and Ladd 1997). The capability to work autonomously, the self-directed academic behaviors, and the collaborative actions are also some characteristics of behavioral engagement (Birch and Ladd 1997; Buhs and Ladd 2001). This engagement is considered essential for achieving positive academic outcomes (Fredricks et al. 2004).

The emotional engagement is associated with students' feelings and their affective reactions, such as interest, boredom, happiness, sadness, and anxiety (Fredricks et al. 2004). In general, this type of engagement involves positive and negative reactions toward teachers, classmates, and educational institutions that influence their willingness to do a specific task. Finn and Voelkl (1993) described it as identification with the school.

In turn, the cognitive engagement can be conceptualized as a psychological investment in learning that promotes improvements in comprehension by using self-regulated strategies (Fredricks et al. 2004). In a theoretical view, this engagement is possible when it involves problem solving, hard work, and ways of coping with perceived failure. At the same time, it is important to highlight the integration of "doing the work," which primarily involves behavioral and mental efforts to deeply understand some theoretical aspect. Students exerting more mental effort create more connections among ideas and may achieve greater understanding of concepts.

Engagement is fundamental to achieve learning. Thus, students need to be actively involved in a meaning-making process wherein they interpret, create, and act on reality. Learning is seen herein as an action by which language (or languages) is transformed (Kress et al. 2001). This perspective is in agreement with the social semiotic background.

This chapter reports on an exploratory study that examines some contributions of video production to chemistry teaching based on the evaluation of videos produced by students and their opinions about the production process. The research question that guided this inquiry was: How can the production of digital videos of experiments foster the undergraduates' engagement and chemistry learning?

Methods

This study involved 31 students enrolled in a general chemistry course at a Brazilian Federal University. The students were divided into groups of two to four members, and each group selected a general chemistry topic for the experiment. The groups subsequently planned, set up, and tested the experimental situation. Each experiment was first presented to the teacher and then videotaped without prompting. A total of 11 videos were produced. The content of the videos and the process of production were analyzed under the students' point of view for this investigation.

The analysis of the videos took into account a research-based analysis model based on the tetrahedron chemistry by Mahaffy (2004) and modified by Sjöström (2013). The analysis model consists basically of four fields: pure chemistry (includes formulae/symbols and safety procedures), applied chemistry (comprises applications and everyday-life aspects), socio-chemistry (involves historical context, risks, and benefits of chemistry, larger cultural milieu), and nature of chemistry (encompasses descriptions, explanations, analysis, synthesis, and knowledge uncertainties). In this perspective, chemistry education includes not only content knowledge in chemistry but also knowledge about chemistry, the nature of chemistry, and its role in society.

Structural components related to aesthetic characteristics were also evaluated based on the film analysis (Vanoye and Goliot-Lété 2013). Film analysis allows interpreting the video production as a cultural production from two different steps: deconstruction and reconstruction. Basically, deconstruction represents a description of the video components, whereas reconstruction presents the interpretation of these components within the video context and its production.

Information about the production process was obtained through a Likert-type questionnaire containing ten items, which also required explanatory answers. The questionnaire was structured following the Technology Acceptance Model, which aims to understand the system's acceptability to users. This model is based on the perceived usefulness, perceived ease of use, and real use of the technology (Davis 1989). Perceived usefulness is related to the users' beliefs in using a technology system to improve their performance in a task. People tend to use or ignore an application to the extent that they believe it can enhance their job/academic quality. Furthermore, even if the potential is recognized by the users, their beliefs about the degree of difficulty can contrast with the usefulness. Accordingly, the perceived ease of use is also a determinant factor in user behavior to accept or not accept a technology. Therefore, the decision of acceptance is concerned with the possibility for the real use of the technology.

Following these statements, the questionnaire items were constructed in order to attend to these definitions and to include questions about perceived usefulness (five questions), perceived ease of use (two questions), and real use (three questions). Explanatory questions were asked to evaluate which aspects of the video production plan were important (from the students' point of view), either in a positive or negative sense. Likert-type questions were counted and the percentage for each one was calculated. Explanatory answers were gathered by similarity generating excerpts that could be associated with engagement's characteristics. This analysis followed qualitative content analysis principles (Bardin 2011).

Results

Video Analysis

The audiovisual production showed a flexible structure in terms of filmic composition. For instance, the videos were filmed in different setups (i.e., six were produced at the university's laboratory, three of them at a home set laboratory, and two were recorded in both places). Movie length varied from 3 min 50 s to 6 min 32 s, and a soundtrack was employed in nine of the 11 videos. Video editing was carried out both linearly and nonlinearly. The videos included narration (except for one video), with two cases having been done in a voice-over. Credits and legends were inserted in ten videos and a making-of in two of them. The making-of exhibited not only a playful environment characterized by happiness and relaxation but seriousness and interest as well. These examples demonstrate both freedom and commitment during video creation.

It is worth noting that two videos produced in the laboratory also employed dramatization resources simulating a chemistry class in which a teacher discussed the experiment with the students. In these videos, jokes, laughs, and a relaxed situation (all elements of emotional engagement) were also identified. One video presented a film scene followed by a problem situation introduced to discuss some chemical aspects. In addition to these results, aesthetic elements (e.g., music, dramatization, picture/image, movie scenes, animation, and paintings, among others) were spontaneously inserted in all video productions. These resources enriched the videos and are related to diverse cultural dimensions. Considering that these elements were included by the students themselves, the variety of cultural aspects demonstrated creativity in that students created situations to present their experiments. These characteristics refer to positive reactions in academic tasks, revealing attitudes like effort and collaborative actions in order to produce material concatenated with students' desires and expectations.

With regard to chemistry aspects, students' concerns were mostly related to pure chemistry and to the nature of chemistry. The main results are summarized in Table 1.

Social issues including historical aspects, risks, and benefits of chemistry and cultural milieu were observed in two situations. For instance, in video 5, the context of the Second World War was presented to discuss chemical ethics, especially the employment of chemical weapons. In the same video, risks versus benefits were highlighted. The dark history of chemical weapons was contrasted with the advances in medicine. In turn, video 1 dealt with the pharmaceutical treatment of stomach acidity by using sodium bicarbonate or calcium hydroxide (milk of magnesia).

Materials found in our everyday life were presented in three of the videos. Examples included electricity generation by means of cells and batteries, food preservation (mainly the role of temperature control in chemical reactions), cooking process, refrigerator use, and the utilization of sodium chloride use to conserve meat in the past. Potassium permanganate application as a chicken pox medicine was also described.

The students mostly failed to explain phenomenological changes in terms of atomic-molecular theory. Most of them described experimental evidences without establishing a correlation with chemical phenomena as the following illustrates:

> This (phenomenological evidence) occurred due to a neutralization reaction between the acid (carbonic gas) and the alkaline (calcium hydroxide) substances.
>
> First, a little naphthalene ball stayed between honey and water. Then, we put a piece of paraffin that was between oil and alcohol. Afterward, we added a coin that was on the bottom because it is more dense than all liquids.

Table 1 Conceptual aspects assessed in the videos

Items analyzed	Video										
	1	2	3	4	5	6	7	8	9	10	11
Presentation of chemistry applications		X			X				X		
Presentation of social and cultural issues	P				X				X		
Correct description of the chemical process	X	X	X	X	X	X	P	X	X	X	P
Correct explanation of the chemical process					X			P			
Adequate use of chemistry nomenclature	X	X		X	X	X		P	X	X	
Correct presentation of experimental procedures	X		X	X	X	X		X	X		X
Correct presentation of chemistry equations		X						P			
Correct use of chemistry formulas		X			X			P			X
Safety recommendations		P									
Presentation of proper waste disposal											

"X" indicates the presence of the item analyzed. "Blank" indicates absence. "P" indicates partial presence

Although the videos offered descriptions, which are an important step in understanding chemistry, explanations were only given in one video (Table 1).

Students' Points of View

The analysis of students' viewpoints demonstrated a positive acceptance on the video production (Table 2). The answers about the usefulness of videos for learning and for stimulating creativity (items 1 and 7) revealed that all the students agreed "partially" or "totally" with the contributions afforded by video production. The positive statements were based on three aspects: the intensive work involved while producing the videos, the cooperative learning atmosphere, and the pleasure:

> This activity was important to foster a closer relationship. We had a lot of fun on our mistakes during the recording and (...) the video production was a new experience. We were free to choose what to do and this enabled us to use our creativity.

Table 2 Students' answers to the questionnaire ($N = 31$) on the video production activity

Statements	I totally agree	I partially agree	I'm un-decided	I partially disagree	I totally disagree
[1] Producing the video has helped me to understand chemistry concepts	10 (32.3%)	21 (67.7%)	0	0	0
[2] Producing the video has helped me to develop my technical skills	10 (32.2%)	19 (61.3%)	0	2 (6.5%)	0
[3] Producing the video has helped me to understand chemistry concepts applied to daily life	6 (19.3%)	21 (67.7%)	4 (13.0%)	0	0
[4] Producing the video was very difficult	0	23 (74.2%)	0	8 (25.8%)	0
[5] Producing the video was a pleasurable activity	21 (67.7%)	8 (25.8%)	2 (6.5%)	0	0
[6] My ability to explain a chemistry phenomenon has improved	6 (19.3%)	21 (67.7%)	0	2 (6.5%)	2 (6.5%)
[7] Producing the video stimulated my creativity	16 (51.6%)	15 (48.4%)	0	0	0
[8] I would like to participate in another activity like this one	15 (48.4%)	15 (48.4%)	1 (3.2%)	0	0
[9] I think video production may be an important teaching tool	16 (51.6%)	15 (48.4%)	0	0	0
[10] I would like using video production in my future pedagogical practice	12 (38.7%)	16 (51.6%)	3 (16.7%)	0	0

> We had to search for a lot of information about the experiment in order to explain what happened. This was a little hard in the beginning but doing it collaboratively helped us in our video.

Freedom and creativity appeared in 11 students' comments. These findings are in concordance to the video analysis that also showed playfulness, commitment, and collaboration actions during video production.

With respect to items 2 (improvement of technical skills), 3 (chemistry application concepts in everyday life), and 6 (explanations of concepts), which focus on usefulness as well, most of the students presented favorable answers, although some restrictions were mentioned regarding the need for someone to help them in technical abilities and chemical explanations:

> Difficulties to edit because my ICT knowledge was restricted. Recording was also hard because we had to combine the camera position, experimental control, time, and light. Sometimes the recording did not meet our expectations. So, we had to repeat it. Even so, we tried to do the best we could. On account of this, we had an enjoyable experience.
>
> The ability to explain a chemical phenomenon requires high cognitive skills. Although I have had other experiences, I was not able to explain some simple phenomena. We can just explain this one because we worked hard, together, and we sought out other sources. Perhaps if I was alone I wouldn't be able to explain the experiment at all.

Concerning the potential for the real use (8, 9, and 10), all the students were in agreement about the usefulness of video production as a teaching tool. Almost all students (96.8%) partially or totally agreed to participate in activities like this one again, and 83.3% of them would use video production as a teaching strategy. The students' justifications were often associated with playful, stimulating atmosphere, and social interaction:

> Video production was a playful activity that motivated me, giving me autonomy, stimulating me to know chemistry concepts, helping me to develop communicative synthesis skills, and promoting chemistry learning.
>
> It was quite interesting to participate, mainly because it was developed in a group in which we can interact more deeply with each other and to learn chemistry in a different and entertaining way. So, we could develop other skills, both social and technical.

Regarding the ease of video production (statements 4 and 5), most students (73%) had difficulties related to editing and recording techniques. In this sense, they suggested courses and specific attendance to solve these issues. Nevertheless, 93.5% of the participants stated that they had fun producing the videos.

Discussion and Conclusions

Video Analysis: Engagement and Learning Possibilities

The flexible filmic structure can be pointed out as an important magnitude of the audiovisual productions, given that such manifestations are concerned with the comprehension by students about the role of videos as a cultural expression. The flexible filmic structure, as can be seen by the different aesthetic elements included in the videos, and the presence of making-of and "bloopers" are related to autonomy, self-directed, and collaborative actions, which are characteristics of behavioral engagement (Fredricks et al. 2004). The climate of video production has also demonstrated interest and happiness, exhibited mainly in the making of the video and from the audio present in some videos.

The interest of students in searching and including cultural elements (e.g., music, photos, slow motion, and painting images) in their videos is related to willingness to do the task. The involvement in schoolwork such as pursuing one activity out of interest or for the pleasure of doing so has been positively associated with behavioral (e.g., participation and work involvement) and emotional (Fredricks et al. 2004) engagements. It can be understood as a psychological investment in learning, unveiling a desire to go beyond the requirements and a preference for challenge. These features are also included in the set of definitions of cognitive engagement (Connell et al. 1995; Newmann et al. 1992).

Considering the science as culture, videos would be a vehicle to converge different scientific dimensions. In different historic times, people have tried to use the knowledge about nature to make artistic representations of the reality or use art to represent scientific knowledge. Art deals with aesthetics, ideas of beauty, feelings, imagination, values, and so forth. Each one of those aspects may be relevant to learning science by performing an abstract idea in a somewhat concrete and even in a pleasing manner. Furthermore, it is possible to bridge the gap between science teaching and art with the aim to provide different dimensions of human knowledge. Meaning and appearance can be combined within an affective appeal, provoking a special feeling of pleasure of understanding (Galili 2013).

Video is not only a product, but it is a part of the process that drives a cultural and didactic production, a result from an action in which a dialogical relationship between product and process is established. The development of critical thinking relies on respecting and encouraging the curious, free, and creative action. This relationship between creativity, freedom, and playfulness can be noted in video production which reveals the three typologies of engagement (i.e., behavior, emotional, and cognitive). However, it is important to underline the importance of mental efforts to create connections among ideas and to achieve a greater understanding of the concepts.

In agreement with the comprehension of the learning as a transformative sign-making process, students should be capable of transforming the concrete world into a different notation. Cultural and aesthetic elements within the videos worked together to form a coherent text through a range of linguistic aspects that is the product from sign-making transformation.

Some videos tried to connect to everyday life, technology, society, chemical research, and history of chemistry, featuring aspects of applied chemistry and debating on the risks and benefits of chemistry. Concerning the perspective of the present work, these aspects can be considered important in the learning process, as videos contextualize chemistry by providing a broader perspective. By doing so, students demonstrated transformative sign-making by which learning can be achieved by the expression of a narrative that includes different dimensions of the tetrahedron, each one contributing to the whole communicative event. However, only three videos included this kind of discussion. A broader and problematized socio-perspective was missing.

In addition, the absence of discussions about safety, proper waste disposal, and its reduction is another aspect in which students need support. Although simple and low-risk materials were used in the experiments, the discussion about this issue cannot be secondary. Vilches and Gil-Pérez (2013: 1869) state that "chemical education is an ethically laden activity that can and must incorporate sustainability as an essential dimension."

Although the videos offered descriptions about the experiments, explanations in terms of atomic-molecular aspects, and chemical representations including reaction equations, were scarce. Chemistry knowledge can be considered multidimensional, and the tangible world (even applied and socio-chemistry) cannot be the only way to capture it. Thus, chemistry learning involves establishing links between macroscopic phenomena and theoretical models, mediated by a specific language (Mortimer and Scott 2003). However, other aims of chemistry, such as analyzing and synthesizing, were present in a significantly lower number of videos.

The acknowledgment of the meanings attributed to the chemical atomic-molecular level is of central importance from a pedagogical perspective. Therefore, focusing solely on descriptions is not enough to develop an understanding on what happens during an experiment. The phenomenon needs to be reinterpreted from a suitable theoretical model, mostly beyond the tangible and visible. Previous studies (Gabel 1999; Talanquer 2011) reported problems faced by students in building bridges between the phenomena and the intellectual tools used in chemistry to describe or explain them.

Chemistry knowledge should be shaped by rhetorical transformation of everyday entities into scientific entities. The teacher's actions need to provide the rhetorical construction of the entities and request the students to see the world in a particular way, through the representations, equations, and formulas. Teacher's action has a central role to shape a dialogical process in which students should participate actively. Taking advantage of the videos as a dialogical product and expressive of the students' interests, the teacher can use them to open the way of seeing the chemistry world.

Just as the audiovisual material, learning is a sort of a flame in which ideas are in constant movement and transformation. This can be seen as potential that has been afforded by video production in a learning process, especially when the video is seen like a cultural production. From this perspective, it is necessary to keep in mind that video production is a complex process through which students discuss, argue, select, (re)descript, construct, and integrate different effects (e.g., sound, image, speech, and feelings, among others) that allows to combine various communicative modes.

Precisely on account of this convergence of features, some difficulties are presented to the measurement and to the development of chemistry learning exclusively from the videos. On the one hand, cultural and aesthetic elements gave evidence of different learning and types of engagement. On the other hand, a requirement for specific chemistry aspects that need to be included in a video by students can hinder freedom, creativity, and, furthermore, engagement. Hence, the discussion of the chemistry aspects, particularly at the atomic level, can be a way to contrast what students have explored (i.e., positive aspects) and what they need to explore and identify gaps in learning. This is the teacher's role in terms of chemistry teaching.

Students' Points of View: Engagement Characteristics

The answers about the usefulness of videos showed that all students "agreed" or "partially agreed" about the contribution of videos for learning and for stimulating creativity. According to students' comments, the approach promoted collaborative engagement during the activities (i.e., experiments and recording), as well as individual organization, in a way which fosters autonomy, playfulness, and as a consequential behavioral, emotional, and cognitive engagement. Especially in the case of comments related to the usefulness and real use, the students attributed importance to the role of peers (e.g., "this activity was important to foster a closer relationship"; "we can interact more deeply").

Some studies have shown that engagement (or disengagement) is linked with peers and interpersonal relationships in classrooms (Marks 2000; Turner et al. 2014). Engagement is enhanced when class members actively discuss ideas, debate points of view, and critique each other's work. Other studies have showed that contexts in which autonomy is stimulated can favor engagement (Connell et al. 1995). In this study, the video production has created happiness and a relaxed environment, both being elements of emotional engagement. These results were corroborated by videos that have also demonstrated the same characteristics in making-of and audio. A possible reason may be associated with the nature of the task, since that video production created opportunities for students to work collaboratively and to freely introduce their ideas. These outcomes are in concordance with previous studies for which engagement can be enhanced in classrooms when the tasks (a) are authentic; (b) provide opportunities for students to assume ownership of their conception, execution, and evaluation; (c) provide opportunities for collaboration; (d) permit diverse forms of talents; and (e) provide opportunities for fun (Fredricks et al. 2004).

In addition to these results, persistence and dedication were identified as behavioral engagement categories in students' comments (e.g., "We had to search for a lot of information about experiments;" "We had to repeat it"). Emotional engagement has also been identified in students' comments when they referred to happiness while recording and, mainly, to identification with the school activity. The acceptance of video production seemed to be influenced by those characteristics of engagement. However, as in all experimental work, learning success is not ensured by simply performing the task, which was revealed by video analysis. In this regard, some aspects of content knowledge in chemistry (especially explanations, analysis, synthesis, and knowledge uncertainties) but also human elements from tetrahedron by Sjöström (2013) need more attention in video discussion. Thus, postproduction activities offer a way to overcome some of these learning difficulties that have been identified during video analysis.

Negative aspects were not identified in explanatory answers. Some students reported technical difficulties while recording and editing, as well as in chemistry comprehension, pointing out the role of collaborative work. Likewise, the results for the perceived ease of use and real use of the technology reinforced those aspects earlier presented. Again, the collaborative work and challenges faced during production operated in a positive way for the emotional (e.g., "it was funny"), behavioral (e.g., "we tried to do the best we could;" "we have repeated it"), and cognitive engagement (e.g., "we worked hard"; "we sought out other sources to explain the experiment"). As ascertained in other studies (Confrey 1996; Pereira et al. 2012), the participants stated that they had fun while producing the videos, which underscores the importance of freedom and social interaction.

Intellectual engagement and creativity have been pointed out as positive characteristics in video production (Goldman 2004). Other trends observed in raising engagement are flexibility and fair tasks (Finn and Voelkl 1993). Feelings of autonomy and competence seem to be strongly connected to engagement (Ryan and Deci 2000). This suggests an important support from social relationships, creativity, and entertaining in fostering the engagement. Social relationships and entertaining can be also related to the feeling of belonging. As described by Polman and Hope (2014), creations based on personally meaningful topics open opportunities for developing identities while fostering participation in critical thinking about science. As earlier mentioned, the students had freedom to choose and set up their experiments and videos. Thus, these aspects may have provided a deep identification with the task.

Some Final Considerations

Although student engagement cannot be seen as a guarantee for learning, it can result in a commitment or investment and, consequently, may be a key factor to decrease student apathy and enhance learning. In this manner, a positive correlation between video production and engagement was observed. The video analysis results were in concordance with the Technology Acceptance Model, and both demonstrated the presence of the three typologies of engagement. A few technical and conceptual difficulties were reported. At the same time, those difficulties worked like an additional incentive for the intellectual engagement. The recording and prerecording activities demonstrated an important role in student engagement during video production. These aspects require further investigations, mainly because they may play an important role during ongoing engagement and consequently in providing bridges between the phenomena and the intellectual tools used in chemistry.

Hence, the creation of a learning environment particularly involving the discussion of misconceptions related to chemistry in the videos would be desirable. This discussion is possible during each step in video production and is essential after its conclusion. Having this in mind, it is possible to connect initial engagement in promoting reflections that can continue over a long time. It will probably improve the learning process by addressing new perspectives for science classrooms.

Overall, the students' engagement has been an important first step. The video can be seen like a cultural production by which students can express different actions and a tangle of effects, resulting in a complex sign-making process. On the other hand, the nature of chemistry knowledge requires ongoing engagement with the aim to achieve links between macroscopic phenomena and theoretical models, as well as applied and socio-chemistry. In this way, students' understanding of chemistry may be improved through a cooperative learning environment after video presentations followed by discursive interactions.

Acknowledgments The author is grateful to CAPES for the financial support by the means of AEX August 2015.

References

Bardin, L. (2011). Análise de conteúdo [Content analysis] (Vol. 70). Lisbon: Edições.

Birch, S., & Ladd, G. (1997). The teacher-child relationship and children's early school adjustment. *Journal of School Psychology, 35*(1), 61–79.

Buhs, E. S., & Ladd, G. W. (2001). Peer rejection as an antecedent of young children's school adjustment: An examination of mediating process. *Developmental Psychology, 37*(4), 550–560.

Confrey, J. F. (1996). Focus on science concepts: Student-made videos zoom in on key ideas. *The Science Teacher, 63*, 16–19.

Connell, J. P., Halpern-Felsher, B. L., Clifford, E., Crichlow, W., & Usinger, P. (1995). Hanging in there: Behavioral, psychological, and contextual factors affecting whether African American adolescents stay in school. *Journal of Adolescent Research, 10*(1), 41–63.

Davis, F. D. (1989). Perceived usefulness, perceived ease of use, and user acceptance of information technology. *MIS Quarterly, 13*(3), 319–339.

Erdmann, M. A., & March, J. L. (2014). Video reports as a novel alternate assessment in the undergraduate chemistry laboratory. *Chemistry Education Research and Practice, 15*(4), 650–657.

Finn, J. D., & Voelkl, K. E. (1993). School characteristics related to school engagement. *The Journal of Negro Education, 62*(3), 249–268.

Fredricks, J. A., Blumenfeld, P. B., & Paris, A. (2004). School engagement: Potential of the concept, state of the evidence. *Review of Educational Research, 74*(1), 59–109.

Gabel, D. (1999). Improving teaching and learning through chemistry education research: A look to the future. *Journal of Chemical Education, 76*(4), 548–554.

Galili, I. (2013). On the power of fine arts pictorial imagery in science education in science education. *Science & Education, 22*(8), 1911–1938.

Goldman, R. (2004). Video perspective meets wild and crazy teens: A design ethnography. *Cambridge Journal of Education, 34*(2), 157–178.

Hodson, D. (1993). Re-thinking old ways: Towards a more critical approach to practical work in school science. *Studies in Science Education, 22*(1), 85–142.

Hofstein, A., & Lunetta, V. N. (2004). The laboratory in science education: Foundation for the 21st century. *Science Education, 88*(1), 28–54.

Kress, G., Jewitt, C., Ogborn, J., & Tsatsarelis, C. (2001). *Multimodal teaching and learning: The rhetorics of the science classroom*. London: Continuum.

Lichter, J. (2012). Using YouTube as a platform for teaching and learning solubility rules. *Journal of Chemical Education, 89*(9), 1133–1137.

Mahaffy, P. (2004). The future shape of chemistry education. *Chemistry Education: Research and Practice, 5*(3), 229–245.

Marks, H. M. (2000). Student engagement in instructional activity: Patterns in the elementary, middle, and high school years. *American Educational Research Journal, 37*(1), 153–184.

Mortimer, E. F., & Scott, P. H. (2003). *Meaning making in secondary science classroom*. Maidenhead: Open University Press/McGraw Hill.

Newmann, F., Wehlage, G. G., & Lamborn, S. D. (1992). The significance and sources of student engagement. In F. Newmann (Ed.), *Student engagement and achievement in American secondary schools* (pp. 11–39). New York: Teachers College Press.

Pereira, M. V., Barros, S. S., Rezende Filho, L. A. C., & Fauth, L. H. A. (2012). Audiovisual physics reports: Students' video production as a strategy for the didactic laboratory. *Physics Education, 47*(1), 44–51.

Polman, J. L., & Hope, J. M. G. (2014). Science news stories as boundary objects affecting engagement with science. *Journal of Research in Science Teaching, 51*(3), 315–341.

Rodrigues, S., Pearce, J., & Livett, M. (2001). Using video analysis or data loggers during practical work first year physics. *Educational Studies, 27*(1), 41–43.

Rouda, R. H. (1973). Student-produced videotapes in a physical chemistry laboratory course. *Journal of Chemical Education, 50*(2), 126–127.

Ryan, R., & Deci, E. (2000). Self-determination theory and the facilitation of intrinsic motivation, social development and well being. *American Psychologist, 55*(1), 68–78.

Sha, L., Shunn, C., Bathgate, M., & Ben-Eliyahu, A. (2016). Families support their children's success in science learning by influencing interest and self-efficacy. *Journal of Research in Science Teaching, 53*(3), 450–472.

Sjöström, J. (2013). Towards *Bildung*-oriented chemistry education. *Science & Education, 22*(7), 1873–1890.

Talanquer, V. (2011). Macro, submicro, and symbolic: the many faces of the chemistry "triplet". *International Journal of Science Education, 33*(2), 179–195.

Turner, J. C., Christensen, A., Kackar-Cam, H. Z., Trucano, M., & Fulmer, S. M. (2014). Enhancing students' engagement: Report of a 3-year intervention with middle school teachers. *American Educational Research Journal, 53*(3), 450–472.

Vanoye, F., & Goliot-Lété, A. (2013). *Ensaio sobre análise fílmica [Essay on the film analysis 7th Ed.]*. Campinas: Papirus Editora.

Vedder-Weiss, D., & Fortus, D. (2013). School, teacher, peers, and parents' goals emphases and adolescents' motivation to learn science in and out of school. *Journal of Research in Science Teaching, 50*(8), 953–988.

Vilches, A., & Gil-Pérez, D. (2013). Creating a sustainable future: Some philosophical and educational considerations for chemistry teaching. *Science & Education, 22*(7), 1857–1872.

Part III
Student Learning and Assessment

Making Sense of 'Making Sense' in Science Education: A Microgenetic Multiple Case Study

Richard Brock and Keith S. Taber

Descriptions of Learning in Science Education

Learning is a complex process (Redish 1999) that is challenging to study as researchers do not have direct accesses to the ideas and thought processes in a learner's mind (Nisbett and Wilson 1977). Descriptions of learning may therefore be thought of as models, that is, as partial descriptions of students' cognitive activities. No model of learning will be entirely free of assumptions, and researchers should be open about the particular conventions that underpin a model (Taber 2013). The incomplete nature of models of learning might suggest that the existence of a range of different descriptions of learning may be seen as a useful strategy for describing a multifaceted phenomenon (Geelan 1997). Different models of learning prioritise different aspects of the process, which might be considered across three dimensions:

(a) *Number of learners involved* – models of learning may vary in the extent to which they focus on individuals or communities of learners. Models that emphasise the personal nature of the learning process (e.g. Rennie and Johnston 2004) may be contrasted with constructions in which social interactions are highlighted (e.g. Jegede and Aikenhead 2006).
(b) *Nature of processes and entities of interest* – given that there is no direct access to cognition (Taber 2013), some models of learning are constructed to focus on students' observable behaviours (Mowrer and Klein 2001: 2) whilst others invoke hypothetical psychological entities, such as the concept, to develop explanations (Hewson 1981).

R. Brock (✉) • K.S. Taber
University of Cambridge, Cambridge, UK
e-mail: rb423@cam.ac.uk

© Springer International Publishing AG 2017
K. Hahl et al. (eds.), *Cognitive and Affective Aspects in Science Education Research*, Contributions from Science Education Research 3,
DOI 10.1007/978-3-319-58685-4_12

(c) *Single entities or the interaction of elements* – models of learning may focus on changes to individual knowledge elements, for example, as in early models of conceptual change (diSessa 2002). More recent models of conceptual change link learning to changes in relationships and contexts of activation of multiple conceptual resources (diSessa and Sherin 1998; diSessa and Wagner 2005).

The construction of these dimensions of difference between models of learning is not intended to critique a particular description of learning but rather to provide a framework for understanding the different emphases of models. In this chapter the focus will be on a loose cluster of types of learning that are seen as desirable outcomes of education. Consider the constructs described below:

• In essence, the deep approach is…an attempt to relate parts to each other, new ideas to previous knowledge and concepts to everyday experiences. There is an internal emphasis where the learner personalises the task, making it meaningful to his or her own experience and to the real world (Chin and Brown 2000: 110).
• Meaningful reception learning is inherently an active process because it requires…some degree of reconciliation with existing ideas in cognitive structure…reformulation of learning material in terms of the idiosyncratic intellectual background and vocabulary of the particular learner (Ausubel 2000: 5).
• Understanding is generally accepted to be an active process in which meaning is constructed. New information is interpreted in the light of currently activated knowledge (Burns et al. 1991: 277).

This cluster of terms (deep learning, meaningful learning and understanding) shares a focus on the interaction of constructed psychological entities of the individual learner. It is assumed that the learner engages in an active process of structuring existing knowledge elements in the context of novel information. Few descriptions of this process exist, and so this chapter presents a fine-grained case study of two students' learning.

Learning as the Personal Act of Structuring Multiple Conceptual Elements

One approach to describe learning is to assume that learners' cognition can be modelled as consisting of multiple conceptual elements of varying types (diSessa 2002; Hammer 2000; Posner et al. 1982). Hammer (2000: 53) has labelled these elements conceptual resources and described them as: '"agents" acting in parallel, sometimes coherently and sometimes not'. Conceptual resources may be related in structures modelled as schemata (Bartlett 1932), mental models (Johnson-Laird 1983) and coordination classes (diSessa 2002) to name just a few. In this research, the preferred term will be conceptual structure, which may be thought of as a network of conceptual resources (Taber 2013). It is assumed that novice learners' conceptual structures related to science topics met in the school curriculum initially consist of knowledge that is poorly differentiated and sparsely interconnected whereas experts possess well-differentiated and richly connected conceptual structure (Ausubel 2000; Reif 2008). Though the development of appropriately interconnected conceptual structure is an educational goal, at the present time, few detailed descriptions of the process of development of conceptual structures in science education exist.

In addition to describing the patterns of conceptual structures, conceptual resource models of learning may focus on understanding why students believe that a particular organisation of elements fits together in a given context. Coherence can be thought of as the '...extent to which category features go together in light of prior theoretical, causal, and teleological knowledge' (Patalano et al. 2006: 408). Coherence is a challenging term to define, in part due to its subjective nature (Hoey 1991); students may possess criteria for coherence which differ from those of scientists (Driver et al. 1985). This difference is significant as cognition may be 'rigged' towards the detection of coherences (Churchland 2004: 50), and coherence is argued to be a driver of changes in conceptual structure (Koponen and Huttunen 2013; Thagard 1989). If learning is modelled as the organisation of elements into structures, then an important facet of such descriptions will be a description of the factors that underlie students' perception of coherences between conceptual resources.

The conceptual resources a student possesses may be thought of as contextually sensitive, that is, they are activated with varying likelihoods in different situations (diSessa 2002; Hammer et al. 2005; Mortimer 1995). Learners at earlier stages of their education may possess a similar set of conceptual resources to experts but differ in their ability to activate combinations of concepts in relevant contexts (Sabella and Redish 2007). This ability to apply learning acquired in one context to another is referred to as transfer (Haskell 2000). It has been suggested that a characteristic of expertise is the ability to apply learning across an appropriate range of contexts (diSessa and Wagner 2005; diSessa 2002; Parnafes 2012). Therefore, assessments of the progress of students towards expertise should focus not only on the development of conceptual structure in a particular context but also seek to examine the application of knowledge across a range of appropriate situations.

Descriptions of learning, such as deep learning, understanding and meaningful learning (see above), model the process as the development of coherent conceptual structure, which can be transferred across appropriate contexts. However, each of the terms invokes additional assumptions: deep learning is linked with a motivational stance (Chin and Brown 2000), understanding has been used in a variety of ways including as a state rather than a process (Zagzebski 2001), and Ausubel's (2000) model of meaningful learning assumes an interaction between a relatively static conceptual structure and novel information. Therefore, the construct of making sense is proposed as a general term to describe the kind of learning discussed above. Making sense is chosen as it is a term used informally to describe learning in science education (e.g. Berland and Reiser 2009; Driver et al. 1994; Wilensky and Resnick 1999) but is not explicitly defined. The related term sense-making is used in organisational psychology to refer to the process of fitting ideas into frameworks and constructing meaning (Weick 1993, 1995). Making sense will be defined as the formation or modification of a conceptual structure in which concepts are related in a coherent system that may be applied to a range of contexts. The empirical study described below set out to investigate how students' conceptual structures related to physics developed. To that end, the research question in this study was:

How do two 16–17-year-old students form or modify conceptual structure in a range of contexts related to forces and dynamics over a 5-week period?

Method

Whilst, in general, conceptual change has been described as a gradual process (Nussbaum 1989; Smith et al. 1993; Vosniadou 2008), some authors report the occurrence of short timescale changes to understand, sometimes labelled as moments of insight (Brock 2015; Chi 1997; Clement 2008). Therefore, a method is required that can capture both short and long timescale changes to conceptual structure. The microgenetic method is an approach to data collection in which probes are applied at a rate which is considered to be high compared to the rate at which the phenomenon of interest changes (Siegler and Crowley 1991). In this case, the participants were interviewed weekly for around 25 min allowing the observation of short timescale changes. In order to observe longer timescale change, 22 sessions were arranged over a period of approximately 6 months. The study reported in this chapter focuses on the first five of the sessions. The participants in the study were five 16–17-year-old students at a secondary school in England. The students were selected via a purposeful sampling strategy to identify extreme cases of learning (Yin 2009): teachers were asked to identify two students who made sense of physics concepts rapidly and with apparent ease, two who struggled to make sense and one student in the middle of those two positions. The chapter is envisaged as a multiple case study (Yin 2009) consisting of two separate cases. Though case studies cannot generate context-independent information, Flyvbjerg (2006) has argued they provide detailed accounts of complex processes, such as learning, that are lost in larger-scale studies. This chapter will focus on two cases: Edward, selected as the intermediate case by his teachers, and Ben, described as an able student. As the study is conceptualised as a multiple case study, no claims to statistical generalisability are intended; it is left to the reader to judge the applicability of the findings to other contexts (Taber 2000). In order to make the case for the trustworthiness of case study research, Bassey (1999) argues that researchers present evidence of prolonged engagement and persistent observation. Within qualitative research, 'thick, rich description' of data and prolonged engagement with the students can be seen as supporting the trustworthiness of the case being made (Creswell and Miller 2000: 126). In a constructivist model of learning, the development of understanding is seen as an idiosyncratic process, and a balance needs to be struck between approaches which describe the detail of individual cases and those which present evidence of more generalisable patterns.

Table 1 Summary of the sessions related to forces and dynamics

Session	Session focus	Date
1	Force concept inventory questions, discussion about attitude to learning	30/9/13
2	Simple pendulum, forces on a car at constant velocity	7/10/13
3	Mass on a spring, forces on an astronaut	14/10/13
4	Loop-the-loop and scales in a lift, beliefs about physics	21/10/13
5	Reflections on making sense, concept map, ball in a bowl	4/11/13

The sessions were recorded, transcribed and analysed using the constant comparative method (Glaser and Strauss 1967).

The students, their parents and the head teacher of the school gave their informed consent to participation in the project, and the students are referred to by pseudonyms. A practical scenario, such as a pendulum, or the oscillations of a ball in a concave bowl, which the student would be asked to discuss in a semi-structured interview, understood in this research as involving the co-construction of meaning with the interviewer, were included in each session. The five sessions that are considered here (shown in Table 1) concerned the topic of forces and dynamics. Later sessions focused on discussions related to electrical circuits.

Emerging Themes of Analysis

The analysis will describe the manner in which Edward and Ben coordinated their conceptual resources in a number of different contexts. Edward displayed evidence of possessing a range of conceptual resources, including both ones that matched and differed from accepted scientific understandings. Making sense is challenging as it is not sufficient to have scientific conceptual resources available; they must be activated in appropriate contexts and linked in suitable combinations. In his first session, Edward showed awareness, in the context of the motion of a car, that an object travels at constant velocity when no resultant force acts. When he came to observe the motion of a simple pendulum in the next interview, Edward perceived the motion as involving an initial acceleration from its stationary state, followed by a period of uniform motion:

> I: And are there any points when it is not accelerating?
> P: Yeah, almost the rest of it.
> I: Because?
> P: It is travelling at a constant velocity.
> (Session 2, 49–52)

Despite developing an understanding that the tension in the string supporting the pendulum's bob varied with displacement, Edward was committed to his perception that the pendulum moves with constant velocity, he nevertheless argued that the 'overall force' on the bob remained constant:

Ah, well for displacement, it goes from being fully left but going back towards the middle and then going into fully right, 'cos the tension of the rope changes direction and pulls it back into the other way. Velocity that same sort of thing and acceleration [pause] well the overall force on from each individual force is always the same, so the acceleration remains constant for most of it, apart from the first bit. (Session 2, 140)

In the third session, Edward was introduced to the oscillations in the vertical axis of a mass on a spring. In this context Edward categorised the motion of the mass as involving changing acceleration with displacement: '…if there's more displacement that means there is more force acting on it therefore higher acceleration'. By linking the changing acceleration to resultant force and displacement, Edward was able to develop a coherent explanation of the motion of the mass on the spring. In the final session, the discussion of a practical situation focused on the oscillations of a marble released from the rim of a large, concave cooking bowl. Edward's initial attempt at explaining the motion activated few resources related to abstract concepts such as force or acceleration and instead focused on physical features of the situation:

Isn't it something to do with the gradient er [pause] and the gradient's like zero at the bottom. I think, yeah it's completely flat, it's zero at the bottom, it's like a large, it's a larger number at the side so it sort of moves down to the centre. (Session 5, 80)

When prompted to explain in more detail, Edward proposed an argument that: '…the gravity acting down on it would be stronger as the further you pull it away the more force acts on it'. This may be a partial reactivation of the force-linked-to-displacement resource that was used in the case of the mass on the spring; however, Edward is unable to describe a mechanism which explains the variation. This difficulty may arise from Edward's weak understanding of reaction force: when asked to describe the forces acting on the ball, he replied: '…probably weight and something to do with lift'. Without the necessary resource, the changing direction of reaction force with displacement, Edward struggled to develop a coherent argument. In a model of making sense as the coordination of conceptual resources, Edward's ability to form a coherent structure in each of the contexts might be summarised in Table 2:

Edward's attempts at making sense of the different contexts shown in Table 2 demonstrate the challenge of coordinating multiple conceptual resources. Edward

Table 2 Summary of Edward's interpretations of three contexts

Pendulum	Mass on a spring	Ball in a bowl
Perceives the bob as moving at constant velocity Is aware of a link between tension and displacement Makes sense by arguing whilst overall force doesn't change, tension does	Perceives motion as involving varying acceleration Is aware of a link between magnitudes of displacement, force and acceleration Makes sense in a manner that matches accepted scientific explanation	Perceives motion as involving speeding up and slowing down Has a weak understanding of reaction force Makes sense by arguing gravitational force varies with displacement but is unable to explain variation

has many of the appropriate resources to develop coherent accounts of the situations, but an erroneous perception of motion and missing resources related to the reaction force lead to the development of alternative accounts. Even when they possess appropriate resources, students may still develop alternative coherences. In the first session, Edward was asked which of two objects with different masses, released above the surface of the Earth, would hit the ground first. His response is shown below:

Hmmm, I would say the one with the heavier mass 'cos, because going back to Newton's law, force equals mass times acceleration. If they are both going the same speed they'd both have a similar acceleration, but if the one has a higher mass then mass times acceleration would bring a higher force behind it than the other...The bigger one would travel further. (Session 1, 32–34)

In this excerpt, conceptual resources that match scientific understandings (Newton's second law and the observation that the masses will accelerate at the same rate) are used to justify an intuition: the larger mass will travel further. In order to develop a coherent argument with these apparently contradictory elements, Edward forms a link between force and distance travelled. Edward's perception of a link between force and velocity (Viennot 1979) may arise from a conflation of velocity and acceleration that appears a number of times in Edward's arguments. Elsewhere in the first interview, in discussing a collision between a lorry and a stationary car, he argued: 'the lorry doesn't just carry a greater mass but there is also more speed, therefore more acceleration acting behind it, so the force would be greater' (Session 1, 40). In the case of Edward's description of the falling objects, despite the availability of appropriate resources to develop a coherence that matches the scientific model, the presence of intuitions regarding the nature of motion leads to the construction of an alternative understanding.

An examination of another case, that of Ben's learning (see Table 3), demonstrates the personal nature of the development of conceptual structures.

Ben's attempts to make sense of the contexts differ from Edward's. He appears to have a more stable link between the concepts of resultant force and acceleration than Edward; however, he struggles to develop explanations in which several forces act together. The nuanced differences in the manner in which the students' conceptual resources are deployed in different contexts highlight the importance of fine-grained descriptions of making sense.

Table 3 Summary of Ben's interpretations of three contexts

Pendulum	Mass on a spring	Ball in a bowl
Argues gravitational force causes acceleration States doesn't understand how the pendulum can move upwards	Links motion to changes in tension Argues tensions decreases as displacement increases Suggests acceleration is zero at maximum displacement	Argues magnitude of gravitational force varies due to changing gradient of bowl Argues at top of slope force is balanced by velocity so ball will be stationary

Making Sense in Science Education

It has been argued that research on conceptual change has developed from a focus on cataloguing domain-specific alternative conceptions to modelling conceptual change as the interaction of multiple conceptual resources (Amin et al. 2014). This kind of multi-element model can begin to account for Edward's various attempts at making sense in different contexts. Even in cases where he is lacking a conceptual resource (e.g. an understanding of reaction force) or has an erroneous perception of a situation (the pendulum moves at constant velocity), he still develops an argument that appears to have some degree of personal coherence. Coherence can be imagined to be a driving force of conceptual change (Thagard 1989), and both scientific and alternative interpretations of coherence may drive change as judgements of fit are subjective (Hoey 1991). The human mind has a tendency to construct patterns in stimuli, even when the information is random (Shermer 2012). Learning about scientific concepts is challenging because multiple, personally coherent arrangements of conceptual resources are possible. Constructions such as the luminiferous aether or phlogiston suggest that coherent collections of assertions can be powerfully compelling even in the face of contradictory evidence (Thagard 1992).

A useful focus for science education research would be to develop an understanding of the factors that lead students to believe a set of conceptual resources fits together. Though students' constructions may sometimes be 'romanced' (Piaget 1979) in response to the interviewer's prompts and have little long-term stability, nonetheless, they are nonarbitrary choices that reflect attempts to produce coherent accounts of a given context with the available conceptual resources. It may be challenging to pick apart students' 'feelings of what's right' (Rohrlich 1996: 1624), as the perceptions may be based on intuitions, pieces of tacit knowledge that are difficult to express in words (Brock 2015). At this early stage of research into systemic conceptual change, studies that can shed light on the subtleties of the process may be useful precursors to research with larger sample sizes that can uncover generalisable patterns. Modelling learning as the coordination of multiple conceptual resources into a coherent framework, making sense, may be a useful model, amongst other alternatives, for examining the dynamics of students' learning across multiple contexts.

References

Amin, T. G., Smith, C., & Wiser, M. (2014). Student conceptions and conceptual change: Three overlapping phases of research. In N. G. Lederman & S. A. Abell (Eds.), *Handbook of research on science education* (pp. 57–81). New York: Routledge.

Ausubel, D. (2000). *The acquisition and retention of knowledge*. Dordrecht: Kluwer Academic Publishers.

Bartlett, F. C. (1932). *Remembering: A study in experimental and social psychology*. Cambridge: Cambridge University Press.

Bassey, M. (1999). *Case study research in educational settings*. Buckingham: Open University Press.

Berland, L. K., & Reiser, B. J. (2009). Making sense of argumentation and explanation. *Science Education, 93*(1), 26–55.

Brock, R. (2015). Intuition and insight: Two concepts that illuminate the tacit in science education. *Studies in Science Education, 51*(2), 127–167. http://doi.org/10.1080/03057267.2015.1049843

Burns, J., Clift, J., & Duncan, J. (1991). Understanding of understanding: Implications for learning and teaching. *British Journal of Educational Psychology, 61*(3), 276–289.

Chi, M. T. H. (1997). Creativity: Shifting across ontological categories flexibly. In B. T. Ward & S. M. Smith (Eds.), *Creative thought: An investigation of conceptual structures and processes* (pp. 209–234). Washington: American Psychological Association.

Chin, C., & Brown, D. E. (2000). Learning in science: A comparison of deep and surface approaches. *Journal of Research in Science Teaching, 37*(2), 109–138. http://doi.org/10.1002/(SICI)1098-2736(200002)37:2<109::AID-TEA3>3.0.CO;2-7

Churchland, P. S. (2004). How do neurons know? *Daedalus, 133*(1), 42–50.

Clement, J. (2008). *Creative model construction in scientists and students.* Dordrecht: Springer.

Creswell, J. W., & Miller, D. L. (2000). Determining validity in qualitative inquiry. *Theory Into Practice, 39*(3), 124–130.

diSessa, A. (2002). Why "conceptual ecology" is a good idea. In M. Limon & L. Mason (Eds.), *Reconsidering conceptual change: Issues in theory and practice* (pp. 29–60). Dordrecht: Kluwer.

diSessa, A., & Sherin, B. (1998). What changes in conceptual change? *International Journal of Science Education, 20*(10), 1155–1191.

diSessa, A., & Wagner, J. F. (2005). What coordination has to say about transfer. In J. P. Mestre (Ed.), *Transfer of learning from a modern multidisciplinary perspective* (pp. 121–154). Greenwich: Information Age Publishing.

Driver, R., Guesne, E., & Tiberghien, A. (1985). *Children's ideas in science.* Milton Keynes: Open University Press.

Driver, R., Rushworth, P., Squires, A., & Wood-Robinson, V. (1994). *Making sense of secondary science: Research into children's ideas.* Abingdon: Routledge.

Flyvbjerg, B. (2006). Five misunderstandings about case-study research. *Qualitative Inquiry, 12*(2), 219–245.

Geelan, D. (1997). Epistemological anarchy and the many forms of constructivism. *Science & Education, 6*(1), 15–28.

Glaser, B. G., & Strauss, A. L. (1967). *The discovery of grounded theory: Strategies for qualitative reserach.* New Brunswick: Aldine Transaction.

Hammer, D. (2000). Student resources for learning introductory physics. *American Journal of Physics, 68*(7), 52–59. http://doi.org/10.1119/1.19520

Hammer, D., Elby, A., Scherr, R. E., & Redish, E. F. (2005). Resources, framing, and transfer. In J. Maestre (Ed.), *Transfer of learning from a modern multidisciplinary perspective* (pp. 89–120). Greenwich: Information Age Publishing.

Haskell, R. E. (2000). *Transfer of learning: Cognition and instruction.* San Diego: Academic Press.

Hewson, P. W. (1981). A conceptual change approach to learning science. *European Journal of Science Education, 3*(4), 383–396. http://doi.org/10.1080/0140528810304004

Hoey, M. (1991). *Patterns of lexis in text.* Oxford: Oxford University Press.

Jegede, O. J., & Aikenhead, G. S. (2006). Transcending cultural borders: Implications for science teaching. *Research in Science & Technological Education, 17*(1), 45–66. http://doi.org/10.1080/0263514990170104

Johnson-Laird, P. N. (1983). *Mental models: Towards a cognitive science of language, inference, and consciousness.* Cambridge: Harvard University Press.

Koponen, I. T., & Huttunen, L. (2013). Concept development in learning physics: The case of electric current and voltage revisited. *Science & Education, 22*(9), 2227–2254.

Mortimer, E. (1995). Conceptual change or conceptual profile change? *Science & Education, 4*(3), 267–285.

Mowrer, R. R., & Klein, S. B. (2001). The transitive nature of contemporary learning theory. In R. R. Mowrer & S. B. Klein (Eds.), *Handbook of contemporary learning theories* (pp. 1–21). New York: Lawrence Erlbaum Associates, Inc..

Nisbett, R. E., & Wilson, T. D. (1977). Telling more than we can know: Verbal reports on mental processes. *Psychological Review, 84*(3), 231–259.

Nussbaum, J. (1989). Classroom conceptual change: Philosophical perspectives. *International Journal of Science Education, 11*(5), 530–540.

Parnafes, O. (2012). Developing explanations and developing understanding: Students explain the phases of the moon using visual representations. *Cognition and Instruction, 30*(4), 359–403.

Patalano, A. L., Chin-Parker, S., & Ross, B. H. (2006). The importance of being coherent: Category coherence, cross-classification, and reasoning. *Journal of Memory and Language, 54*(3), 407–424.

Piaget, J. (1979). *The childld's conception of the world.* (J. Tomlinson & A. Tomlinson, Trans.). Plymouth: Rowman & Littlefield Publishers, Inc.

Posner, G., Strike, K. A., Hewson, P. W., & Gertzog, W. A. (1982). Accommodation of a scientific conception: Toward a theory of conceptual change. *Science Education, 66*(2), 211–227.

Redish, E. F. (1999). Millikan lecture 1998: Building a science of teaching physics. *American Journal of Physics, 67*(7), 562–573. http://doi.org/10.1119/1.19326

Reif, F. (2008). *Applying cognitive science to education.* Cambridge, MA: The MIT Press.

Rennie, L. J., & Johnston, D. J. (2004). The nature of learning and its implications for research on learning from museums. *Science Education, 88*(1), 4–16. http://doi.org/10.1002/sce.20017

Rohrlich, F. (1996). The unreasonable effectiveness of physical intuition: Success while ignoring objections. *Foundations of Physics, 26*(12), 1617–1626. http://doi.org/10.1007/BF02282125

Sabella, M. S., & Redish, E. F. (2007). Knowledge organization and activation in physics problem solving. *American Journal of Physics, 75*(11), 1017–1029. http://doi.org/10.1119/1.2746359

Shermer, M. (2012). *The believing brain: From spiritual faiths to political convictions-how we construct beliefs and reinforce them as truths.* London: Constable & Robinson Ltd.

Siegler, R. S., & Crowley, K. (1991). The microgenetic method: A direct means for studying cognitive development. *American Psychologist, 46*(6), 606–620.

Smith, J. P., diSessa, A. A., & Roschelle, J. (1993). Misconceptions reconceived: A constructivist analysis of knowledge in transition. *The Journal of the Learning Sciences, 3*(2), 115–163.

Taber, K. S. (2000). Case studies and generalisability: Grounded theory and research in science education. *International Journal of Science Education, 22*(5), 469–487.

Taber, K. S. (2013). *Modelling learners and learning in science education: Developing representations of concepts, conceptual structure and conceptual change to inform teaching and research.* Dordrecht: Springer.

Thagard, P. (1989). Explanatory coherence. *Behavioral and Brain Sciences, 12*(03), 435–467.

Thagard, P. (1992). *Conceptual revolutions.* Princeton: Princeton University Press.

Viennot, L. (1979). Spontaneous reasoning in elementary dynamics. *European Journal of Science Education, 1*(2), 205–221. http://doi.org/10.1080/0140528790010209

Vosniadou, S. (2008). Conceptual change research: An introduction. In S. Vosniadou (Ed.), *International Handbook of research on conceptual change* (1st ed., pp. xiii–xxviii). New York: Routledge.

Weick, K. E. (1993). The collapse of sensemaking in organizations: The Mann Gulch disaster. *Administrative Science Quarterly, 38*, 628–652.

Weick, K. E. (1995). *The nature of sensemaking. Sensemaking in organizations.* Thousand Oaks: SAGE Publications.

Wilensky, U., & Resnick, M. (1999). Thinking in levels: A dynamic systems approach to making sense of the world. *Journal of Science Education and Technology, 8*(1), 3–19. http://doi.org/10.1023/A:1009421303064

Yin, R. K. (2009). *Case study research: Design and methods.* Thousand Oaks: Sage Publications Incorporated.

Zagzebski, L. (2001). Knowledge, truth and duty. In M. Steup (Ed.), *Knowledge, truth and duty* (pp. 235–253). Oxford: Oxford University Press.

Student Difficulties with Graphs in Different Contexts

Lana Ivanjek, Maja Planinic, Martin Hopf, and Ana Susac

Introduction

Graphs are very important in all areas of science, and they are an essential part of university, secondary school, and even primary school curricula around the world. Students are introduced to graphs through different subjects but mostly through mathematics and physics. As students are usually introduced to graphs quite early, the ability to interpret them is often assumed by university faculty to be fully developed by the time students enroll at university. Previous physics and mathematics education studies (McDermott et al. 1987; Beichner 1990; Beichner 1994; Forster 2004; Araujo et al. 2008; Nguyen and Rebello 2011; Christensen and Thompson 2012; Dreyfus and Eisenberg 1990; Leinhardt et al. 1990; Graham and Sharp 1999; Kerslake 1981; Hadjidemetriou and Williams 2002) showed that this assumption does not hold and that students still have many difficulties with graph interpretation at university level, as well as at earlier levels.

This paper summarizes the results of two large-scale studies, conducted by Planinic et al. (2013) and Ivanjek et al. (2016), on university students' understanding of graphs in different contexts and presents the main student strategies and difficulties concerning graph interpretation. In addition, it introduces the comparison of the results of students at the University of Zagreb and the students at the University of Vienna. Student ability to interpret graph slopes and areas under the graph changes across three different domains, mathematics, without context (M domain), physics or kinematics (P domain), and mathematics in contexts other than physics (C domain) are discussed, and the students' main strategies as well as reasoning

---

L. Ivanjek (✉) • M. Hopf
University of Vienna, Vienna, Austria
e-mail: lana.ivanjek@univie.ac.at

M. Planinic • A. Susac
University of Zagreb, Zagreb, Croatia

© Springer International Publishing AG 2017   167
K. Hahl et al. (eds.), *Cognitive and Affective Aspects in Science Education Research*, Contributions from Science Education Research 3,
DOI 10.1007/978-3-319-58685-4_13

difficulties expressed through student explanations are identified. In addition, difficulties of different domains for students at the University of Zagreb and University of Vienna are compared and discussed.

In previous research McDermott et al. (1987) found that students have difficulties with discriminating between the slope and height of a graph, relating one type of graph to another, interpreting the area under a graph, distinguishing the shape of a graph from the shape of the trajectory of the body, and understanding the meaning of the sign of velocity and acceleration. Leinhardt et al. (1990) classified student difficulties with graphs as interval-point confusion (focusing on a single point instead of on an interval), slope-height confusion (focusing on the height of the graph instead of the slope), and iconic confusion (interpreting graphs as literal pictures of motion). Although these difficulties were found both in physics and in mathematics education research, the prevalence of these difficulties in both domains was rarely compared. The latest study from Wemyss and van Kampen (2013) investigated the difficulties encountered by first year university students when determining the value of speed from distance time graphs, the rate of change of water level from water level vs. time graph, or the slope in context-free graph. They found that when answering conceptually equivalent questions in different contexts, students deployed different problem-solving strategies and that the students performed poorer on physics problems, which was attributed to students' reliance on learned procedures in physics (e.g., use of formulas). Michelsen (2005) suggested that it is not just the mathematical formalism that presents the barrier in learning physics, but there is also the missing link between mathematics and physics. He suggested that the mathematical domain should be expanded by using examples from physics and from everyday life contexts in mathematics teaching, in order to solve the problem of domain specificity. In such an expanded domain, modeling of real-life situations could be a way of bridging the gap between mathematics and physics.

The previous studies from Planinic et al. (2012, 2013) also showed that the interpretation of graphs in physics questions was more difficult for students than the same task in parallel mathematics questions. The current study attempts to investigate this issue further through adding other context questions (questions situated in contexts other than physics, which did not require any special content knowledge) in order to investigate if the higher difficulty of physics questions primarily originated from students' lack of physics knowledge or from the introduction of context in the problem.

Theoretical Background

There is still some debate going on among researchers in physics education about the nature and origin of student difficulties in physics and the structure of students' naive knowledge. The two opposing views are sometimes described (Özdemir and Clark 2007) as knowledge as theory (Chi 2005; Vosniadou 1994) and knowledge as elements (diSessa 1993; Linn et al. 2004; Hammer et al. 2005). The first one sees naive knowledge as highly organized and interconnected, containing firm beliefs or ideas inconsistent with the accepted physics knowledge, known as misconceptions or alternative conceptions, and the other one sees naive knowledge as a set of relatively loosely connected knowledge elements, whose activation is very context dependent.

In the knowledge-as-elements perspective, the displayed difficulties reflect students' inappropriate or simplified reasoning patterns that originate from using basic reasoning elements called phenomenological primitives (p prims), which are in themselves neither correct nor incorrect (diSessa 1993). Which knowledge or reasoning elements will be activated in a certain situation depends largely on the context of the problem and on students' framing of the problem (Hammer et al. 2005). Framing means that students consciously or unconsciously make choices as to what knowledge to activate and use, based on their perception of the situation and on the social and cultural expectations. Student responses are dynamically created in response to their perception of the task. The question of transfer of knowledge, which is usually defined as the ability to extend what has been learned in one context to new contexts (Bransford et al. 1999), is then replaced with the question of which cognitive resources will be considered appropriate by the student in a given problem and, therefore, activated. That will largely be determined by the context and domain of the problem.

Data Collection and Analysis

In order to investigate and compare student understanding of graphs in different contexts, eight sets of parallel mathematics, physics, and other context questions about graphs were developed by the authors (Planinic et al. 2013). Five sets of questions referred to the concept of graph slope and three to the concept of area under the graph. Besides giving answers to the questions in the test, students were also required to provide explanations and procedures that accompanied their answers. Each set of questions referred to the same concept and required the same mathematical procedure in different contexts—one question was a direct mathematical question, one was situated in the context of physics (kinematics), and one was in some context other than physics (e.g., variation of total cost of a phone call changes with the call duration). An example of one set of parallel slope items is given in Fig. 1, and the whole test can be found in the Supplemental Material of the previous paper of Planinic et al. (2013).

The test consisting of 24 questions was administered to 385 first year students at the Faculty of Science, University of Zagreb, Croatia, who were either prospective physics or mathematics teachers or prospective physicists or mathematicians. Students were tested at the beginning of the first semester, and the allocated time for taking the test was 60 min. In addition, the test was administered to 417 first year physics students at the University of Vienna, Austria, who were either prospective physics teachers or prospective physicists. Students at the University of Vienna

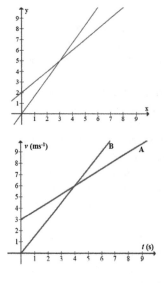

M –S2 Straight lines p and q are shown on the graph.

Compare their slopes at $x = 1$.

a) The slope of p is less than the slope of q.
b) The slopes of p and q are equal.
c) The slope of p is greater than the slope of q.

P –S2 Motions of objects A and B are represented
by a v vs. t graph. Compare accelerations
of the objects at $t = 2$ s.

a) Acceleration of A is less than the acceleration of B.
b) Accelerations of A and B are equal.
c) Acceleration of A is less than the acceleration of B.

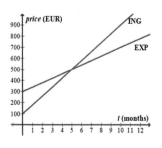

C –S2 Variation of prices of stocks ING and EXP
with time are shown on the graph. Compare
the growth rates of prices at $t = 3$ months.

a) The growth rate of the price of stock ING
is less than the growth rate of the price of stock EXP.
b) The growth rates of the prices of stocks ING and EXP
are equal.
c) The growth rate of the price of stock ING
is larger than the growth rate of the price of stock EXP.

Fig. 1 Questions P-S2, M-S2, and C-S2: an example of a set of parallel slope items

were also tested at the beginning of the first semester, and the allocated time for tak-ing the test was the same. The test was scored by the authors and Rasch analysis (Bond and Fox 2001) was conducted with the WINSTEPS software (Linacre 2006) to obtain linear measures for item difficulties. WINSTEPS software preforms logis-tic transformation on the raw scores of persons and items and in this way transforms the raw scores in linear measures of student ability and item difficulty. From the individual item difficulties, the average difficulties of each domain were calculated. After the analysis of the Croatian students was carried out, the difficulty of nine questions was anchored to the analysis of the data from the Austrian students, in order to be able to compare these two groups of students and to express all results on the same logit scale.

Each student explanation provided with the answers was also carefully analyzed and categorized. The categorization process followed the general guidelines for analyzing verbal data (Chi 1997). The data were always coded and categorized by both authors separately. The categories were different for different questions, although some of them appeared in multiple questions. Different categories reflect different student strategies (some correct, some incorrect). Based on the frequency of students' use of different categories, several incorrect strategies have been identified.

Results

Analysis of Item Difficulties

The functioning of the test as a whole on the University of Zagreb sample was satisfactory with high item reliability (0.99); somewhat lower, but satisfactory, person reliability (0.85); and Cronbach alpha (0.88). Overall, the data seemed to fit the Rasch model. The analysis of the students at the University of Vienna also gave high item reliability (0.99), somewhat lower person reliability (0.86), and Cronbach alpha (0.90).

In order to compare the difficulties of items in each investigated context, the average values of item difficulties over three different domains (mathematics, physics, and other contexts) and two investigated concepts (slope, area) were calculated. The average difficulty of all items in the test is usually set to zero in Rasch analysis, so negative item difficulties indicate items easier than the average, and positive item difficulties indicate items more difficult than the average. The results suggest that the concept of graph slope in mathematics (M) domain is easier for students than the same concept in physics (P) domain, but difficulties of the concept of slope in physics and other contexts (C) domains, as well as of the same concept in M and C domains, cannot be clearly distinguished from one another due to large uncertainty of the average difficulty in the C domain. Altogether, the concept of slope appears rather homogenous in difficulty (Fig. 2). Error bars indicate the level of dispersion of item difficulties from the average difficulty.

The differences among domains are much more pronounced when the concept of an area under a graph is analyzed. The concept of area under a graph in M domain appears to be much easier than the same concept in P and C domains but also much easier than the concept of slope in any of the domains. The concept of area under a graph appears to be of similar difficulty in P and C domains and of much higher difficulty than the concept of slope in any of the domains.

When looking at the comparison of students at the University of Zagreb and the University of Vienna, the quantitative data show small discrepancies between these two populations (Fig. 3). The only place where there is discernible difference between these two groups of the students is the concept of area under a graph in mathematics domain. Students at the University of Zagreb were on average better in calculating area under a graph. On the other hand, when analyzing students' explanations, it was noticed that when calculating slope, students at the University of

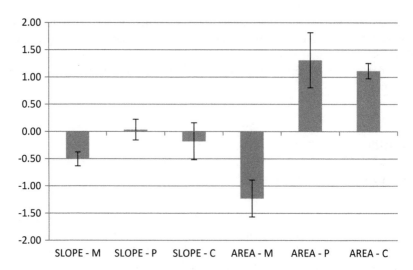

Fig. 2 Average difficulties of slope and area items in three different domains (*M* stands for mathematics without context, *P* for physics, and *C* for other contexts). *Error bars* indicate the combined uncertainties of each average value computed as CE = (SEM2 + SE)

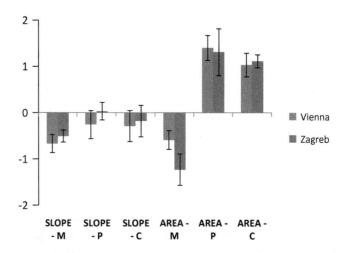

Fig. 3 Average difficulties of slope and area items in three different domains for two groups of students—students from the University of Zagreb and from University of Vienna (*M* stands for mathematics without context, *P* for physics, and *C* for other contexts)
Error bars indicate the combined uncertainties of each average value computed as $CE = (SEM^2 + SE^2)^{1/2}$, where SEM is standard error of the mean and SE is standard Rasch error

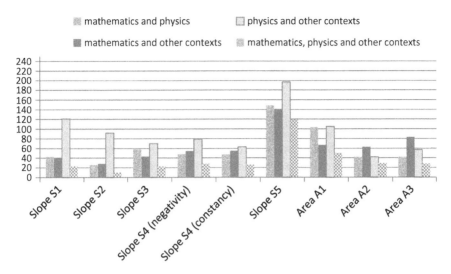

Fig. 4 Number of students in each set of questions who used same strategies in different combinations of domains

Vienna predominantly used rise over run (*Steigungsdreieck* in German) and were on average somewhat more successful in determining slope than students at the University of Zagreb who mostly relied on formulas.

Analysis of Student Explanations

In addition to the analysis of item difficulties, student reasoning patterns and their frequencies were also analyzed. The summary of the findings is described here.

Strategies Used in Parallel Questions Are Often Context Dependent and Domain Specific

The data presented in Fig. 4 indicate that only a small fraction of students used the same strategy in all three questions of the same set. It seems that many students perceived the parallel questions as different and used different approaches when solving them. In all sets of questions, except one, at most 15% of students used the same strategy in all three parallel questions, which suggests that most students probably saw parallel questions as different. The only exception was the slope set of questions S5, where more students used similar strategies in different domains. The reason for that could be that the graphs from set S5 were curved and stood out among line graphs and enabled students to recognize similarities in the questions.

The Preferred Strategy in Physics Seems to Be the Use of Formulas (Often Incorrect Ones)

Among Croatian students the preferred strategy for solving the physics questions tended to be the use of formulas. On all area problems and on some slope problems, the students chose to use a formula for solving physics problems although that strategy was not very productive. The formulas that were used were often incorrect or inappropriate, resulting in wrong solutions. When asked to calculate acceleration, the students often applied incorrect formula $a = v/t$. In the question presented in Fig. 1, where students needed to compare the accelerations from the two linear graphs and where calculations were not necessary, the application of incorrect formula led students to incorrect conclusions. The very extensive use of formula $a = v/t$ (41% of all University of Zagreb students in the question where they were asked to determine the acceleration from the graph and 22% of all University of Zagreb students in the question where they were asked to compare accelerations) led to many incorrect solutions. The same was true in area questions, where incorrect and misapplied formulas were quite common in physics questions. Some examples of incorrect and inappropriate formulas include: $\Delta v = \Delta a \Delta t$ (18% of Croatian students in question P – A2), $\Delta v = v_2 - v_1 = a_2 t_2 - a_1 t_1$ (20% of Croatian students in question P – A2), and $v = at$ (6% of Croatian students in question P – A2). These findings suggest that students mostly do not see formulas as mathematical models in physics situations. Translating physics into mathematics is not an easy task for students, as some previous studies on modeling in physics and student understanding of physics equation have already shown (Erickson 2006; Sherin 2001).

Students Use More Creative Strategies on Other Context Problems Than on Physics Problems

On other context problems students used different strategies than on physics questions. Many of those approaches originated from physics (e.g., dimensional analysis) and could have helped them in solving physics questions, but their reliance on formulas seemed to prevent students from using those strategies on physics questions. For example, in area questions, twice as many students came to the idea to calculate area under the graph in other context questions than in parallel physics questions, where they mostly used incorrect formulas.

Physics Seems to Provoke More Naive Reasoning Than Mathematics or Other Contexts (Interval-Point and Slope-Height Confusion)

Interval-point and slope-height confusion were present in all three domains, but they were much more pronounced in physics questions. In this study we have observed interval-point and slope-height confusion, while iconic confusions were very rare. The slope-height confusion was mostly pronounced in the slope question

S4, where the slope of the line was negative. The visual cue in that question was very strong (the decreasing height of the graph), and it seems that students relied on that when incorrectly concluding that the slope, or some slope-related quantity such as acceleration, was decreasing. In physics, 29% of all students concluded incorrectly that the acceleration is decreasing because the height of the graph is decreasing, while 20% of students said that the slope was decreasing in the parallel mathematics question. Interval-point confusion was mostly pronounced in the question where the students were asked to compare the accelerations, where they reasoned on the basis of the incorrect formula $a = v/t$, which suggests that they do not distinguish point from interval. In this particular question, many more students expressed interval-point confusion in the physics question when asked to compare accelerations from the velocity-time graphs (20% of students) than in mathematics and other context questions (only 1% of students).

Slope Is a Vague Concept for Many Students: Calculating Slope Seems to Be the Most Difficult Aspect of the Concept

Although the students often referred to slope in their explanations in physics and in other contexts, one should not automatically assume that students therefore fully understood the meaning of the concept. The students often gave unclear or no explanations on all mathematical slope questions, which suggests that for many of them the slope may not be more than a vague notion of how steep the straight line is, sometimes identified with an angle that the straight line forms with one of the coordinate axes. That kind of reasoning was often sufficient to produce the correct answer in the qualitative comparison of slopes. However, when it came to calculating slope, these vague ideas no longer helped. Slope calculation was required on the first set of slope questions. This was the most difficult slope question in physics and other context domain and the second most difficult in mathematics. Only 54% of students calculated slope correctly in mathematics domain, which is similar to what was found in other studies (Beichner 1994; Wemyss and van Kampen 2013).

Interpretation of the Meaning of Area Under a Graph Is Very Difficult for Students

As one can notice in Figs. 2 and 3, most of the students know how to determine the area under a graph when asked directly. But on physics and other context problems in order to answer the question correctly, students first needed to interpret the area under a graph as the quantity of interest. In contrast to difficulty of area questions in mathematics, the much higher difficulty of the questions from the two other domains suggests that the interpretation of the meaning of area under a graph is very difficult for students. That is also confirmed by the explanations that students gave on area questions. It was noticed that only few students can interpret areas under graphs in situations other than those previously encountered. Typically only 10–25% of students calculated area in physics and other context questions.

Discussion and Conclusion

This study attempted to analyze and compare student reasoning on mathematically similar problems, which were situated in different domains and contexts. The analysis of student explanations suggested that student reasoning about problems is often very much bound by the contexts and the disciplines in which their knowledge was acquired. The observed dependence of student strategies on the domain and context of the questions seems to support the knowledge-in-pieces framework, which explains this dependence through context-dependent activation of cognitive resources and the importance of framing. For example, it is possible that when students identified (framed) a question as a mathematics question, they felt obliged to answer it in a standard "mathematical" way. Students seemed to think more freely and creatively in problems in which their perception probably did not fall in the category of either discipline (other context problems). The students' use of physics formulas was often superficial and uncritical—students often did not seem to understand the suitability or even correctness of the formula they are using. Students' almost exclusive reliance on formulas in physics presents, in our opinion, a significant obstacle for the development of students' deeper reasoning in physics and sometimes even an obstacle for the application of their already existing knowledge and reasoning developed in other domains.

The results of the study also reveal some differences between student understanding of the concept of the graph slope and the concept of the area under the graph. Although we can say generally that the concept of slope seems to be better understood than the concept of the area under a graph, we have still noticed that for many students the concept of slope may be quite vague and needs refinement and strengthening. The area under a graph seems to be very difficult for students to interpret. An important implication for physics teaching is that we should work more on building student understanding of such interpretations and not only give ready-made interpretations for specific cases in physics.

Students seemed to think more freely and creatively and to transfer more of their knowledge in other context problems. This finding is in accordance to the findings of Wemyss and van Kampen (2013), who found that the number of students' correct answers to a problem involving a water level vs. time graph (which students had not encountered in the formal educational setting before) was much higher than the number of correct answers to the supposedly more familiar kinematics problem. This suggests that other context problems may have a potential to expose and develop student reasoning more than the standard domain-specific mathematics and physics questions. They should be used more, in both mathematics and physics teaching. We should work more on establishing links between common concepts and procedures in mathematics and physics and to promote their integration in students' minds to a much larger extent than is the case now.

This study confirms once again that human knowledge is very complex and multifaceted. Whenever we are trying to probe or assess student knowledge, we should be aware of that fact. The problem lies not only in the quantity and quality of knowledge but also in its accessibility. It is possible to pose basically the same problem three times, as we have done in this study, and to get three very different answers from the same student. The context and formulation of the question, the students' framing of the question, the procedures and conventions of the domain in which a certain piece of knowledge was first acquired, and the existing or missing links between the domains, all that and much more contribute to the form and content of a student's answer. Using many contexts during teaching, and constantly building links between different domains, could be a good way to building stronger student knowledge.

Acknowledgments This research is part of the Lise Meitner Project M1737-G22 "Development of Graph Inventory".

References

Araujo, I. S., Veit, E. A., & Moreira, M. A. (2008). Physics students' performance using computational modeling activities to improve kinematics graphs interpretation. *Computers & Education, 50*, 1128–1140.

Beichner, R. J. (1990). The effect of simultaneous motion presentation and graph generation in a kinematics lab. *Journal of Research in Science Teaching, 27*, 803–815.

Beichner, R. J. (1994). Testing student interpretation of kinematics graphs. *American Journal of Physics, 62*, 750–762.

Bond, T. G., & Fox, C. M. (2001). *Applying the Rasch model: Fundamental measurement in the human sciences.* Mahwah: Lawrence Erlbaum.

Bransford, J. D., Brown, A. L., & Cocking, R. R. (1999). *How people learn: Brain, mind, experience, and school.* Washington, DC: National Academy Press.

Chi, M. T. H. (1997). Quantifying qualitative analyses of verbal data: A practical guide. *The Journal of the Learning Sciences, 6*(3), 271–315.

Chi, M. T. H. (2005). Commonsense conceptions of emergent processes: Why some misconceptions are robust. *The Journal of the Learning Sciences, 14*(2), 161–199.

Christensen, W. M., & Thompson, J. R. (2012). Investigating graphical representations of slope and derivative without a physics context. *Physical Review Special Topics – Physics Education Research, 8*, 023101.

diSessa, A. A. (1993). Toward an epistemology of physics. *Cognition and Instruction, 10*(2 & 3), 105–225.

Dreyfus, T., & Eisenberg, T. (1990). On difficulties with diagrams: Theoretical issues. In G. Booker, P. Cobb, & T. N. De Mendicuti (Eds.), *Proceedings of the fourteenth annual conference of the International Group for the Psychology of mathematics education* (Vol. 1, pp. 27–36). Oaxtepex: PME.

Erickson, T. (2006). Stealing from physics: Modeling with mathematical functions in data-rich contexts. *Teaching Mathematics and its Applications: An International Journal of the IMA, 25*, 23–32.

Forster, P. A. (2004). Graphing in physics: Processes and sources of error in tertiary entrance examinations in Western Australia. *Research in Science Education, 34*, 239–265.

Graham, T., & Sharp, J. (1999). An investigation into able students' understanding of motion graphs. *Teaching Mathematics Applications, 18*, 128–135.

Hadjidemetriou, C., & Williams, J. S. (2002). Children's graphical conceptions. *Research in Mathematics Education, 4*(1), 69–87.

Hammer, D., Elby, A., Scherr, R. E., & Redish, E. F. (2005). Resources, framing, and transfer. In J. Mestre (Ed.), *Transfer of learning from a modern multidisciplinary perspective* (pp. 89–120). Greenwich: Information Age Publishing.

Ivanjek, L., Susac, A., Planinic, M., Andrasevic, A., & Milin-Sipus, Z. (2016). Student reasoning about graphs in different contexts. *Physical Review Special Topics – Physics Education Research, 12*, 010106.

Kerslake, D. (1981). Graphs. In K. M. Hart (Ed.), *Children's understanding of mathematics: 11–16* (pp. 120–136). London: John Murray.

Leinhardt, G., Zaslavsky, O., & Stein, M. K. (1990). Functions, graphs, and graphing: Tasks, learning, and teaching. *Review of Educational Research, 60*(1), 1–64.

Linacre, J. M. (2006). *WINSTEPS Rasch measurement computer program*. Available at http://www.winsteps.com

Linn, M. C., Eylon, B., & Davis, E. A. (2004). The knowledge integration perspective on learning. In M. C. Linn, E. A. Davis, & P. Bell (Eds.), *Internet environments for science education* (pp. 29–46). Mahwah: Lawrence Erlbaum Associates.

McDermott, L. C., Rosenquist, M. L., & van Zee, E. H. (1987). Student difficulties in connecting graphs and physics: Examples from kinematics. *American Journal of Physics, 55*, 503–513.

Michelsen, C. (2005). Expanding the domain – Variables and functions in an interdisciplinary context between mathematics and physics. In A. Beckmann, C. Michelsen, & B. Sriraman (Eds.), *Proceedings of the 1st International symposium of mathematics and its connections to the arts and sciences* (pp. 201–214). Schwäbisch Gmünd: The University of Education.

Nguyen, D. H., & Rebello, N. S. (2011). Students' understanding and application of the area under the curve concept in physics problems. *Physical Review Special Topics – Physics Education Research, 7*, 010112.

Özdemir, G., & Clark, D. B. (2007). An overview of conceptual change theories. *Eurasia Journal of Mathematics, Science & Technology Education, 3*(4), 351–361.

Planinic, M., Milin-Sipus, Z., Katic, H., Susac, A., & Ivanjek, L. (2012). Comparison of student understanding of line graph slope in physics and mathematics. *International Journal of Science and Mathematics Education, 10*(6), 1393–1414.

Planinic, M., Ivanjek, L., Susac, A., & Milin-Sipus, Z. (2013). Comparison of university students' understanding of graphs in different contexts. *Physical Review Special Topics – Physics Education Research, 9*, –020103.

Sherin, B. L. (2001). How students understand physics equations. *Cognition and Instruction, 19*(4), 479–541.

Vosniadou, S. (1994). Capturing and modeling the process of conceptual change. *Learning and Instruction, 4*(1), 45–69.

Wemyss, T., & van Kampen, P. (2013). Categorization of first-year university students' interpretations of numerical linear distance-time graphs. *Physical Review Special Topics – Physics Education Research, 9*, 010107.

Students' Mental Models of Human Nutrition from a Literature Review

Aurelio Cabello-Garrido, Enrique España-Ramos, and Ángel Blanco-López

Introduction

Over the past three decades, a large body of literature has developed around the topic of mental models. Although definitions and ideas about mental models vary widely, a generally accepted view is that these constructs refer to dynamic and generative representations which can be manipulated mentally to provide causal explanations of physical phenomena and make predictions about the state of affairs in the physical world (Vosniadou 1994). Mental models are built both by scientists and learners to interpret their experiences and to make sense of the physical world, and they may serve a number of purposes and functions when it comes to providing explanations and justifications (Coll and Treagust 2003). A number of authors (Johnson-Laird 1983; Norman 1983; Hafner and Stewart 1995), however, consider mental models to be incomplete, unstable, unscientific and lacking in firm boundaries. Norman (1983), on the other hand, argues that mental models are held over long periods of time and are relatively stable.

Although mental models represent personal constructs, there is consensus that they are subject to social influence. As Coll and Treagust (2003) point out, mental models are of interest for two reasons. First, such models influence cognitive functioning, and second, they can provide science education researchers and teachers with valuable information about learners' conceptual frameworks, that is, their underlying knowledge structures (Vosniadou 1994). Information of this kind could be used to improve the design of school curricula.

In the particular case of mental models related to human nutrition, knowledge of how they evolve may be useful for encouraging healthy eating habits. Furthermore,

A. Cabello-Garrido (✉) • E. España-Ramos • Á. Blanco-López
University of Malaga, Malaga, Spain
e-mail: aureliocabello@uma.es

© Springer International Publishing AG 2017
K. Hahl et al. (eds.), *Cognitive and Affective Aspects in Science Education Research*, Contributions from Science Education Research 3,
DOI 10.1007/978-3-319-58685-4_14

understanding the mental models that people build in relation to nutrition is especially relevant at a time when overweight and obesity are becoming a serious public health issue worldwide, and schools are being asked to contribute to raising students' awareness of this problem. Recent research (Zeyer and Dillon 2014) has also highlighted that health and the environment are increasingly important contexts for science education because they are close to students' interests and needs and, therefore, can help to link science education to personally relevant questions. In this regard it is worth noting that both these areas have often been neglected in science education research.

In light of the above, our aim here was to review the literature in an attempt to answer this question: Is it possible to identify and describe students' mental models of human nutrition and to consider how they evolve as students age and gain experience?

Method

The methodology proposed by Prieto-Ruz et al. (2002) and by Talanquer (2009) was employed. The study involved three phases (Table 1).

Phase 1 A comprehensive inventory of students' conceptions about human nutrition was built based on an analysis of the extensive literature published over the last three decades on students' ideas in this domain. Particular attention was paid to longitudinal and cohort studies that have explored students' core ideas at different stages of their learning (e.g. Turner 1997; Bullen and Benton 2004), as well as to monographs and reviews focused on identifying students' conceptions and mental models about human nutrition and how these change with age and instruction (e.g. Francis and Hill 1993; Núñez and Banet 1997). The review of the literature was performed together since the models are defined in this paper from the review in its whole. Altogether, we have analysed 43 works with the characteristics indicated.

Phase 2 Having built the inventory we were then able to identify the variables (aspects) most frequently cited in the literature as being those through which students' ideas about human nutrition seem to evolve with learning. These variables were food, nutrients, food energy, food groups and diet (Cabello-Garrido et al. 2009).

Table 1 An outline of the methodology used

Phase 1	Building a comprehensive inventory of students' conceptions about human nutrition
Phase 2	Identification of the variables (aspects) through which students' ideas about human nutrition seem to evolve with learning: food, nutrients, food energy, food groups and diet
Phase 3	Definition of models by grouping the key characteristic of the ideas expressed by students of similar ages or with similar level of experiences regarding the variables identified

Table 2 Main features of the three proposed models

Model	Characterised by	Main features
1. Vitalist	Vitalist thinking about food. It is believed that they are preset and ideas concerning food are based on information obtained directly from the sensory organs	Vitalism
		Predeterminism
		Sensoriality
2. Dualist	Categorical thinking splitting food and nutrients into those that are good vs. bad for one's health	Duality
		'Monotonic mind' belief
		Dose insensitivity
		Bijectivity
3. Dynamic	Flexibility in thinking about food, hence its name	Dynamic ideas
		Abstract reasoning
		Dose sensitivity
		Multicausal thinking

Phase 3 We defined different types of mental model by grouping the key characteristics of the ideas expressed by students of similar ages or with similar levels of experience regarding the variables identified.

Results

As shown in Table 2, the method described above led to the identification of three mental models of human nutrition, which we labelled 'vitalist', 'dualist' and 'dynamic'. In what follows, we describe the features of each model and illustrate them with examples or quotations taken from the literature.

Vitalist Model

This model is characteristic of younger people. Its main features are:

- Vitalism: Foods are considered to contain something natural that gives vitality or life force to the person who eats them: "Bread and milk have natural energy" (Francis and Hill 1993: 83). This characteristic is neither physical nor chemical but is something that transmits wellbeing and dynamism (Brophy et al. 2001). Physiological effects on the body are attributed to the foods themselves. Sometimes, this 'strength' obtained from food is called 'energy', but only in the colloquial sense, perhaps because an *energetic person* is seen as someone healthy.
- Predeterminism: This is the belief that what makes something a food is established in advance, because foods are composed of what people need to be healthy. Thus, if you eat what people have always eaten, then it is impossible to suffer any food-related disorder. The following quotations illustrate this idea:

  Food is food because it is edible (Lee and Diong 1999: 83).

> We eat things because they are food and food is food because we eat it (Brophy et al. 2001: 170).
>
> *En mi casa siempre se comió así* [In my house, that's what we always ate] (Rivarosa and De Longhi 2006: 543).

From this perspective, human foods are preset and cannot be changed: they contain what a person needs and they are eaten and have always been eaten in that person's environment (Brophy et al. 2001). This viewpoint leads to certain foods not being considered as food:

> Highly processed foods are not really food (Francis and Hill 1993: 83).
> If you can't chew it, it is not a food. Liquids are not food (Francis and Hill 1993: 83).

- Sensoriality: Ideas concerning food are based on information obtained directly from the sensory organs. The taste, shape, colour and texture of food are the main qualities which young children consider when assessing and classifying foods (Turner 1997). Concepts that cannot be perceived do not feature in this model. For instance, nutrients may solely be understood as visible or palatable parts of food, as illustrated in the following quotations:

> Fat is on bacon and it's all white and it's all wobbly (Turner 1997: 499).
> I would consider corn beef high in fibre because it is often stringy (Francis and Hill 1993: 81).

It is frequently thought that vitamins and minerals are pills. Turner (1997) notes that the belief that vitamins in the form of tablets/pills equated with health, or were good for you, was widespread among the children she studied. Likewise, the children in the study by Oakley et al. (1995) mentioned vitamins and iron as examples of healthy food. McKinley et al. (2005) also reported that children associated the word 'vitamins' with supplements (pills) more than with food.

Dualist Model

As young people gain knowledge about food and nutrients and their causal thinking improves, a more sophisticated model replaces the vitalist one. This model largely coincides with the heuristics described by Rozin et al. (1996), who found it to be very common among the American adult population. Oakes and Slotterback (2004) claim that the general public have acquired the belief that some foods promote healthfulness while others cause disease and death. Outlined below are the main features of the proposed model, along with some quotes that illustrate them.

- Duality. The defining characteristic of this model is categorical thinking, whereby food and nutrients are dichotomised into those that are good for one's health (such as fruit and vegetables) and those that are bad (e.g. high in salt, sugar and fat). Categorising foods as good or bad was first described in people with eating disorders in the early 1980s, but a decade later the phenomenon was observed to be widespread (Oakes and Slotterback 2004), with numerous studies documenting it in people of all ages. Chapman and Maclean (1993) characterised the two types as follows: bad foods, sometimes referred to as fast or junk food, contain a lot of fat, sugar, cholesterol, calories, salt and additives. Many people are unable to stay away from these foods because of their good taste, convenience and affordability. Conversely, healthy food is seen as something natural, providing nutrients that the body needs and as not containing the aforementioned 'bad' substances. These 'good' foods are often perceived as healthy but as not very tasty, and they are only eaten as a means of improving health or of achieving weight goals (Birch 1998). In this context, the word 'energy' is used to denote some positive feature of a food, whereas the term 'calorie' is associated with negative characteristics such as weight gain.

People's use of this mental model appears to take into account the latest research findings as reported by the media. For example, the students investigated by Bledsoe (2013) tended to divide carbohydrates into 'good' and 'bad' categories, which would reflect the latest research on the relationship between metabolic disease and carbohydrates that are quickly absorbed versus those which are gradually absorbed in the intestine.

- 'Monotonic mind' belief. This type of reasoning, when applied to food, leads to the belief that if something is harmful at high levels (e.g. dietary fat), then it is also harmful at low levels. A consequence of this belief is the magical principle of contagion (Rozin et al. 1996), whereby the presence in food of an ingredient considered 'bad' (e.g. fat) will, whatever the dose, *infect* all of the food:

> Sugar is bad for you and watermelon is good, so it gives you more energy (Francis and Hill 1993: 83).
> Foods bad for you like fats and oils don't have fibre (Francis and Hill 1993: 83).
> ...a few students classified sodium as a fat because they associated it with unhealthful foods (Bledsoe 2013: 28).

A corollary of this belief is that if a nutrient is healthful, then it is good for any function. This is particularly the case for vitamins and proteins:

> Proteins were universally viewed in a positive light as a source of energy and strength for muscles (Bledsoe 2013: 28).

- Dose insensitivity (Rozin et al. 1996). A major problem for many people is the inability to gauge correctly a certain amount of food (e.g. the amount recommended by the authorities or the amount ingested on a given day):

> ...the overestimation of calories in perceived unhealthy foods and the underestimation of calories in perceived healthy foods appears to be a pervasive error in the general population (Carels et al. 2007: 457).

Several studies have also examined the difficulties that many adults have in understanding the amounts stated on food labels:

> Almost a third of the recipients thought that "mg" means a millionth of a gramme (Abbott 1997: 47).

- Bijectivity. In mathematics, a bijection or one-to-one correspondence is a function between the elements of two sets, where every element of one set is paired with exactly one element of the other set and every element of the other set is paired with exactly one element of the first set. This kind of bijective or one-to-one correspondence can be observed in this mental model, since a single function is attributed to each type of nutrient (Francis and Hill 1993). Likewise, it is considered that a certain food can only provide a single nutrient, or it is believed that each nutrient can only be found in one food group.

This is a unicausal view which suggests that certain foods must be eaten to avoid deficiency diseases or that a particular food which is known to contain a certain nutrient cannot also provide other nutrients:

> Milk contains calcium, fruit has vitamins, veggies have minerals and meat has protein (Francis and Hill 1993: 83).
> He doesn't drink much milk so he's short (Wellman and Johnson 1982: 141).
> *Quien no bebe mucha leche, en general tiene bastante flojos los huesos* [People who don't drink much milk tend to have weak bones] (España-Ramos 2008: 272).

Dynamic Model

In the above-mentioned study by Rozin et al. (1996), 48% of American adults showed categorical thinking, similar to that described in our proposed dualist model. This would appear to highlight the difficulty that people experience in moving beyond this second model towards a more complex conception of human nutrition. In the third or dynamic model, the main emphasis is placed on diet rather than on individual foods, which are no longer seen simply as good or bad. In this model, nutritional value is associated with other factors, such as how foods are cooked, the amounts eaten, the other types of food in the diet, etc.

Recommendations and messages in the media frequently reinforce the dichotomy of good and bad foods, and this is one reason why it is important for young people to begin building more complex mental models. Wellman and Johnson (1982) observed many years ago that some children needed to create categories other than good or bad to refer to some foods. Similarly, Rawlins (2008) describes how some high-school students argued that whether or not a food was good or bad depended on how it was cooked and eaten.

> The boiled egg is in the centre cos ... it's basically healthy and unhealthy as well (Rawlins 2008: 141).

The third model is the closest to the one that people would ideally achieve, a model that would enable them to understand nutrition messages and be able to apply them in order to improve their diet and health. The dynamic model is labelled as such because it is characterised by flexibility in thinking about food, and its main features are as follows:

- Dynamic ideas. This refers to the capacity for flexible thinking and the recognition that concepts related to food are constantly changing, not only through the emergence of new scientific knowledge but because our physiological balance is never static, its elements interacting to generate different situations over time.

In this model, therefore, nutrients are no longer classified in absolute terms as good or bad, but rather they are evaluated in relation to the person's overall diet and circumstances and even according to their interactions with one another:

> ...increased amounts of oat bran may inhibit the uptake of iron (Schibeci and Wong 1994: 292).
>   Vitamin C... assists in iron absorption (Schibeci and Wong 1994: 292).

- Abstract reasoning. The ability to make inferences about unseen aspects of reality enables us to understand complex ideas, such as food being the source of nutrients and energy. Ideas of this kind form the basis of the food groups and most recommendations in nutrition guides.

> [to form groups of food] ...younger children used nutrient-specific criteria less frequently than older children (Bullen and Benton 2004: 361).

- Dose sensitivity. Acquiring this skill is what allows us to understand the meaning of the quantities of food and nutrients recommended in nutrition guides (Francis and Hill 1993), in prescriptions and on food labels.

> ...only 32% of patients could correctly calculate the amount of carbohydrates in a 20-oz bottle of soda that had 2.5 servings in the bottle (Rothman et al. 2006: 393).
>   One participant gave ice cream as an example of junk food but said that it does have some nutritional value so 'it depends on how much you use it' (Chapman and Maclean 1993: 111).

- Multicausal thinking. This is the ability to understand the multiple relationships between nutrients and the functions they perform, as well as between different foods and the nutrients they provide. Individuals now understand that there are many types and combinations of foods that can satisfy the body's needs, which in turn leads to the idea of a 'balanced diet' (Freeland-Graves and Nitzke 2002).

> Fay said that a high BMI value meant that people were either eating too much or not exercising enough (Schibeci and Wong 1994: 291).

Discussion and Conclusions

By examining the existing literature in the domain of human nutrition, we were able to organise students' ideas about food into three mental models that follow one another chronologically as young people develop their knowledge.

The first model, which we labelled 'vitalist', is typical of young children and results from the combination of their intuitive beliefs, social interactions and early educational interventions. In this model only readily perceptible aspects are considered, such as the organoleptic properties of food (flavour, shape, colour, etc.), where or when it is eaten and obvious physiological consequences of eating (satiation, growth, fattening). Abstract concepts, such as nutrients, cannot yet be understood and therefore do not feature in this model.

As young people gain knowledge about food and nutrients and their causal thinking improves, a more sophisticated 'dualist' model replaces the vitalist one.

Gradually, through educational instruction and social experiences, young people become familiar with the names of different types of nutrients and the value ascribed to them. Although not all nutrients will be known about to the same extent (Turner 1997; Bullen and Benton 2004), the second, dualist model represents an important step forward in people's ability to understand that foods are composed of chemical compounds, each of which performs important functions in the body.

According to Rozin et al. (1996), a sizeable proportion of the adult population continues to show categorical thinking, similar to that described in our dualist model. This kind of thinking is often reinforced by the media and advertising (Girón-Gambero et al. 2015), and even some influential nutritional authorities seem to support a good versus bad dichotomy concerning food (Oakes and Slotterback 2004). Consequently, there are few opportunities for people to move beyond such beliefs about food. In addition, the progression to a more complex model is subject to a number of cognitive constraints, many of which will be familiar to teachers of chemistry, for example, difficulties understanding the idea of chemical energy or the concepts of compound and chemical change, among others (Talanquer 2009). All this could explain why people find it difficult to progress to the third of our proposed mental models.

In what we have labelled the 'dynamic' model, ideas about human nutrition are no longer categorical or absolute, and the value of food and nutrients is seen as being dependent on a person's circumstances at a particular time. Nutrition is understood in a holistic way, and what matters is diet, not individual foods. Furthermore, people who have progressed to this kind of mental model are able to understand nutritional messages expressed in quantitative terms. Therefore, a goal for any society that wishes to improve levels of health in the population is to ensure that the majority of people are able to develop this dynamic mental model about food and nutrition. Hence, this model reflects the level of skill and knowledge that students will ideally have acquired by the time they leave school.

It is important to note that because learning is a nonlinear process, a given individual may show an overlap of ideas from different models at a particular point in time (Blanco-López and Prieto-Ruz 2004; Talanquer 2009), depending on which aspect of knowledge is being considered (food groups, nutrients, etc.). New ideas may replace some but not all of the old ones, and thus a student may simultaneously show characteristic features of two or even three models across different aspects of his or her nutrition knowledge. It must also be stressed that the three models we have proposed here should be considered as hypotheses to explain progress in learning in the domain of human nutrition. If they are confirmed through further empirical research, then our findings could be useful not only in studies of conceptual progression but also for informing curriculum sequencing.

With respect to nutrition education in schools, the current consensus is that this should not merely provide students with scientific knowledge but also aim to foster the development of skills and behaviours relevant to the buying, preparation and conservation of food, to achieving a balanced family diet and to the adoption of a healthy lifestyle (España-Ramos et al. 2014; Cabello-Garrido et al. 2016). Nonetheless, knowledge of food and nutritional processes provides an important foundation and cannot be separated from the attitudes and other psychosocial factors that may influence people's eating habits (Worsley 2002; McCullough et al. 2004). In the school context, therefore, an understanding of the mental models that students may have about food could help teachers to select the most suitable teaching strategies. This could play a key role in encouraging young people to develop healthy eating habits, a goal of considerable importance given that overweight and obesity now constitute a serious public health problem in Western societies.

Acknowledgements This work is part of the R + D Project "Development and evaluation of scientific competences through context-based and modeling teaching approaches. Case studies" (EDU2013-41952-P), funded by Spanish Ministry of Economy and Finance through its 2013 research call.

References

Abbott, R. (1997). Food and nutrition information: A study of sources, uses, and understanding. *British Food Journal, 99*(2), 43–49.

Birch, L. L. (1998). Development of food acceptance patterns in the first years of life. *Proceedings of the Nutrition Society, 57*(4), 617–624.

Blanco-López, A., & Prieto-Ruz, T. (2004). Un esquema para investigar el progreso en la comprensión de los alumnos sobre la naturaleza de la materia [A framework for investigating students' progress in understanding the nature of matter]. *Revista de Educación, 335*, 445–465.

Bledsoe, K. E. (2013). "Starch is very fatty": Understanding the logic in undergraduate student conceptions about biological molecules. *Electronic Journal of Science Education, 17*(2).

Brophy, J., Alleman, J., & O'Mahony, C. (2001). *Primary-grade students' knowledge and thinking about food as a cultural universal.* Michigan State University. Retrieved from http://files.eric.ed.gov/fulltext/ED451124.pdf

Bullen, K., & Benton, D. (2004). Moving to senior school: An under-exploited opportunity to teach nutrition? *International Journal of Science Education, 26*(3), 353–364.

Cabello-Garrido, A., España-Ramos, E., & Blanco-López, Á. (2009). Una hipótesis de trabajo para investigar el progreso en la comprensión de la alimentación humana por parte de los alumnos [A working hypothesis for investigating students' progress in understanding human nutrition]. *Enseñanza de las Ciencias,* Número extra, 1729–1735.

Cabello-Garrido, A., España-Ramos, E., & Blanco-López, Á. (2016). *La competencia en alimentación [Food competence].* Barcelona: Octaedro.

Carels, R. A., Konrad, K., & Harper, J. (2007). Individual differences in food perceptions and calorie estimation an examination of dieting status, weight, and gender. *Appetite, 49*(2), 450–458.

Chapman, G., & Maclean, H. (1993). "Junk food" and "healthy food": Meanings of food in adolescent women's culture. *Journal of Nutrition Education, 25*(3), 108–113.

Coll, R. K., & Treagust, D. F. (2003). Learners' mental models of metallic bonding: A cross-age study. *Science Education, 87*(5), 685–707.

España-Ramos, E. (2008). *Conocimiento, actitudes, creencias y valores en los argumentos sobre un tema sociocientífico relacionado con los alimentos* [Knowledge, attitudes, beliefs and values in discussions about a social and scientific topic related to food]. Doctoral dissertation, University of Malaga, Spain.

España-Ramos, E., Cabello-Garrido, A., & Blanco-López, Á. (2014). La competencia en alimentación. Un marco de referencia para la educación obligatoria [Food competence: A reference framework for compulsory education]. *Enseñanza de las Ciencias, 32*(3), 611–629.

Francis, R., & Hill, D. (1993). Developing conceptions of food and nutrition. *Research in Science Education, 23*(1), 77–84.

Freeland-Graves, J., & Nitzke, S. (2002). Position of the American dietetic Association: Total diet approach to communicating food and nutrition information. *Journal of the American Dietetic Association, 102*(1), 100–108.

Girón-Gambero, J. R., Blanco-López, Á., & Lupión-Cobos, T. (2015). Uso de la publicidad de un producto alimenticio para aprender un modelo sobre las defensas en el intestino humano. Un estudio en 3° de ESO [Using food advertising to teach a model of human intestinal defences: A study with year-three secondary school students]. *Revista Eureka sobre Enseñanza y Divulgación de las Ciencias, 12*(2), 278–293.

Hafner, R., & Stewart, J. (1995). Revising explanatory models to accommodate anomalous genetic phenomena: Problem solving in the "context of discovery". *Science Education, 79*(2), 111–146.

Johnson-Laird, P. N. (1983). *Mental models: Towards a cognitive science of language, inference, and consciousness (No. 6).* Cambridge, MA: Harvard University Press.

Lee, Y. J., & Diong, C. H. (1999). Misconceptions on the biological concept of food: Results of a survey of high school students. In M. Wass (Ed.), *Enhancing learning: Challenge of integrating thinking and information technology into the curriculum* (pp. 825–832). Singapore: Education Research Association.

McCullough, F. S., Yoo, S., & Ainsworth, P. (2004). Food choice, nutrition education and parental influence on British and Korean primary school children. *International Journal of Consumer Studies, 28*(3), 235–244.

McKinley, M. C., Lowis, C., Robson, P. J., Wallace, J. M. W., Morrissey, M., Moran, A., & Livingstone, M. B. E. (2005). It's good to talk: Children's views on food and nutrition. *European Journal of Clinical Nutrition, 59*(4), 542–551.

Norman, D. A. (1983). Some observations on mental models. *Mental Models, 7*(112), 7–14.

Núñez, F., & Banet, E. (1997). Students' conceptual patterns of human nutrition. *International Journal of Science Education, 19*(5), 509–526.

Oakes, M. E., & Slotterback, C. S. (2004). Prejudgments of those who eat a "healthy" versus an "unhealthy" food for breakfast. *Current Psychology, 23*(4), 267–278.

Oakley, A., Bendelow, G., Barnes, J., Buchanan, M., & Husain, O. N. (1995). Health and cancer prevention: Knowledge and beliefs of children and young people. *British Medical Journal, 310*(6986), 1029–1033.

Prieto-Ruz, T., Blanco-López, Á., & Brero-Peinado, V. (2002). La progresión en el aprendizaje de dominios específicos [Learning progressions in specific domains]. *Enseñanza de las Ciencias, 20*, 3–14.

Rawlins, E. (2008). Citizenship, health education and the obesity 'crisis'. *ACME: An International E-Journal for Critical Geographies, 7*(2), 135–151.

Rivarosa, S., & De Longhi, A. L. (2006). La noción de alimentación y su representación en alumnos escolarizados [Ideas about food and their representation in school children]. *Revista Electrónica de Enseñanza de las Ciencias, 5*(3), 534–552.

Rothman, R. L., Housam, R., Weiss, H., Davis, D., Gregory, R., Gebretsadik, T., & Elasy, T. A. (2006). Patient understanding of food labels: The role of literacy and numeracy. *American Journal of Preventive Medicine, 31*(5), 391–398.

Rozin, P., Ashmore, M., & Markwith, M. (1996). Lay American conceptions of nutrition: Dose insensitivity, categorical thinking, contagion, and the monotonic mind. *Health Psychology, 15*(6), 438–447.

Schibeci, D. R., & Wong, D. K. Y. (1994). Have you got any cholesterol? Adults' views of human nutrition. *Research in Science Education, 24*(1), 287–294.

Talanquer, V. (2009). On cognitive constraints and learning progressions: The case of "structure of matter". *International Journal of Science Education, 31*(15), 2123–2136.

Turner, S. A. (1997). Children's understanding of food and health in primary classrooms. *International Journal of Science Education, 19*(5), 491–508.

Vosniadou, S. (1994). Capturing and modeling the process of conceptual change. *Learning and Instruction, 4*(1), 45–69.

Wellman, H. M., & Johnson, C. N. (1982). Children's understanding of food and its functions: A preliminary study of the development of concepts of nutrition. *Journal of Applied Developmental Psychology, 3*(2), 135–148.

Worsley, A. (2002). Nutrition knowledge and food consumption: Can nutrition knowledge change food behaviour? *Asia Pacific Journal of Clinical Nutrition, 11*(s3), 579–S585.

Zeyer, A., & Dillon, J. (2014). Science|Environment|Health—Towards a reconceptualization of three critical and inter-linked areas of education. *International Journal of Science Education, 36*(9), 1409–1411.

The PISA Science Assessment for 2015 and the Implications for Science Education: Uses and Abuses

Jonathan Osborne, Magnus Oskarsson, Margareta Serder, and Svein Sjøberg

Introduction

For many national governments, the outcome of the triennial OECD PISA tests matter. For instance, in a survey conducted of the impact of PISA for 17 countries, PISA was seen to be "very influential", 11 others identified it as "moderately influential" and only 5 countries saw PISA as "not very influential" (Breakspear 2012). The director of PISA, Andreas Schleicher, sees PISA as a tool for identifying poor performance in any countries' educational system. Indeed performance on PISA has been shown to correlate with economic growth (Hanushek and Woessmann 2012). Along with the results of the TIMSS study, these tests have become an international benchmark that enable a country to judge the performance of its education system against that of other countries. Germany, for instance, suffered a severe blow to its sense of self-esteem when the 2000 results showed that their performance was merely mediocre (Breakspear 2012). As a result, both Germany and Switzerland initiated significant programs of reform in response to lower than expected performance. However, PISA is not without its critics. In a series of articles, Meyer and colleagues argue that PISA has become part of "a pervasive normalizing discourse, legitimizing historic shifts from viewing education as a social and cultural project to

J. Osborne (✉)
Stanford University, Stanford, California, USA
e-mail: osbornej@stanford.edu

M. Oskarsson
Mid Sweden University, Härnösand, Sweden

M. Serder
Malmö University, Malmö, Sweden

S. Sjøberg
University of Oslo, Oslo, Norway

© Springer International Publishing AG 2017   191
K. Hahl et al. (eds.), *Cognitive and Affective Aspects in Science Education Research*, Contributions from Science Education Research 3,
DOI 10.1007/978-3-319-58685-4_15

an economic one engendering usable skills and 'competences'" (Meyer et al. 2014). Labaree has argued that PISA assesses what nobody teaches (Labaree 2014) which Münch details in his exploration of how an Anglo-Saxon model of education has been forced on the German system (Münch 2014). To what extent are these criticisms justified? As science is the major focus of the tests in 2015, the findings will be particularly salient for the science education community. This chapter therefore offers a summary of a set of papers presented at a symposium at ESERA 2015 and seeks to explore the value of PISA and the legitimacy of such criticisms.

The PISA Science Assessment Framework: Advancing What It Means to Teach and Learn Science?

Jonathan Osborne

The goal of PISA is to define a set of competencies in reading, mathematics and science. Competencies are seen as "more than just knowledge and skills" requiring the ability "to meet complex demands, by drawing on and mobilising psychosocial resources (including skills and attitudes) in a particular context" (Rychen and Salganik 2003: 4). Thus the assessment framework for PISA offers an opportunity to define what constitutes a leading-edge conception of what the outcomes of formal education might achieve. In the case of PISA, the operationalization of what should be assessed in these programs is a product of a dialogue between the OECD directorate, the PISA governing body, and small panels of experts who draft the framework for consideration. For science, these outcomes are defined by the frameworks written for assessment in 2000, 2006 and 2015. The 2015 framework (OECD 2012) is the first revision since 2006 and, hence, can be seen as an important contribution to defining an international perspective on what the outcomes of formal science education should currently be.

The PISA framework draws on the view of many countries that an understanding of science is so important that it should be a feature of every young person's education (Confederacion de Sociedades Cientificas de España 2011, Millar and Osborne 1998, National Research Council 2012, Sekretariat der Ständigen Konferenz der Kultusminister der Länder in der Bundesrepublik Deutschland, KMK 2005). Many of these documents and policy statements give pre-eminence to an education for citizenship. Likewise, the emphasis in the PISA frameworks is on science for citizenship seeking to assess the competency of 15-year-old students to become informed critical consumers of scientific knowledge – a competency that all individuals are expected to need during their lifetimes. The particular focus of the PISA science framework is on scientific literacy which is defined for 2015 as the competency to:

1. Explain phenomena scientifically.
2. Evaluate and design scientific enquiry.
3. Interpret data and evidence scientifically.

These competencies are seen to lie at the heart of what it means to reason scientifically and require a knowledge of science or what is commonly called *content knowledge*. PISA defines this very vaguely in terms of what it might be reasonable to expect a 15-year-old student to know. To do more would be impossible given that over 70 countries participate in the test. The second and third competencies, however, require more than a knowledge of what we know. Rather, they depend on an understanding of how scientific knowledge is established and the degree of confidence with which it is held. Historically, specific calls have been made for teaching what has variously been called "the nature of science" (Lederman 2007), "ideas about science" (Millar and Osborne 1998) or "scientific practices" (National Research Council 2012). Within PISA, the 2006 framework operationalized this aspect of science using the term "knowledge about science". The major innovative feature of the PISA framework for 2015 has been to demarcate a knowledge of the standard procedures of the diverse methods and practices used to establish scientific knowledge – what is commonly called *procedural knowledge* from what is called *epistemic knowledge*. The latter is needed to understand the rationale for the common practices of scientific enquiry, the status of the knowledge claims that are generated and the meaning of foundational terms such as theory, hypothesis and data.

Procedural and epistemic knowledge are necessary to identify questions that are amenable to scientific inquiry, to judge whether appropriate procedures have been used, to ensure that the claims are justified and to distinguish scientific issues from matters of values or economic considerations. What then might be the constructs of procedural knowledge and epistemic knowledge? How might they be defined and how might they be demarcated from each other? The first major contribution of the PISA assessment framework for 2015 has been to clarify these forms of knowledge required for science literacy. In addition, it is the first document to bring into being the construct of epistemic knowledge as an explicit feature of assessment and, by inference, an explicit feature of teaching and learning.

Second, the new framework has introduced a means of assessing the cognitive demand of items using a definition which defines the depth of knowledge required for any task – a feature which was absent from the previous frameworks (Webb 2007). Finally, 2015 will be the first year in which the assessment will be undertaken using a computer-based platform. Computer-based assessment not only offers some adaptive testing but also a wider and more diverse form of assessment of student competency – producing a more valid assessment of student competency.

However, this view of the positive value of PISA is challenged in the following section by an argument that there are negative policy implications for educators, teachers and students.

PISA: A Global Educational Arms Race?

Svein Sjøberg

Since the first publication of PISA results in 2001, the results have become a kind of global "gold standard" for educational quality – a single measure of the quality of the entire school system. An OECD report on the policy impact of PISA proudly states that "PISA has become accepted as a reliable instrument for benchmarking student performance worldwide, and PISA results have had an influence on policy reform in the majority of participating countries/economies" (Breakspear 2012: 4).

Similarly, Andreas Schleicher (2012), director of PISA and recently also of Directorate of Education and Skills in OECD, in a TED talk starts his presentation by stating that PISA is "really a story of how international comparisons have globalized the field of education that we usually treat as an affair of domestic policy".

The intentions of PISA are, not surprisingly, related to the overall political aims OECD and the underlying concern for economic development in a competitive global free market economy. PISA is constructed and intended for the 30+ industrialized and wealthy OECD countries but has later been joined by a similar number of other countries with developing "economies". When the PISA results are presented, they are seen as an indicator for future competitive edge in a global economy (Sjøberg 2016). Governments are blamed for low scores, and governments are quick to take the honour when results are improving. In many countries, educational reforms have been launched as direct responses to the PISA results. While some try to copy the PISA winners, others do just the opposite of what high-achieving countries actually do.

The PISA undertaking is also a well-funded multinational "techno-scientific" exercise, undoubtedly the world's largest and costliest empirical study of schools and education. Given the size and importance, PISA has to be understood not just as a study of student learning but also as a "social phenomenon" in its wider political, social and cultural context, as also acknowledged by people who played a key role in the OECD preparations of PISA. As chair of Centre for Educational Research and Innovation (CERI) in the OECD, Professor Ulf P Lundgren had until 2000 a key role in the preparation of PISA. Ten years later, he writes:

> The outcomes of PISA we hoped could stimulate a debate on learning outcomes not only from an educational perspective but also a broad cultural and social perspective. Rarely has a pious hope been so dashed. (Lundgren 2011: 27)

PISA rankings create anxiety and discomfort in practically all countries, even in high-scoring countries (Alexander 2012). This produces an urge for politicians and bureaucrats to do "something" to rectify the situation. But since PISA cannot tell us much about cause and effect, creativity blossoms and educational reforms that are not empirically founded are introduced, often overnight. National curricula, cultural values and priorities are pushed aside.

Consequently, in many countries new curricula have been introduced, caused by "PISA shocks", (e.g. Norway, Denmark, Sweden, Germany and Japan). In many countries new national standards as well as new systems of obligatory national testing have been introduced. Some of these are directly influenced by PISA documents, as also proudly noted in a comprehensive report by the OECD itself (Breakspear 2012).

Many countries publish their own national test scores as league tables using them to rank school districts and schools. Some countries have introduced incentives such as salary systems that use test scores for teachers and (in particular) principals. Free choice of schools further exacerbates the importance of the rankings, often widening the gap between schools, as well as creating ways to "improve" test rankings. Such rankings have several consequences, like the obvious "teaching to the test", but the rankings also influence the price of neighbourhood housing, thereby widening socioeconomic gaps between districts.

The strive for better test scores also serves commercial interests. Companies deliver products such as tests and teaching materials that are supposed to increase scores, and cramming schools make substantial profit from preparing students to achieve higher test scores. The largest PISA contractor is the US-based non-profit assessment and measurement institution ETS. Maybe more important is that the world's largest commercial educational company, Pearson Inc., was involved in PISA 2015 and won the bid to develop the framework for PISA 2018. The joint press release from OECD and Pearson explains:

> Pearson, the world's leading learning company, today announces that it has won a competitive tender by the Organisation for Economic Co-operation and Development (OECD) to develop the Frameworks for PISA 2018. [...]. The frameworks define what will be measured in PISA 2018, how this will be reported and which approach will be chosen for the development of tests and questionnaires. (OECD and Pearson 2014)

The partnership with PISA/OECD is also a strategic door opener for Pearson "with 40,000 employees in more than 70 countries" into the global educational market. In company with the OECD, Pearson also produces "The Learning Curve", a ranking of nations according to a set of test-based indicators. PISA leader Andreas Schleicher sits on the Advisory Panel of The Learning Curve. These rankings get media coverage and further create anxiety among politicians and policymakers. The result is a further pressure towards doing "something" to climb in the league tables.

PISA is now used to legitimize neoliberal policies and reforms that are duly labelled New Public Management (Møller and Skedsmo 2013). The PISA outcomes are also leading to an emerging global governance and standardization of education, as also noted by key educational experts (Ball 2012). The process is also called "governing with numbers" and the "PISA effect in Europe" (Grek 2009).

The PISA testing framework (OECD 2012) is a most interesting document that could be used to inspire discussions about the purpose and contents of science curriculum and teaching. However, problems arise when the brave intentions of the PISA framework are translated to concrete test items to be used in a great variety of languages, cultures and countries. It is, of course, impossible to construct a test that in a fair and objective way can be used across countries and cultures to assess the quality of learning in "real-life" situations with "authentic texts". The requirement of "fair testing" implies by necessity that local, current and topical issues must be excluded. This runs against most current thinking in, e.g. science education, where "science in context" and "localized curricula" are ideals promoted by, e.g. UNESCO, science educators as well as in national curricula.

The use of PISA for political purposes is very selective. While the rankings of nations get a lot of attention, other results are ignored. It seems, for instance, that pupils in high-scoring countries develop the most negative attitudes to the subject.

It also seems that PISA scores are unrelated to educational resources, funding, class size, etc. PISA scores also seem to be negatively related to the use of active teaching methods, the inquiry-based instruction and the use of ICT. The fight to improve PISA rankings may conflict with the work to make science education relevant, contextualized, interesting and motivating for young learners. Whether one "believes in PISA" or not, such intriguing results need to be discussed.

As a contribution to that discussion, the next two sections draw on data from Sweden – one country whose performance on PISA has declined in recent years. These papers explore to what extent there is, or is not, a positive value to the outcomes of PISA and the comparisons that are made.

School Science in a Market-Driven School System

Magnus Oskarsson

Sweden's results in PISA have shown the largest drop of all countries in all three subjects over the last 12 years. From results which were above the OECD average and with a high degree of equity in PISA 2000, these have fallen by 2012 to results which were below average in science as well as in math and reading and with a sharp increase of low-performing students and low-performing schools. One important reason behind this seems to be the decentralization and market adaptations of the Swedish school since the mid-1990s. A free choice of school and a voucher system were introduced together with new legislation that allowed private schools to be fully financed by public means through the vouchers. This was followed in the mid-2000s by new control and steering mechanisms with an expanded grading system, a vast increase in the number and occasions of national tests, and a school inspectorate.

Sweden like the other Nordic countries has a long history of successful efforts to create a comprehensive school system with good results and a high degree of equity. The differences between high and low achievers and the difference between different schools have been smaller than in many other countries, and the same has been true for the impact of social background. The first PISA study 2000 showed that Swedish students were above mean in all subjects (OECD 2001).

From the 1990s Sweden has developed a more decentralized school system with a new curriculum and a benchmarked grading system. Influenced by New Public Management theories, a school voucher system was established and the students were free to choose their school. Private schools were allowed, fully financed by public means. The effects of the reforms were minor during the 1990s and during the first years of the following decade. From around 2003, however, changes have become more noticeable (Lundahl et al. 2013).

The majority of the schools are still public, but PISA shows that Sweden has had the fastest growth in the proportion of private schools between 2003 and 2012 of all

OECD countries. One effect is that differences between schools have increased. A number of schools seem to have been abandoned by the most ambitious students, and segregation both by socioeconomic background and ethnic background has increased – a finding which supports Sjøberg's critique.

Since 2006 many new school reforms have been introduced. Examples of reforms are a formation of a national school inspectorate, increased quality control, more national testing and again an expanding grading system. A new mandatory national test in science for grade nine was piloted in 2009 and fully operational in 2010. A number of partly government-funded school development and in-service programmes have been offered to schools, and several of these programmes have been directed to math and science.

When the PISA 2012 results were presented, it showed that the performance of Swedish students was below the international average in all tested domains. The drop was the largest in OECD in all three domains since the start of PISA. A closer look at the science results from 2006 and onwards reveals not only a drop in mean result but also an increasing difference between low and high achievers (OECD 2014a). The decline has been more rapid since 2006 and especially after 2009 despite all efforts by the government.

There was, however, no significant change either in the number of top-performing schools or in the number of top-performing students in science. Rather, it is the number of low achievers that has increased, and the same is true for the number of low-achieving schools, where the proportion of schools with a mean below 450 in PISA science has increased from less than 5% in 2006 to 20% in 2012. Results for both boys and girls have dropped, but there was a larger deterioration in boys' results.

Ambitious students choose schools with high reputation, while other students are left behind in less advantaged schools. And as PISA shows, there are few winners and many losers. Several reports show that the school choice system and the voucher system are two important reasons (Skolverket 2012). A study by Östh et al. (2013) shows that the cause of increasing differences between schools is school choice rather than increasing residential segregation. Another recent study shows covariance between increased between-school variance and decreasing PISA science results in a number of countries (Davidsson et al. 2013).

While the Swedish science results have dropped not only in PISA but also in TIMSS, the same is not true, however, for national grading or the national testing. Data for national test results and for final grades in science neither show increasing numbers of failing students nor increasing differences between boys and girls, and there is growing evidence that neither the national tests nor grades are stable over time (Lundahl et al. 2013).

A recent report has pointed out that participation in school development programmes varies greatly between different regions. Larger cities and towns with universities have participated in a majority of these programs, while smaller communities in more remote areas only have a participation rate at around 15% or less in the national school development programmes.

These large differences between schools and regions also have the effect that teachers' employment choices are more diverse. For instance, it becomes more attractive to work at schools with not only high-achieving students but also good in-service training and strong support for professional development. This means that schools, that for a variety of reasons do not or cannot take part in development programmes, are likely to show decreasing results and increasing difficulties in recruiting teachers. This is exactly what has been pointed out in a recent report from OECD (2014b) where Sweden is one of the countries where low-performing schools report both lack of resources and the largest difficulties in recruiting competent teachers.

Moreover, there are not only increasing differences in students' results but also in students' attitudes. Some students feel more motivated while others feel increasing social exclusion, and several of these background factors have strong correlations with students' results. More control and testing seem to have increased the extrinsic motivation among some students, while others show more negative response to the greater pressure.

The value of PISA, however, is that it can reveal this kind of important information about national school systems, such as the Swedish – a point which Sjøberg does not consider. In a period with many reforms, the pros and cons of such change are often unknown when the next reforms are introduced. The Swedish example shows that when a nation's assessment system is not stable over time, international comparisons give invaluable data to educators and policymakers about the state of the educational system.

The final section looks not at the national data but at how students interpret the assessment tasks and their cultural specificity arguing that the assessments are too culturally specific to offer valid cultural comparisons.

An Interpretation of PISA Results from a Science Classroom Perspective

Margareta Serder

In Sweden, a series of educational reforms have been launched in the last 10 years, often explicitly addressing the decreasing results of the Swedish students in international assessments and PISA (Ringarp and Rothland 2010). A few recent examples of reforms intended to strengthen the outcomes of Swedish students are a new teacher education, a new grading system, a new curriculum and 1 h more of mathematics for all students. Still, the negative Swedish trend has not shifted.

In the academic conversation about PISA, several concerns with the assessment have been raised. Those address, for instance, the increasing impact that OECD/PISA has on educational policy (Sellar and Lingard 2013), methodological weaknesses (Allerup 2007) and translation issues that affect international comparability (Arffman 2010). Meanwhile, in science education, researchers have argued that the concept of scientific literacy as articulated in PISA might have a positive effect in offering a good example of the goals and emphases of science education (Fensham 2009). Other scholars argue that the role that literacy in its more fundamental, linguistic sense (Norris and Phillips 2003) plays in the PISA results needs to be more clearly emphasized in the scientific literacy framework (Olsen 2012).

While the reforms above are uses of PISA on a policy level, this paper leaves behind the statistical assumptions that inform policy to offer, instead, an empirical investigation of the effects of PISA from a pedagogical (classroom) perspective. More specifically, it describes a study that seeks to understand the interaction between the students and the items used for testing. It assumes that observation of problem-solving in action can give us information such as: What impediments or difficulties develop in students' encounters with specific test questions? What meanings are offered by and produced from the science problems as presented in the test? The focus of these questions is more on the validity of the data.

A Design to Explore Test Items in Scientific Literacy

To explore these questions, a study was designed in which 15-year-old students collaboratively answered 11 PISA test questions in scientific literacy (OECD 2007) during a science lesson (this design is thoroughly described in, e.g. Serder and Jakobsson 2015). Three PISA units were included: Greenhouse, Acid Rain and Sunscreens (PISA units S114, S447, S485). In total, 21 groups of 3–4 students were formed. This collaborative design was chosen based on a sociocultural understanding of knowledge construction (Wertsch 1998) which emphasizes that knowledge/knowing is shaped in action and obtains meaning from real-life situations. The interactions were video recorded producing 16 h of video data for further observation and semantic analysis (Mäkitalo et al. 2009). The focus for the analysis presented here was what specific problems the students experienced with the test questions.

A Stereotyped Portrayal of Science to Resist and Terms with Hybrid Meanings

In the analysis of the data, two main themes were discerned: (1) how the students discussed science as it was portrayed in the test items and (2) how words and formulations of the test items were used in the student conversations. PISA scientific literacy test items are required to address "real-life issues" (OECD 2012: 102) which infer that the problems are contextualized in everyday life situations. According to the analysis, this condition seems to affect how the students in this study approached the problems. A common difficulty for the groups was to identify the intended meaning of various words. Words with hybrid meaning, that have different meanings in different contexts, were found to be frequently negotiated by the students (Serder and Jakobsson 2016). The Swedish words, e.g. *pattern, factor, reference, constant* and *better* are examples from the study, all with differing meanings depending upon whether they are used in an everyday, science or mathematical context. In order for the groups to respond successfully in the test, it appeared to be crucial to ignore all alternative, possible contexts – including the inferred everyday context. Two examples from the observed student conversations are the word "pattern" that could be discussed as a mathematical term (in the sense *regularity*) instead of as (intended) a result of a scientific experiment and the word "factor" used to denote a "sun cream", instead of a scientific variable. The results also indicate that some meaning is likely to be added, or lost, in translation (Serder and Jakobsson 2016). The second analytic theme concerned meaning in a different sense, namely, the meaning of science itself. The students tended to discuss the manner in which the fictive characters of the test items were speaking and approaching the scientific problems presented to them in their imaginary everyday life. While so doing, the students often expressed resistance towards the image of science that they were presented and interacted with. However, in order to approach the test questions productively, the students need to accept the artificial aspects of the "real-life" problems, as well as the authoritative, highly stereotyped way in which science was (implicitly) portrayed in those items (Serder and Jakobsson 2015).

Use or Misuse?

The group situations used for the purpose of this study were different from individual testing situations. However, the design permits an insight in the interactions that possibly take place in the test situation, of which very little is known. The reasons for the decreasing Swedish results are likely to be various. This research poses questions about comparability between different national versions of the PISA test because the hybrid meanings of the words that were negotiated by the student groups are unlikely to be overlapping in different languages (see Arffman 2010). This is a finding that merits further exploration. As for the testing of "real-life skills", the work shows that the everyday aspects of the test questions invite a number of various discourses and meanings. The conclusions support previous research that suggests that scientific competence is intimately linked to students' discursive knowledge (Norris and Phillips 2003; Olsen 2012). Moreover, the inferred everyday contexts in PISA seem to enhance the way science gets portrayed as a very particular – and peculiar – human culture with certain norms and values (cf. Aikenhead 1996). For instance, the students questioned "everyday" situations inferred in the test items such as conducting experiments outside school, putting marble chips in vinegar or discussing graphs in a library (Serder and Jakobsson 2015).

Sweden may be an exceptional case, both in the decline in its PISA scores and the number of rapid education reforms in recent years. Unfortunately, there might be little correspondence between national school reforms launched to address the political anxieties of *PISA shock* and the actual problems lived by students and teachers in the school system that need addressing. In order to reduce misuse of the PISA scores, these findings raise the question of whether there is a need for the scholarly community to stress less the need to attend to the general outcomes and superficial interpretations of the results and attend more to a fine-tuned analysis of the results and their implications. This study, in which the interactions of students and test items were studied, was an attempt to show what such analyses reveal and why they might be valuable.

Postscript

What conclusions then can be drawn from these somewhat disparate perspectives? The first is to recognize that there are both positive and negative aspects to PISA. The positive is that this international test attempts to measure competency – not just knowledge and understanding but the ability to use scientific knowledge to undertake science-specific tasks which might reasonably be expected of a science literate 15-year-old. These competencies are defined in a manner which represents the best thinking about what should be the outcomes of a contemporary science education. Moreover, the testing for PISA is undertaken in 72 countries in a manner which is as rigorous and systematic as can be given the constraints of producing one test in multiple languages in a short time scale. As a consequence, it produces an enormous quantity of data which raises issues for individual countries rather than definitive answers. Embedded within the data are some clear trends and patterns. For instance, Canada, Estonia, Germany and Hong Kong (China) have all attained high levels of performance with high or improving levels of equity in the 2015 results. What these countries would appear to share to a greater or lesser degree is (OECD 2016: 7):

1. A clear education strategy to improve performance and equity.
2. Rigorous and consistent standards are applied across all classrooms.
3. Teacher and school leader capacity has been improved.
4. Resources are distributed equitably across schools – preferentially to those schools and students that need them most.
5. At-risk students and schools are proactively targeted.

   If such findings succeed in raising questions or providing pointers to where we might concentrate our efforts to improve the quality of the education that each country might offer its young people, then PISA has value. If, in contrast, PISA is simply seen as part of an international educational arms race from which little can be learnt, then all will have been in vain.

References

Aikenhead, G. S. (1996). Science education: Border crossing into the sub-culture of science. *Studies in Science Education, 27,* 1–52.

Alexander, R. (2012). Moral panic, miracle cures and educational policy: What can we really learn from international comparison*? Scottish Educational Review, 44*(1), 4–21.

Allerup, P. (2007). Identification of group differences using PISA scales – Considering effects of inhomogenous items. In S. T. Hopmann & G. Brinek (Eds.), *PISA zufolge PISA – PISA according to PISA* (pp. 175–201). Vienna: Universität Wien.

Arffman, I. (2010). Equivalence of translations in international reading literacy Studies. *Scandinavian Journal of Educational Research, 54*(1), 37–59.

Ball, S. J. (2012). *Global education Inc: New policy networks and the neo-liberal imaginary*. London: Routledge.

Breakspear, S. (2012). *The policy impact of PISA: An exploration of the normative effects of international benchmarking in school system performance, OECD Education working papers, No. 71*. Paris, France: OECD Publishing. Retrieved from http://dx.doi.org/10.1787/5k9fdfqffr28-en

Confederacion de Sociedades Cientificas de España. (COSCE). (2011). Informe ENCIENDE: Enseñanza de las Ciencias en la Didáctica Escolar para edades tempranas en España. [*Report ENCIENDE: Teaching science in school education for early ages in Spain*]. Madrid, Spain: COSCE. Retrieved from http://www.cosce.org/pdf/Informe_ENCIENDE.pdf

Davidsson, E., Karlsson, K.-G., & Oskarsson, M. (2013). Trender och likvärdighet: Svenska elevers resultat på PISA naturvetenskap i en internationell jämförelse [trends and equivalence: Swedish students' performance in the PISA science in an international comparison]. *Utbildning & Demokrati 2013, 22*(3), 37–52.

Fensham, P. J. (2009). Real world contexts in PISA science: Implications for context-based science education. *Journal of Research in Science Teaching, 46*(8), 884–896.

Grek, S. (2009). Governing by numbers: The PISA 'effect' in Europe. *Journal of Education Policy, 24*(1), 23–37.

Hanushek, E. A., & Woessmann, L. (2012). Do better schools lead to more growth? Cognitive skills, economic outcomes, and causation. *Journal of Economic Growth, 17*(4), 267–321.

Labaree, D. F. (2014). Let's measure what no one teaches: PISA, NCLB, and the shrinking aims of education. *Teachers College Record, 116*(9), 1–14.

Lederman, N. G. (2007). Nature of science: Past, present and future. In S. Abell, & N. G. Lederman (Eds.), Handbook of research on science education (pp. 831–879). Mawah: Lawrence Erlbaum.

Lundahl, L., Erixon Arreman, I., Holm, A.-S., & Lundström, U. (2013). Educational marketization the Swedish way. *Education Inquiry, 4*(3).

Lundgren, U. P. (2011). PISA as a political instrument. In M. A. Pereyra, H. G. Kotthoff, & R. Cowen (Eds.), *PISA under examination. Changing knowledge, changing tests, and changing schools*. Rotterdam: Sense publishers.

Mäkitalo, Å., Jakobsson, A., & Säljö, R. (2009). Learning to reason in the context of socioscientific problems. Exploring the demands on students in 'new' classroom activities. In K. Kumpulainen, C. Hmelo-Silver, & M. Cesar (Eds.), *Investigating classroom interaction. Methodologies in action* (pp. 7–26). Rotterdam: Sense Publishers.

Meyer, H.-D., Tröhler, D., Labaree, D. F., & Hutt, E. L. (2014). Accountability: Antecedents, power, and processes. *Teachers College Record, 116*(9), 1–12.

Millar, R., & Osborne, J. F. (Eds.). (1998). *Beyond 2000: Science education for the future*. London: King's College London.

Møller, J., & Skedsmo, G. (2013). Modernising education: New public Management reform in the Norwegian education system. *Journal of Educational Administration and History, 45*(4), 336–353.

Münch, R. (2014). Education under the regime of PISA & co.: Global standards and local traditions in conflict—The case of Germany. *Teachers College Record, 116*, 1–16.

National Research Council. (2012). *A framework for K-12 science education: Practices, crosscutting concepts, and core ideas*. Washington, DC: National Academies Press.

Norris, S. P., & Phillips, L. (2003a). How literacy in its fundamental sense is central to scientific literacy. *Science Education, 87*, 224–240.

OECD. (2001). *Knowledge and skills for life: First results from the OECD Programme for International Student Assessment (PISA) 2000*. Paris: OECD Publishing.

OECD. (2007). *PISA 2006: Science competencies for tomorrow's world. Volume 1: Analysis*. Paris: OECD Publishing.

OECD. (2012). *PISA 2012 assessment and analytical framework: Mathematics, reading, science, problem solving and financial literacy*. Paris: OECD Publishing.

OECD. (2014a). *PISA 2012 results: What students know and can do*. (Vol. I, Rev. edn, February 2014) [Electronic resource]: OECD publication.

OECD. (2014b). *Resources, policies and practices in Sweden's schooling system: an in-depth analysis of Pisa 2012 results*. Retrieved from http://www.lcis.com.tw/paper_store/paper_store/86952f1a-2015226205027484.pdf

OECD. (2016). *Country note: Key findings from PISA 2015 for the United States*. Paris, OECD Publishing.

OECD & Pearson. (2014). (Joint press release) *Pearson to develop PISA 2018 Student Assessment 21st Century Frameworks for OECD*. https://www.pearson.com/news/announcements/2014/december/pearson-to-develop-pisa-2018-student-assessment-21st-century-fra.html

Olsen, R. V. (2012). An exploration of cluster structure in scientific literacy in PISA: Evidence for a Nordic dimension? *Nordic Studies in Science Education, 1*(1), 81–94.

Oskarsson, M., Eliasson, N., & Karlsson, K. (in progress). *School science in a market-driven school system*.

Östh, J., Andersson, E., & Malmberg, B. (2013). School choice and increasing performance difference: A counterfactual approach. *Urban Studies, 50*(2), 407–425.

Ringarp, J., & Rothland, M. (2010). Is the grass always greener? The effect of the PISA results on education debates in Sweden and Germany. *European Educational Research Journal, 9*(3), 422–430.

Rychen, D. S., & Salganik, L. H. (Eds.). (2003). *Definition and selection of key competencies: Executive summary*. Göttingen: Hogrefe.

Schleicher, A. (2012). *Use data to build better schools*. TEDGlobal. Available at http://www.ted.com/talks/andreas_schleicher_use_data_to_build_better_schools?language=en

Sekretariat der Ständigen Konferenz der Kultusminister der Länder in der Bundesrepublik Deutschland (KMK). (2005). *Bildungsstandards im Fach Biologie für den Mittleren Schulabschluss* (Jahrgangsstufe 10). Berlin, Germany.

Sellar, S., & Lingard, B. (2013). PISA and the expanding role of the OECD in global educational governance. In H. Meyer & A. Benavot (Eds.), *PISA, power and policy* (pp. 185–206). Oxford: Symposium Books Ltd..

Serder, M., & Jakobsson, A. (2015). "Why bother so incredibly much?": Student perspectives on PISA science assignments. *Cultural Studies of Science Education, 10*(3), 833–853.

Serder, M., & Jakobsson, A. (2016). Language games and meaning as used in students encounters with scientific literacy test items. *Science Education, 100*(2), 321–343.

Sjøberg, S. (2016). OECD, PISA, and globalization. The influence of the international assessment regime. In C. H. Tienken & C. A. Mullen (Eds.), *Education policy perils. Tackling the tough issues*. New York: Routledge.

Skolverket. (2012). *Educational equity in the Swedish school system? A quantitative analysis of equity over time*. Retrieved from http://www.skolverket.se/

Webb, N. L. (2007). Issues related to judging the alignment of curriculum standards and assessments. *Applied Measurement in Education, 20*(1), 7–25. doi:10.1207/s15324818ame2001_2.

Wertsch, J. V. (1998). *Mind as action*. New York: Oxford University Press.

Part IV
Language in Science Classrooms

Part IV
Language in Science Instruction

Chemistry Teachers' Professional Knowledge, Classroom Action, and Students' Learning: The Relevance of Technical Language

Holger Tröger, Elke Sumfleth, and Oliver Tepner

Introduction

Research on quality of instruction is an important facet of educational research (Abell 2007; Arnold 2007; Clausen 2002). Teaching is considered as a highly complex activity and several different models have been developed to describe the relation between different elements of teaching, as there are teachers' and students' characteristics and attitudes, teachers' knowledge in different forms, general conditions of teaching and classes, as well as students' learning achievement and motivation (Bauer 2011; Baumert et al. 2010; Berry et al. 2015).

Though these models vary in complexity and focus, teachers' characteristics are regarded as fundamentally important for students' learning outcome (Krauss et al. 2008). Findings of recent empirical studies on teachers' professional knowledge (e.g., COACTIV or MT-21) document a connection between teachers' professional knowledge and students' learning achievement (Kunter et al. 2013).

H. Tröger • E. Sumfleth
University of Duisburg-Essen, Essen, Germany

O. Tepner (✉)
University of Regensburg, Regensburg, Germany
e-mail: oliver.tepner@chemie.uni-regensburg.de

© Springer International Publishing AG 2017

207

K. Hahl et al. (eds.), *Cognitive and Affective Aspects in Science Education Research*, Contributions from Science Education Research 3, DOI 10.1007/978-3-319-58685-4_16

Theoretical Framework

Professional Knowledge

Following Shulman's (1987) initial operationalization, teachers' professional knowledge comprises at least seven dimensions: *content knowledge* (1), *general pedagogical knowledge* (2), *curriculum knowledge* (3), *pedagogical content knowledge* (4), *knowledge of learners and their characteristics* (5), *knowledge of educational contexts* (6), and *knowledge of educational ends, purposes, and values* (7). Recent research focuses on three distinct dimensions of professional knowledge:

- Pedagogical knowledge (PK)
- Content knowledge (CK)
- Pedagogical content knowledge (PCK)

Following Shulman's description of pedagogical knowledge (PK), this type of knowledge comprises strategies and principles of classroom management and classroom organization and is generally regarded as not subject-specific (König et al. 2011; Shulman 1987).

Content knowledge (CK) comprises the in-depth knowledge of a particular subjects' content (Kleickmann et al. 2012; Riese and Reinhold 2012). Findings in recent studies in mathematics education indicate that teachers' repertoire of actions and explanations is related to the depth and range of their CK (Kunter et al. 2013).

Over time, PCK has been described by many approaches, and while there is no consistent conceptualization of PCK (Abell 2007, 2008; Berry et al. 2015; Loughran et al. 2012; Park and Oliver 2007), it is considered to be important for teaching. Several national and international studies focus on and discuss the meaning of teachers' PCK for teaching (Park et al. 2010) and the development of PCK (Abell et al. 2009; de Jong and van Driel 2005). PCK can be described in various models like the model of teacher professional knowledge and skill (TPK&S) which integrates the construct PCK in a very complex structure of teaching and learning (Gess-Newsome 2015). Research operationalizes PCK in various forms and nuances. General professional knowledge and topic-specific professional knowledge (TSPK) can be seen as a knowledge base and are quite static. In a PCK summit, the conception of PCK was extended taking classroom action into account. Personal PCK is described as "the *knowledge* of, *reasoning* behind, and *planning* for teaching a particular *topic* in a particular *way* for a particular *purpose* to particular students for enhanced *student outcomes* (Reflection *on* Action, Explicit)" (Gess-Newsome 2015, p. 36). Personal PCK&S is operationalized as "the *act of teaching* a particular *topic* in a particular *way* for a particular *purpose* to particular *students* for enhanced *student outcomes* (Reflection *in* Action, Tacit or Explicit)" (Gess-Newsome 2015, p. 36).

These descriptions suggest that teachers' PCK has manifold aspects and facets. As PCK is a highly complex construct, any investigation requires an emphasis on certain aspects of PCK. In this study, the focus is on pedagogical content knowledge about technical language (PCK_{TL}).

Technical Languages

Language is of central importance for learning. Language is not only the primary medium of interpersonal communication used to transfer information like thoughts, beliefs, or knowledge (Bußmann 2002; Eunson 2012), but it is also deeply connected to cognition and thinking (Childs et al. 2015; Johnstone and Selepeng 2001). Technical languages (TL) are an artificial form of language (Bußmann 2002) used with the specific purpose to communicate on a subject-specific level (Grucza 2012; Roelcke 2005; Schmölzer-Eibinger 2013). Although technical languages are not clearly defined in detail, their common characteristics can be derived. Technical languages have a specific lexis, commonly known by the expression *technical terms* (Özcan 2013; Taber 2015) and more complex syntactic structures. In addition, more compositions and nominalizations are used than in everyday language (Schmölzer-Eibinger 2013). In chemistry, technical language is central for a meaningful communication as it also comprises formal and pictorial elements of language like symbolic expressions (e.g., structural formulas, chemical equations) which are chemistry immanent. These are considered as of fundamental importance to communicate about chemistry (Childs et al. 2015; Taber 2015).

Students' knowledge of and proficiency in language and technical language are deeply and inseparably connected to their subject learning as well as their development of scientific literacy (Özcan 2013; Taber 2015; Yore 2012; Yore et al. 2003). Both are regarded as central aims of present science education. Deficiencies in this field might lead to the development of misconceptions, inadequate representations of scientific concepts, as well as misunderstandings (Barke 2015; Johnstone and Selepeng 2001; Taber 2015).

Teachers have to deal with their own and their students' technical language in class as it is an integral component of subject-specific teaching. Therefore, the handling of technical language is seen as an important aspect of PCK is this study.

Research Objective

Studies in the field of educational research consider the connection between teachers' professional knowledge and their students' learning outcome as well as the development of teachers' or student teachers' PCK. However, research lacks findings on the relation between teachers' professional knowledge, their actual action in class, and their students' learning. This study is part of the joint research project ProwiN (**Pro**fessions**wi**ssen in den **N**aturwissenschaften) [Professional Knowledge in Sciences]. The joint research project explores teachers' professional knowledge and its connection to students' achievement and classroom action in biology, chemistry, and physics (e.g., Cauet et al. 2015; Förtsch et al. 2016). This study's objective is to shed light on the aforementioned desideratum. It focuses on the investigation of the relation between chemistry teachers' subject-specific professional knowledge regarding technical language, their classroom action, and students' learning achievement.

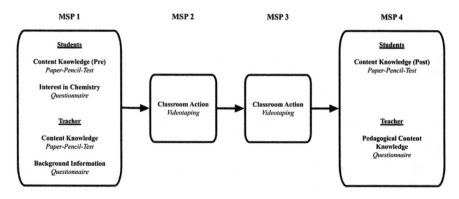

Fig. 1 Course of testings

Research Design

Chemistry classes of German secondary school teachers (German *Gymnasium*) in North Rhine-Westphalia and Bavaria took part in this study. The lessons cover the topic *atomic structure and periodic table of the elements*. In general, this topic is taught in the eighth grade at German Gymnasium (upper secondary school).

There are four points of measurement (MSP) in order to collect all required data (see Fig. 1).

The first testing had taken place before the teaching sequence on the aforementioned topic began. During the first testing, students' prior content knowledge and their general interest in chemistry as well as teachers' CK and background information had been ascertained. Then, two successive lessons were videotaped. At the end of the teaching sequence, students' post content knowledge and teachers' pedagogical content knowledge regarding the handling of technical language (PCK_{TL}) were assessed.

Test Instruments

Students' learning achievement was examined using a multiple-choice single-select test in a pre-post design. The applied test was specifically developed in order to measure students' content knowledge regarding the topic *atomic structure and the periodic table of the elements* and it has been evaluated in a pilot study. The piloted test comprised 40 items in two test booklets using a multi-matrix design and was answered by a total of 149 students. The items were analyzed using Rasch measurement. Evaluation shows good values for item (.92) and person (.77) reliability. Thirty items were selected based on item discrimination (>.75), item fit (.80 < mean square (MNSQ) > 1.2), and criteria of validity and were included in the final test instrument.

Students' interest in chemistry was assessed by a questionnaire. Students answered to several statements on a 7-point Likert scale (subscales, e.g., regarding their personal interest in chemistry, the perceived importance of chemistry for their personal future).

Teachers' CK was examined using an evaluated multiple-choice-single-select test (Tepner and Dollny 2014). The test instrument was carefully developed in a previous study (Dollny 2011) and validated in a subsequent assessment. This test comprised 29 items covering distinct aspects of the subject's content taught in school.

Teachers' PCK_{TL} was assessed by a questionnaire which was specifically developed for this project (Strübe et al. 2014). Fifteen items have been constructed and evaluated in a pilot study. Each item comprised a fictional dialogue between a student teacher and one or more of his students and four statements describing possibilities of action or judgments of the dialogue. All dialogues were constructed based on theoretical assumptions and video data of a prior study. The participants were asked to rate each statement on a Likert scale from 1 (very appropriate) to 6 (not appropriate at all). All ratings were treated as answers to a dichotomous knowledge test and scored on the basis of expert ratings. Nine university professors for chemistry education answered the same statements as the teachers. Teachers' answers were scored based on their accordance with the experts' opinion on the statement. Experts showed a good agreement on the statements. In consensus with expert notes and discussions, statements showing less satisfying item characteristics were adapted and 12 items were selected to form the final test instrument.

Teachers' classroom action regarding their handling of technical language was examined using video coding. Each video was analyzed in two coding procedures. A low inferent basis rating provides insights on the general structure of the teaching and distribution of shares of speech in class. Following this, a highly inferent coding covers specific elements of teachers' and students' use of technical language in class. The coding manual was developed based on theory and video data at hand and comprised categories regarding the technical language and the content (e.g., *correctness of utterance* or *content-related complexity* of the turn). Subcategories regarding the language include, e.g., the language-related complexity (*form of lexis, composition of syntactic structure*) and the inducement for the utterance. Subcategories regarding the content comprise the content-related complexity and the correctness of the utterance. Using the coding manual, teachers' and students' use of technical language should be described from a perspective of subject-specific education.

The category *correctness of utterance* addresses the bare subject-specific correctness of the utterance (right, wrong, indecisive, e.g., with questions or hypothesis), while the *content-related complexity* addresses the content in relation to cognition and differentiates whether the utterance is on a level of mere facts (e.g., "Hydrochloric acid is an acid") or refers to a relation (e.g., "Hydrochloric acid is an acid because it is a proton donor") or even a major concept in chemistry (e.g., "Hydrochloric is an acid according to the Arrhenius theory of acids and bases"). This differentiation is supposed to shed light on possible difficulties to comprehend the utterance as studies found a connection between the content-related complexity and, for example, the item difficulty in paper-pencil tests on content knowledge (Ropohl 2010). The category *language-related complexity* is operationalized closely to the *content-related complexity* as a connection between the complexity of used language and comprehension can be assumed. Additionally, each utterance is rated regarding the *form of lexis* (*technical term(s)* or *everyday language*) and its syntactic complexity (*Word*, *Sentence(s)*, or *Dependent Sentences*). Using these codings, structure- and content-related reasons for students' difficulties in comprehension could be identified.

Moreover, utterances are rated regarding their *technical language-related correctness* as video data at hand shows that utterances might be correct from a subject-related perspective but wrong from the perspective of technical language (e.g., a student could describe hydrochloric acid as *HCl* instead of as the correct *HCl (aq)*. A teacher might accept *HCl* as an answer but the answer remains wrong).

Using these aforementioned categories, ten videos of chemistry lessons from a prior video project were rated by two trained raters. Each utterance of the teacher or his students was rated on the basis of the coding manual. Evaluation shows a very good interrater reliability (on average $\kappa = .88$), ranging from nearly perfect ($\kappa = 1.0$) to good agreement ($\kappa = .72$). Based on experts' opinion and theoretical assumptions, the manual can be regarded as valid.

Results

All reported performance measures were analyzed using Rasch measurement, taking item difficulty into account. According to this, for example, results for teachers' CK do not reflect their raw scores but their ability in CK. Raw scores highly correlate to Rasch-scaled measurements (e.g., CK: $r = .97$, $p < .001$). Since Rasch measurement takes into account varying item difficulty and the test subjects' ability, it is considered as an appropriate way to measure ability (Boone et al. 2014). It should be mentioned that a negative score for a person's ability does not reflect "negative knowledge" but indicates a lower than 50% probability for the person to answer an item of average difficulty correctly (Boone et al. 2014).

Teachers

Twenty-eight (28) chemistry teachers participated ($♀ = 50\%$) in this study. The teachers are, on average, between 42 and 43 years old and have spent between 12 and 13 years in school service ($SD = 11.52$). The teachers show very good content knowledge and variance in their PCK_{TL}. A comparison of this study's teacher sample (ProwiN 2 sample) to the prior study's and subsequent assessment's sample of teachers of the same school form (validation sample) indicates that the ProwiN

Table 1 Teachers' CK and PCK$_{TL}$

Scale	$N_{Teachers}$	N_{Items}	Ability (M)	SD	Person reliability	Item reliability
CK	28	29	2.62	1.35	.58	.58
PCK$_{TL}$	28	35	1.42	1.14	.78	.84

Table 2 Students' development of content knowledge

Scale	$N_{Students}$	N_{Items}	Ability (M)	SD	Person reliability	Item reliability
Pretest	764	30	−.77	.60	.40	.97
Posttest	764	30	.27	.90	.73	.99

Table 3 Predictors for students' learning achievement

	B	SE	β	R^2	ΔR^2	p
Student variables						
Prior knowledge	.63	.05	.43	.22	–	<.001
Interest in chemistry	.01	.00	.24	.02	.02	<.001
Teacher variables						
CK	.07	.02	.25	.01	.01	<.001
PCK$_{TL}$	.08	.03	.26	.01	.01	.004

2 teachers have significantly better CK than the validation sample ($t(201) = 3.32$, $p < .001$, $g = 1.08$). This fact might explain the mediocre person reliability of the CK test items. The reliability of the PCK$_{TL}$ test is good (see Table 1).

Students

Reported results are based on the complete datasets of 764 students ($\female = 49.1\%$). The 764 students come from 34 classes taught by the aforementioned 28 teachers and are, on average, between 13 and 14 years old. Students show a significant development of their personal ability regarding their content knowledge on the topic *atomic structure and periodic table of the elements* with a strong effect size $t(763) = 36.14$, $p < .001$, $d = 1.4$. This reflects their actual development in content knowledge (see Table 2). Students show only little prior knowledge, explaining the poor person reliability of the pretest. Posttest person reliability is good.

Regression analyses indicate that students' prior knowledge is the strongest predictor for students' learning, followed by their interest in chemistry. Teachers' CK and their PCK$_{TL}$ contribute to variance explanation (see Table 3). Under consideration of all four aforementioned predictors, a total variance of 26% can be explained.

Comparative analyses of students' learning achievement with high and low prior knowledge show a distinct change in the variable predictive power and amount of

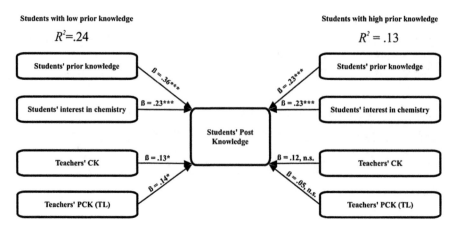

Fig. 2 Predictors for students' post knowledge differentiated by prior knowledge

explained variance. For this purpose, the student sample was differentiated by their prior knowledge. Only the 25% with the highest and the 25% with the lowest prior knowledge make up the data basis for the following analyses. For students with low prior ability, 24% of variance can be explained, while for students with high prior ability, explained variance decreases remarkably to 13%. Analyses also show that the significance and predictive power of each predictor changes considerably (see Fig. 2). Results for students with low prior ability retain the results of the entire sample; prior knowledge remains the strongest predictor followed by interest in chemistry. Findings indicate that students with low prior knowledge benefit more from teachers' CK and PCK_{TL} than students with high prior knowledge. For students with high prior knowledge, only their prior knowledge itself and their interest in chemistry are significant predictors and even equally strong.

Teachers' Acting in Class

The reported results of qualitative analyses focus on extreme classes: the class with the highest post knowledge (class A) and the class with the lowest post knowledge (class B). Both classes are taught by trained chemistry teachers (teacher A and teacher B) and comprise 20 students each. Qualitative analysis is conducted based on the aforementioned evaluated overall rating questionnaire and covers both video-taped lessons of each teacher. Though their PCK_{TL} does not differ strongly, analysis of videotaped lessons shows a distinct difference in their actual handling of techni-cal language in class. Both teachers show a strong use of technical terms, complex syntactical structures when giving monologue or explanations and rather simple syntactical structures with ellipses and other characteristics which can be expected in spoken language when confronted with spontaneous situations (e.g., questions).

Although both teachers show a strong use of technical terms, teacher A uses fewer and more specific technical terms, while teacher B shows a high frequency and variance of technical terms. Teacher A fosters a more dialogic structure in class. Confronted with spontaneous situations, teacher B reacts faster than teacher A but uses technical terms less precisely. On the contrary, teacher A takes more time to answer and uses his technical language very precisely. In addition to that, teacher A pays attention to his students' correct use of technical language, while teacher B barely reacts to inadequate utterances. Students in both classes use technical language sometimes wrong, but while teacher A corrects students' use of technical language and additionally pays attention to the correctness of the utterances from a content-related perspective, teacher B pays more attention to his students' content-related correctness of utterances and seems indifferent to their technical language-related correctness. Teacher B seems to anticipate what his students want to say but not what they actually say as several utterances are right from a content-related perspective but wrong from a technical language-related perspective. Although the aforementioned results are based on the comparison of the two classes differing most in learning achievement, subsequent qualitative analyses of the entire sample sustain the reported observations. Teachers of varying successful learners differ strongly in their technical language-related behavior in class.

Summary and Perspectives

Teachers' professional knowledge is a central subject to educational research and regarded as an important aspect of quality of instruction. National and international studies indicate a positive relation of teachers' professional knowledge and students' learning outcome albeit most studies focus on theoretical aspects of PCK and types of knowledge. Teaching is a highly complex and dynamic activity which makes it necessary to consider teachers' actual acting in class when examining the connection of teacher traits and students' learning.

This study uses paper-pencil tests, questionnaires, and video analyses to investigate the relation between teachers' content knowledge (CK), pedagogical content knowledge regarding technical language (PCK_{TL}), teachers' acting in class, and students' learning achievement in chemistry. The sample consists of 28 German secondary school chemistry teachers teaching the topic *atomic structure and the periodic table of the elements* in eighth grade.

The findings show a significant development in students' content knowledge and confirm a significant relevance of teachers' CK and PCK_{TL} for students' learning outcome. Especially students with low prior knowledge benefit from high teacher CK and PCK_{TL}. Preliminary results of video analyses indicate a disparity between teachers' theoretical professional knowledge and their actual acting in class. Albeit teachers' theoretical knowledge about the handling of technical language differs only slightly for the entire sample, teachers of the classes with the highest and lowest learning outcome show a divergent handling of technical language in class and their students' learning achievement varied strongly.

Further investigations will focus on the quantitative analyses of classroom action in order to consider teachers' actual acting in an explanation model for students' learning outcome. Classroom action could mediate between teachers' CK and PCK$_{TL}$ and students' learning achievement. The study's final results might contribute to a better understanding of the relevance of technical languages for teaching and learning and the complex construct of PCK. In addition, the connection between PCK, teachers' actual acting in class, and students' learning will be examined, and findings will make a contribution to this complex and foremost important relation for teaching.

References

Abell, S. K. (2007). Research on science teacher knowledge. In S. K. Abell & N. G. Lederman (Eds.), *Handbook of research on science education* (pp. 1105–1149). Mahwah: Erlbaum.

Abell, S. K. (2008). Twenty years later: Does pedagogical content knowledge remain a useful idea? *International Journal of Science Education, 30*(10), 1405–1416. doi:10.1080/09500690802187041.

Abell, S. K., Rogers, M. A. P., Hanuscin, D. L., Lee, M. H., & Gagnon, M. J. (2009). Preparing the next generation of science teacher educators: A model for developing PCK for teaching science teachers. *Journal of Research in Science Teaching, 20*(1), 77–93. doi:10.1007/s10972-008-9115-6.

Arnold, K.-H. (Ed.). (2007). *Unterrichtsqualität und Fachdidaktik [quality of instruction and subject didactics]*. Bad Heilbrunn: Verlag Julius Klinkhardt.

Barke, H.-D. (2015). Learners ideas, misconceptions, and challange. In J. García-Martínez & E. Serrano-Torregrosa (Eds.), *Chemistry education* (pp. 395–420). Weinheim: Wiley-VCH.

Bauer, K.-O. (2011). Modelle der Unterrichtsqualität [models of quality of instruction]. In K.-O. Bauer & N. Logemann (Eds.), *Unterrichtsqualität und Fachdidaktische Forschung. Modelle und Instrumete zur Messung fachspezifischer Lernbedingungen und Kompetenzen* [Quality of Instruction and Subject Didactis Research. Models and Instruments for Measuring Subject-Specific Learning Conditions and Competencies] (pp. 51–74). Münster: Waxmann.

Baumert, J., Kunter, M., Blum, W., Brunner, M., Voss, T., Jordan, A., et al. (2010). Teachers' mathematical knowledge, cognitive activation in the classroom, and student progress. *American Educational Research Journal, 47*(1), 133–180. doi:10.3102/0002831209345157.

Berry, A., Friedrichsen, P., & Loughran, J. (Eds.). (2015). *Re-examining pedagogical content knowledge in science education*. New York: Routledge.

Boone, W. J., Staver, J. R., & Yale, M. S. (2014). *Rasch analysis in the human sciences*. Dordrecht: Springer.

Bußmann, H. (2002). *ikon der Sprachwissenschaft* [Dictionary of linguistics] (3rd ed.). Stuttgart: Körner.

Cauet, E., Liepertz, S., Borowski, A., & Fischer, H. E. (2015). Does it matter what we measure? Domain-specific professional knowledge of physics teachers. *Revue suisse des sciences de l'éducation, 37*(3), 462–479.

Childs, P. E., Markic, S., & Ryan, M. C. (2015). The role of language in the teaching and learning of chemistry. In J. García-Martínez & E. Serrano-Torregrosa (Eds.), *Chemistry education* (pp. 395–420). Weinheim: Wiley-VCH.

Clausen, M. (2002). *Unterrichtsqualität: Eine Frage der Perspektive? Empirische Analysen zur Übereinstimmungs-, Konstrukt- und Kriteriumsvalidität* [Quality of instruction: A question of perspective? Empirical analyses regarding congruity-, construct- and criteria validity]. Münster: Waxmann.

de Jong, O., & van Driel, J. H. (2005). Exploring the development of student teachers' PCK of the multiple meaning of chemistry topics. *International Journal of Science and Mathematics Education, 2*, 477–491.

Dollny, S. (2011). *Entwicklung und Evaluation eines Testinstruments zur Erfassung des fachspezifischen Professionswissens von Chemielehrkräften* [Development and evaluation of a testinstrument for measuring chemistry teachers' subject specific professional knowledge]. Berlin: Logos.

Eunson, B. (2012). *Communicating in the 21st century* (3rd ed.). Milton: Wiley.

Förtsch, C., Werner, S., Dorfner, T., von Kotzebue, L., & Neuhaus, B. J. (2016). Effects of cognitive activation in biology lessons on students' situational interest and achievement. *Research in Science Education.* doi:10.1007/s11165-016-9517-y.

Gess-Newsome, J. (2015). A model of teacher professional knowledge and skill including PCK: Results of the thinking from the PCK Summit. In A. Berry, P. Friedrichsen, & J. Loughran (Eds.), *Re-examining pedagogical content knowledge in science education* (pp. 28–42). New York: Routledge.

Grucza, S. (2012). *Fachsprachenlinguistik* [Linguistics of technical languages]. Frankfurt am Main: Lang.

Johnstone, A. H., & Selepeng, D. (2001). A language problem revisited. *Chemical Education Research and Practice, 2*(1), 19–29.

Kleickmann, T., Richter, D., Kunter, M., Elsner, J., Besser, M., Krauss, S., & Baumert, J. (2012). Teachers' content knowledge and pedagogical content knowledge: The role of structural differences in teacher education. *Journal of Teacher Education, 64*(1), 90–106. doi:10.1177/0022487112460398.

König, J., Blömeke, S., Paine, L., Schmidt, W. H., & Hsieh, F. J. (2011). General pedagogical knowledge of future middle school teachers: On the complex ecology of teacher education in the United States, Germany, and Taiwan. *Journal of Teacher Education, 62*(2), 188–201. doi:10.1177/0022487110388664.

Krauss, S., Brunner, M., Kunter, M., Baumert, J., Blum, W., Neubrand, M., & Jordan, A. (2008). Pedagogical content knowledge and content knowledge of secondary mathematics teachers. *Journal of Educational Psychology, 100*(3), 716–725. doi:10.1037/0022-0663.100.3.716.

Kunter, M., Baumert, J., Blum, W., Klusmann, U., Krauss, S., & Neubrand, M. (Eds.). (2013). *Cognitive activation in the mathematics classroom and professional competence of teachers – Results from the COACTIV project.* New York: Springer.

Loughran, J., Berry, A., & Mulhall, P. (2012). *Understanding and developing science teachers' pedagogical content knowledge* (2nd ed.). Rotterdam: Sense Publishers.

Özcan, N. (2013). *Zum Einfluss der Fachsprache auf die Leistung im Fach Chemie: Eine Förderstudie zur Fachsprache im Chemieunterricht* [The influence of technical language on performance in the subject chemistry. A study to foster technical language in chemistry classes]. Berlin: Logos.

Park, S., & Oliver, J. S. (2007). Revisiting the conceptualisation of Pedagogical Content Knowledge (PCK): PCK as a conceptual tool to understand teachers as professionals. *Research in Science Education, 38*(3), 261–284. doi:10.1007/s11165-007-9049-6.

Park, S., Jang, J.-Y., Chen, Y.-C., & Jung, J. (2010). Is Pedagogical Content Knowledge (PCK) necessary for reformed science teaching?: Evidence from an empirical study. *Research in Science Education, 41*(2), 245–260. doi:10.1007/s11165-009-9163-8.

Riese, J., & Reinhold, P. (2012). Die professionelle Kompetenz angehender Physiklehrkräfte in verschiedenen Ausbildungsformen. [The professional competency of perspective physics teachers in different forms of qualification]. *Zeitschrift für Erziehungswissenschaft, 15*(1), 111–143. doi:10.1007/s11618-012-0259-y.

Roelcke, T. (2005). *Fachsprachen* [Technical languages] (2nd ed.). Berlin: Schmidt.

Ropohl, M. (2010). *Modellierung von Schülerkompetenzen im Basiskonzept chemische reaktion. Entwicklung und analyse von testaufgaben* [Modelling of students' competencies in the basic concept chemical reaction. Development and analysis of test-items]. Berlin: Logos.

Schmölzer-Eibinger. (2013). Sprache als medium des Lernens im Fach [language as medium of learning in subjects]. In M. Becker-Mrotzek (Ed.), *Sprache im Fach. Sprachlichkeit und fachliches Lernen* [Language as medium of learning in subjects] (pp. 25–41). Münster: Waxmann.

Shulman, L. S. (1987). Knowledge and teaching: Foundations of the new reform. *Harvard Education Publishing Group, 57*(1), 1–23.

Strübe, M., Tröger, H., Tepner, O., & Sumfleth, E. (2014). Development of a pedagogical content knowledge test of chemistry language and models. *Educatión Química, 25*(3), 380–390.

Taber, K. S. (2015). Exploring the language(s) of chemistry education. *Chemical Education Research and Practice.* doi:10.1039/c5rp90003d.

Tepner, O., & Dollny, S. (2014). Measuring chemistry teachers' content knowledge: Is it correlated to pedagogical content knowledge? In C. Bruguière, A. Tiberghien, & P. Clément (Eds.), *Topics and trends in current science education. 9th ESERA conference selected contributions* (pp. 243–254). Dordrecht: Springer.

Yore, L. D. (2012). Science literacy for all: More than a slogan, logo, or rally flag! In K. C. D. Tan & M. Kim (Eds.), *Issues and Challenges in science education research* (pp. 5–24). Dordrecht: Springer.

Yore, L. D., Bisanz, G. L., & Hand, B. M. (2003). Examining the literacy component of science literacy: 25 years of language arts and science research. *International Journal of Science Education, 25*(6), 689–725. doi:10.1080/09500690305018.

Analyzing Teachers' Discursive Participation in Co-taught Science-and-English CLIL Classrooms

Laura Valdés-Sánchez and Mariona Espinet

Introduction

Our society and schools are multilingual and multicultural contexts embedded in a globalized world. European educational institutions are promoting new teaching approaches related to bilingual education in order to encourage every citizen to acquire a command of at least three languages through compulsory education (Eurydice 2006, UNESCO 2003). An understanding of Science is considered a competence that also needs to be promoted so that everyone can participate actively and critically in the functioning of our society (Eurydice 2011, COSCE 2011).

Within the context of a multicultural and globalized world, the European Commission has promoted Content and Language Integrated Learning (CLIL) projects to improve the learning of a foreign language (Eurydice 2006). The acronym CLIL is used as "a generic term to describe all types of provision in which a second language (a foreign, regional or minority language and/or another official state language) is used to teach certain subjects in the curriculum other than languages lessons themselves" (Eurydice 2006: 8). In Catalonia, a bilingual autonomous community of Spain, English is taught as a foreign and a third language. CLIL projects in this region often aim to integrate the teaching and learning of Science and English.

The teaching of Science in English has been widely studied by the international educational research community (Escobar-Urmeneta et al. 2011; Gajo 2007; Valdés–Sánchez and Espinet 2013b), which has documented the benefits to both language and Science teaching and learning. In the case of Science teaching, the approach encourages teachers to strengthen the focus on language through increased dialogue and negotiation (Moate 2011). This enhances the development of scientific competences, since learning Science means learning to use the language associated

L. Valdés-Sánchez (✉) • M. Espinet
Universitat Autònoma de Barcelona, Barcelona, Catalonia, Spain
e-mail: lauravaldessanchez@gmail.com

© Springer International Publishing AG 2017   219
K. Hahl et al. (eds.), *Cognitive and Affective Aspects in Science Education Research*, Contributions from Science Education Research 3,
DOI 10.1007/978-3-319-58685-4_17

with school Science (Edwards and Mercer 1987; Espinet et al. 2012; Lemke 1990; Mortimer and Scott 2003; Sanmartí 2002).

However, from the perspective of Science Education, it should be noted that, in order to initiate this kind of project, teachers must go far beyond simply explaining the subject in English. Indeed, it highlights the need for new professionals who are capable of integrating the teaching of Science and English without compromising the educational quality of either discipline (Escobar-Urmeneta et al. 2011). It also involves collaboration between teachers in different specialized disciplines, and reflection on how best to achieve real integration between the learning objectives of the two subjects (Dalton-Puffer 2007; Valdés–Sánchez and Espinet 2012).

Within this context, co-teaching emerges as a strategy with great potential to help build discipline integration projects. Understood as a process in which two or more teachers plan, instruct, and evaluate together (Davis-Wiley and Crespo 1998), it involves collaborative work based on dialogue between disciplines that occurs inside and outside the classroom, enriching learning environments through the knowledge and experience of two experts. Davis-Wiley and Crespo have compiled some of the benefits of this practice, as documented by several researchers. Roth and Tobin (2004) advocate this model as a tool for teachers' professional development. They argue that teachers who work together in the classroom expand their identities through cooperation based on established goals and interests. Siry and Martin understand it as a "method for learning how to teach, as well as a methodological approach to learn about teaching" (Siry and Martin 2010: 57).

This chapter presents part of a broader study that explores the potential of co-teaching as a strategy to build a Science-and-English CLIL project in public primary schools. By analyzing the discourse of several pairs of co-teachers, the broader study aims to shed light on certain aspects of the interaction that takes place in such CLIL classrooms. Its objectives are to reflect on how the integration of Science teaching and English teaching occurs in the classroom and how the co-teachers' discursive collaboration develops.

In the following pages, we present the methodology developed to analyze the interaction between co-teachers and the findings from the analysis of their discursive participation while co-teaching Science and English. These findings give us some insights into (a) how CLIL co-teachers manage their participation in a shared classroom, (b) how CLIL co-teachers' participation changes, and (c) how diverse co-teachers' participation is.

Methodology

Study Design

This exploratory qualitative study is based on assumptions shared by the discourse studies community in relation to Language and Science Education. We are interested in contributing to the understanding of "how learning occurs through language, how access to knowledge derives from participating in the social and symbolic worlds, and how disciplinary knowledge is constructed through language" (Kelly 2007). The methodological approach draws on Science Education discourse studies from a sociocultural perspective (Edwards and Mercer 1987; Lemke 1990; Mortimer and Scott 2003).

We have studied the discursive interaction that occurs between co-teachers from different disciplines while they try to integrate foreign-language teaching and Science Education. We understand discursive participation in the classroom as a social act that implies the use of spoken language (or the choice to remain silent) under certain social rules that all the participants situate and continually negotiate within the interaction. Although we specifically focus on spoken language for the purposes of our study, it is not our intention to underestimate non-spoken language or other means of interaction, as we understand that participation occurs through multimodal means.

There is a long tradition of classroom Science Education discourse studies from a sociocultural perspective, though they usually refer to classrooms with only one teacher interacting with pupils. The interaction of two teachers in a classroom has been analyzed within different contexts, such as classrooms with one teacher and one special educator supporting certain pupils or classrooms where co-teaching has been used as a tool for preservice teacher training. Within our context, however, two expert teachers collaborate in an attempt to integrate the teaching of two disciplines.

The data collection strategies included video recordings of full classroom interactions. In 2012/2013, we recorded the classroom interactions of four primary school teachers (one English teacher and three Science teachers) that were grouped into three pairs of co-teachers. Each pair was formed by the English teacher and one of the Science teachers, and they all did the same activity in the third grade (8–9-year-olds), within the subject entitled Knowledge of the Natural and Social Environment. The recorded material related to a hands-on activity about the properties of materials used for a given technological purpose (building a mud-brick house) in a historical context (prehistory). In addition, we collected video recordings of a Science workshop on the topic of air, which the co-teachers who initiated the project held in 2010, when they began working together.

We performed two types of analysis: (a) a longitudinal study of one pair of co-teachers (from 2010 to 2013) and (b) a cross-sectional analysis of three pairs of co-teachers doing the same activity in 2013. Data analysis was performed in two phases: macroanalysis and microanalysis.

The ***macroanalysis phase*** started with a selection of activity fragments in which the co-teachers were constructing a collaborative discourse, that is to say, those sections of the video when the co-teachers were interacting with each other and with the whole class. The second step was the transcription of the selected fragments. Finally, we segmented the transcriptions by their semantic patterns (Lemke 1990) in order to create our units of analysis: episodes. An episode is a transcribed interactional sequence in which the co-teachers collaborated to complete a specific task with the whole class. In the longitudinal study, 85 episodes were identified: 36 episodes correspond to the workshop that the initial pair of co-teachers (*Pair 1*) developed during their first year of collaboration (2011), and 49 episodes belong to the activity that the same pair did in 2013. In the cross-sectional study, 132 episodes were analyzed, 49 from the co-teachers *Pair 1*, 49 from *Pair 2*, and 34 from *Pair 3*.

Having completed the macroanalysis and identified the units of analysis, we began the ***microanalysis phase***. Through an iterative process of analysis and discussion of certain episodes, we developed an inductive-deductive analytical tool. This tool was validated by other researchers in the field of Science Education and Language Education through several data sessions held at the Autonomous University of Barcelona and the University of Luxemburg. This tool enabled us to perform a qualitative analysis of each episode and to generate a coding system using ATLAS.ti software.

Analytical Tool

We developed an inductive-deductive analytical tool that guided the process of coding. Based on the analysis performed in a preliminary study (Valdés–Sánchez and Espinet 2012, 2013a), this tool was subsequently expanded during data analysis as a result of new theoretical and empirical contributions. The analytical tool characterized the episodes into three dimensions: (1) semantic pattern, (2) participation pattern, and (3) linguistic pattern. It also characterized how alternations among different categories occurred within each pattern (Table 1).

In the analysis of the semantic pattern, the teacher's interventions were classified, in terms of their objectives, into three categories: (a) objectives related to Science teaching, (b) objectives related to English teaching, and (c) objectives related to the management of the group. The participation pattern emphasized and characterized moments of individual or shared leadership and the role of each teacher in the participation. Finally, in the analysis of the linguistic pattern, we looked for the moments when the co-teachers changed the agreed linguistic pattern, and we analyzed the types of linguistic alternation and their causes.

Table 1 Analytical tool for the qualitative analysis of episodes

EPISODE:		Code:		
GOAL:		Co-teacher's position:		
TURNS:		Number of turns:		
QUALITATIVE DESCRIPTION:				
		Content	Participation pattern	Linguistic pattern
Semantic pattern	Science			
	English			
	Management			
Alternations				

Results

The preliminary study focusing on the co-teachers' questioning interactions (Self-references 1; 2; 3) showed that co-teaching generated reciprocal learning between teachers. Such learning promoted change in the structure of the activity, shared responsibility in the classroom, and transfer between teachers of the ability to formulate scientific questions. The study presented in this chapter provides a more in-depth analysis of co-teachers' participation models and their development.

Co-teachers' Discursive Participation Models

The first objective of our discursive participation analysis was to describe how two co-teachers – and experts in different disciplines – having the same responsibility within the classroom interacted with the pupils in a Science and English integrated learning context. The analysis of the participation patterns revealed three main models of participation that all pairs of co-teachers enacted in the classroom: (1) *passive*, (2) *supporting,* and (3) *co-constructing*. We identified these models and characterized the most relevant participation strategies associated with each of them (Table 2).

The *passive* model identifies a co-teaching method in which one co-teacher interacts with the pupils and the other decides to remain silent. The model in which one co-teacher leads the teaching and the other offers her/him support is called the *supporting* model. The co-teacher who is not leading can help with the management of the classroom or contribute with repetitions, reformulations, extensions, and feedback. We have exemplified this model with the classroom episode extract shown in Fig. 1, in which the English teacher (ET) is leading the dialogue and the Science teacher (ST) is helping with the management of the classroom. As the Science teacher and certain pupils (A) are speaking in Catalan, the translation into English is included in gray.

Finally, in the *co-constructing* model, both co-teachers lead the classroom together, acting as if they were one teacher, that is to say, both of them decide how the classroom discourse should be developed, and they collaborate on the construction of a fluent dialogue with the pupils. This shared leadership was constructed using different discursive strategies, such as dialogue co-construction with the pupils, talking to each other as teachers or acting as if one were a teacher and the other were a pupil. An example of a dialogue co-construction strategy is shown in Fig. 2. In this episode extract, both teachers are trying to encourage pupil A4 to use English in order to communicate to a visitor about the materials they are going to collect to build a mud-brick house.

Table 2 CLIL Science and English co-teachers' discursive participation strategies associated with discursive participation models

Participation model	Participation strategy	Description
Passive	Silence	The co-teacher who is not leading classroom interaction remains silent
Supporting	Repetition	The co-teacher who is not leading repeats what the other co-teacher has said with the same words, either in the same language (English or Catalan) or a different language (English or Catalan)
	Reformulation	The co-teacher who is not leading repeats what the other co-teacher has said using other words
	Extension	The co-teacher who is not leading adds some new information to what the other co-teacher has said
	Feedback	The co-teacher who is not leading gives some feedback on the interventions of the pupils or on what the other co-teacher has said
	Management assistance	The co-teacher who is not leading helps the other co-teacher with the management of the classroom, e.g., attracting the pupils' attention, offering help with the materials, etc.
Co-constructing	Dialogue co-construction	Both co-teachers decide how the classroom discourse should be developed, and they collaborate on the construction of a fluent dialogue with the pupils
	Open dialogue	The co-teachers talk to each other in order to communicate something to the pupils
	Open dialogue with teacher-pupil roles	Co-teachers talk to each other while one of them acts as if he/she were a pupil

47	ET	- And today, is a very special day
48	Ax	Aaah!
49	Ax	Special!
50	ST	Yago, per què no enlloc de estar amb això per què no escoltes? Yago, could you stop doing that and listen?
51	A4	Jo ja ho sé. I already know that
52	ET	Because, we, our class, we are going walking to the...
53	Ax	Mountain
54	ET	a little mountain, a little eh

Fig. 1 Example of the supporting participation model with a management assistance strategy

Changes in Co-teachers' Discursive Participation Models

Pair 1 of the co-teachers taught a Science-and-English CLIL classroom for 3 years, and changes were expected after this long period of collaboration. Table 3 allows us to compare the distribution of the different discursive participation models in their

103	A4	Però li explico en català o...?
		But should I tell her in Catalan or...?
104	ST	Noo
		No!
105	ET	I think she speaks English not Catalan
106	A4	Llavors no
		In that case I'm not going to
107	ST	Va, una mica segur que en saps.
		Come on, I'm sure you can
108	ET	Yes
109	A22	Paula, li dius: "the materials..."
		Paula, tell him: "the materials…"
110	ST	Clar, "the materials..."
		Right, "the materials..."
111	ET	Materials...

Fig. 2 Example of the co-constructing participation model with a dialogue co-construction strategy

Table 3 Longitudinal study of co-teachers' participation models

| | Participation model | | | |
Stage of collaboration	Being (Science teacher) (%)	Being (English teacher) (%)	Supporting (%)	Co-constructing (%)
1st year of co-teaching (2011)	0	15	58	27
3rd year of co-teaching (2013)	5	6	45	44

initial stage of collaboration with the distribution of these models 3 years later. The percentages indicate the number of discursive turns associated with each model with respect to the total number of turns for each activity. The total number of turns was 1703 for the first year activity and 1792 for the third year activity.

From the beginning of their co-teaching, these teachers spent most of their time collaborating through discourse, enacting *Supporting* or *Co-constructing* participation models. In fact, sequences of turns in which only one co-teacher talks and the other remains silent were less frequent. After 3 years of collaboration, as shown in Table 3, the distribution of participation models had developed in two ways.

On the one hand, there is an important change regarding who remains silent. In the first year, the *passive* model was only enacted by the English teacher, i.e., while the Science teacher talked to the classroom, the English teacher remained silent. In contrast, after 3 years of collaboration, the two co-teachers enacted the *passive* model with a similar frequency, i.e., both teachers assumed full responsibility in guiding the dialogue with pupils at different moments of their interactions. However, within the classroom interactions occurring in the third year, the *Passive* model was enacted less frequently when compared to the first year.

On the other hand, there is an evolution with respect to how these co-teachers collaborated through discourse, which corresponds to an increase in the *Co-constructing* model and a decrease in the *Supporting* model. In addition, in the third year, they incorporated new strategies like the one shown in Fig. 3. In this episode extract, the Science teacher (ST) is explaining to the English teacher (ET) that mud bricks were used in prehistory to build houses instead of huts.

The episode extract exemplifies how the co-teachers talk to each other in order to communicate something to the pupils. Although talking to each other, they use a rhythm that is normally used with the pupils rather than between them. That is what we call an *Open dialogue*. What makes this *Open dialogue* special is the fact that they assume different roles, i.e., while the Science teacher acts as a teacher, the English teacher assumes a pupil role. In this activity, we located various moments when the co-teachers used the Open dialogue with teacher-pupil roles strategy, with either acting as teacher or pupil.

91	ST	Però el que, el que farem, els tovots, amb els tovots construïen cabanes? But what, what we are going to do with the mud bricks ... were mud bricks used to build huts?
92	As	No
93	ST	No.
94	ET	Ah, no?
95	ST	[Talking to ET:] Mira, construïen això Look, that's what they used to build [ST points to a photo of a prehistoric house on the classroom wall].
96	ET	What's this?
97	ST	Això es vist des de dalt. This is seen from above
98	ET	Mmh
99	ST	Això és, una... casa This is, a... house
100	Ax	Casa House
101	ST	pre-històrica prehistoric
102	Ax	de tovots made from mud bricks
103	ST	que els tovots els tenen col·locats aquí. Això són les parets, però que es veuen des de dalt. The mud bricks are located here. These are the walls, but seen from above
104	ET	Okay
105	ST	Val? Okay?
106	ET	So it's not a House.
107	ST	Is not a House
108	ET	Not hut, House
109	ST	Però el, el teulat és com a "hut". És en forma de cabana But the roof is like a hut. It's hut-shaped
110	ET	Okay.

Fig. 3 Example of the open dialogue with teacher-pupil roles strategy associated with the co-constructing participation model

To sum up, after 3 years of collaboration, these co-teachers used more developed co-teaching models, as they interacted through various participation models and they incorporated new co-construction discursive strategies. The changes observed were (a) the *passive* model was less frequent and was used by both teachers as opposed to by the English teacher only in the first year, (b) the *co-constructing* model gained in importance over the *supporting* model, and (c) they incorporated new co-construction strategies.

Diversity of Co-teachers' Participation Models

In 2013, the initial pair of co-teachers started their third year of collaboration, and two new Science teachers joined the project. They formed two new pairs of co-teachers with the same English teacher (*Pair 2* and *Pair 3*). As the school had three third grade classrooms, the activity consisting of building a mud-brick house was carried out by the same English teacher co-teaching with three different Science teachers in three different classrooms.

The distribution of the participation models of the three different pairs of co-teachers is shown in Table 4. If we look at the distribution of the models of co-teacher *Pairs 2* and *3*, we can see that although they were in their first year of co-teaching, they had some features that are typical of a developed co-teaching model. We must bear in mind that these Science teachers started co-teaching in 2013, but the English teacher they co-taught with was an expert co-teacher in her third year of practice.

The features that these new pairs of co-teachers shared with the developed co-teaching model were the low frequency of the *Passive* model and its distribution between co-teachers. In contrast, the distribution of the *Supporting* and *Co-constructing* participation models for the new co-teacher *Pairs 2* and *3* was typical of an initial collaboration, and they did not use all the co-construction strategies that the English teacher was able to use in *Pair 1* after 3 years of co-teaching.

Table 4 Cross-sectional study of co-teachers' participation models

Pair of co-teachers	Participation model			
	Passive (Science teacher) (%)	Passive (English teacher) (%)	Supporting (%)	Co-constructing (%)
Pair 1 (2013 – 3rd year of co-teaching)	5	6	45	44
Pair 2 (2013 – 1st year of co-teaching)	4	2	66	28
Pair 3 (2013 – 1st year of co-teaching)	7	1	74	18

Discussion and Conclusions

Co-teaching in Science-and-English CLIL classrooms is not a common teaching strategy in primary education. The study presented in this chapter is based on data obtained from the unique, stable, and diverse co-teaching experiences under scrutiny. The interaction between two co-teachers trying to teach primary Science and English is complex and diverse. In fact, co-teaching can be understood as a situated social act that occurs inside the classroom, which is continually negotiated by the participants in the interaction, both the two teachers and the pupils. Within our specific context, co-teaching implied the collaborative orchestration of the language used (Catalan, Spanish, or English), the discipline to support the construction of meaning (Social Sciences, Natural Sciences, Technology, and English), and, finally, the way teachers participated, sharing their leadership within the classroom.

This study focused solely on one of these three dimensions: the participation pattern of co-teachers' discursive collaboration. The teachers in our study used a variety of discursive participation strategies that can be classified into three types of discursive participation models ranging from less to more collaborative: *passive*, *supporting*, and *co-constructing*. However, those teachers seemed to feel more comfortable creating and enacting participation strategies for discursive collaboration by either *supporting* the leading teacher or sharing leadership by *co-constructing* a dialogue with the pupils. This was possible thanks to two types of process: (a) the learning of one pair of co-teachers through 3 years of shared practice (*Pair 1*) and (b) the transfer of one expert co-teacher's discursive participation competence to new co-teachers (*Pairs 2* and *3*).

The first conclusion drawn from this study was that co-teachers' participation strategies progressed, through practice, toward the acquisition of a more co-constructed discourse. This result is in line with the findings from a study by Roth and Tobin (2004), which emphasized an improvement in Science teachers' co-teaching through practice, an expansion of their agency, and a change in their praxis. The knowledge and experience of two experts from different disciplines, in this case Science teaching and English teaching, led to a more co-constructed dialogue that enriched the classroom (Davis-Wiley and Crespo 1998). One of the most interesting co-construction participation strategies that we identified was the way in which teachers managed their expertise and positioned themselves in CLIL classrooms. In fact, the most expert co-teachers were able to share their lack of knowledge by positioning the English teacher as a Science apprentice, and the Science teacher as a foreign-language learner. By openly expressing and sharing their lack of knowledge, the co-teachers created a context in which pupils felt safe to express their doubts or needs and were able to observe models of how to manage their learning and their relationship with the teacher. In addition, it created a context within which teachers also felt safe to express their lack of knowledge of a discipline outside their respective fields of expertise. We consider this discursive participation strategy to be an indicator of a more developed collaborative discourse through which pupils are able to expand their learning, and co-teachers are able to expand both their learning and their professional development opportunities.

The second conclusion drawn from this study was that more developed discursive participation strategies could be transferred to some extent from expert to novice co-teachers. The comparison between different pairs of novice co-teachers' interactions showed that they used the same participation models and similar strategies. However, their interactions were characterized by a lower presence of less collaborative participation models such as the *Passive* participation model. This could be interpreted as an indicator of transfer from the side of a co-teacher with 3 years of experience in co-teaching. Another feature of novice co-teachers' interactions was the distribution of *Supporting* and *Co-constructing* participation models, which were typical of an initial collaboration characterized by a lower presence of co-construction participation strategies. This might indicate that even though expert teachers could easily establish discursive collaboration in the classroom with the help of an expert co-teacher, the elaboration of more sophisticated co-construction strategies may require more time teaching together.

The authors of this study advocate the need to include co-teaching as a beneficial approach to the teaching of Science-and-English CLIL programs in primary education. However, this alone might not be sufficient to support the dynamics of change that primary schools currently need. The establishment of stable and interdisciplinary partnerships could provide a context to support innovation either in preservice or in-service primary Science teacher training (Espinet et al. 2016). These partnerships could involve actors from different institutions such as local authorities, universities, and schools, as well as from different disciplines of reference such as foreign-language education and Science Education. Today, Science teacher training requires practitioners to be able to work through difference and come up with boundary-crossing practices that facilitate authentic partnerships for change.

Acknowledgments The study was partially funded by grant 2014SGR1492 from AGAUR (Government of Catalonia), grant EDU2015-66643-C2-1-P (Spanish Ministry of Economy and Competitiveness), and the ESERA Early Career Researcher Travel Award 2016. We would like to express our thanks for the help and invaluable feedback received from our coaches Manuela Welzel-Breuer and Dimitris Stavrou at the ESERA Summer School 2014 and our colleagues Vicky Wong, Leanne P. Hinch, Lotta Leden, Katrin Weber, Alex Dawes, and Manou Leonidas. It was amazing to have another opportunity to discuss our research at the Helsinki ESERA Conference 2015. We would also like to thank Vicente Mellado for his ideas from the Spanish doctoral school *Escuela de doctorado de los XXVI EDCE*. Finally, thanks go to the researchers that helped us validate our analysis through their participation in data sessions: Emilee More, Jennifer Adams, Christina Siry, and her research team.

English-language editing by Steven Norris, Member of the Institute of Translation & Interpreting.

References

COSCE Confederación de Sociedades Científicas de España. (2011). *Informe ENCIENDE. Enseñanza de las Ciencias en la Didáctica Escolar para edades tempranas en España* [ENCIENDE report. Science teaching in early years education in Spain]. Retrieved from http://www.cosce.org/pdf/Informe_ENCIENDE.pdf

Dalton-Puffer, C. (2007). *Discourse in Content and Language Integrated Learning (CLIL) classrooms*. Philadelphia: John Benjamins.

Davis-Wiley, P., & Crespo, A. (1998, November). *Are two instructors better than one? Planning, teaching and evaluating à Deux*. Paper presented at the annual meeting of the mid-south educational research association. New Orleans, Louisiana. Retrieved from http://eric.ed.gov/?id=ED428038

Edwards, D., & Mercer, N. (1987). *Common knowledge: The development of understanding in the classroom*. London: Methuen/Routledge.

Escobar-Urmeneta, C., Evnitskaya, N., Moore, E., & Patiñio, A. (Eds.). (2011). *AICLE-CLIL-EMILE: Educació plurilingüe. Experiencias, research & polítiques* [AICLE-CLIL-EMILE: Plurilingual education. Experiences, research and policies]. Bellaterra: Servei de Publicacions UAB.

Espinet, M., Izquierdo, M., Bonil, J., & Ramos, S. (2012). The role of language in modeling the natural world: Perspectives in science education. In K. Tobin, B. Fraser, & C. McRobbie (Eds.), *Second international handbook of science education* (pp. 1385–1405). New York: Springer.

Espinet, M., Valdés-Sánchez, L., Carrillo, N., Farró, L., Martínez, R., López, N., & Castillón, A. (2016). Promoting the integration of inquiry based science and English learning in primary education through triadic partnerships. In A. W. Oliveira & M. H. Weinburgh (Eds.), *Science teacher preparation in content-based second language acquisition*. Dordrecht: Springer. Retrieved from http://www.springer.com/us/book/9783319435145.

Eurydice. (2006). *Content and language integrated learning (CLIL) at school in Europe*. Retrieved from http://ec.europa.eu/languages/documents/studies/clil-at-school-in-europe_en.pdf

Eurydice. (2011). *Science education in Europe: National policies, practices and research*. Retrieved from http://eacea.ec.europa.eu/education/eurydice

Gajo, L. (2007). Linguistic knowledge and subject knowledge: How does bilingualism contribute to subject development? *International Journal of Bilingual Education and Bilingualism, 10*(5), 563–581.

Kelly, G. J. (2007). Discourse in science classrooms. In S. Smith, S. Abell, K. Appleton, & D. Hanuscin (Eds.), *Handbook of research on science teaching*. Mahwah: Lawrence Erlbaum.

Lemke, J. (1990). *Talking science: Language, learning and values*. Stamford: Ablex Publishing Corporation.

Moate, J. (2011). Using a sociocultural CLIL pedagogical model to develop CLIL. In C. Escobar-Urmeneta, N. Evnitskaya, E. Moore, & A. Patiñio (Eds.), *AICLE-CLIL-EMILE educació plurilingüe: Experiencias, research & polítiques* [AICLE-CLIL-EMILE: Plurilingual Education. Experiences, research and policies] (pp. 101–111). Bellaterra: Servei de Publicacions UAB.

Mortimer, E., & Scott, P. (2003). *Meaning making in secondary science classrooms*. Maidenhead: Open University Press.

Roth, W. M., & Tobin, K. (2004). Coteaching: From praxis to theory. *Teachers and Teaching: Theory and Practice, 10*(2), 161–179.

Sanmartí, N. (2002). *Didáctica de las ciencias en la educación secundaria obligatoria* [Science teaching in compulsory secondary education]. Madrid: Editorial Síntesis.

Siry, C., & Martin, S. (2010). Coteaching in science education courses: Transforming teacher preparation through shared responsibility. In C. Murphy & K. Scantlebury (Eds.), *Coteaching in international contexts: Research and practice* (pp. 57–78). Dordrecht: Springer.

UNESCO. (2003). *Education in a multilingual world* (UNESCO education position paper). Paris: UNESCO. Available at http://unesdoc.unesco.org/images/0012/001297/129728e.pdf

Valdés-Sánchez, L., & Espinet, M. (2012). *Construint l'aprenentatge integrat de ciències i anglès a Primària. Retrat d'una evolució fruit del coteaching i la pràctica autoreflexiva.* [Building science and English integrated learning in primary education. Portrait of an evolution as a result of coteaching and reflective practice.] Post-graduate dissertation. Autonomous University of Barcelona.

Valdés-Sánchez, L., & Espinet, M. (2013a). La evolución de la co-enseñanza de las ciencias y del inglés en educación primaria a partir del análisis de las preguntas de las maestras. [The evolution of science and English coteaching in primary education from the analysis of teachers' questioning] *Enseñanza de las ciencias. Número Extra, 2013,* 3588–3594.

Valdés-Sánchez, L., & Espinet, M. (2013b). Ensenyar ciències i anglès a través de la docència compartida. [Teaching science and English through coteaching] *Ciències. Revista del professorat de Ciències d'Infantil, Primària i Secundària, 25,* 26–34.

Multilingual Contexts: A New Positioning for STEM Teaching/Learning

Philip Clarkson and Lyn Carter

Introduction

Multiple cultures and languages are represented in most classrooms worldwide. Hence the majority of teachers may now expect to work with at least some pupils from ethnic, linguistic, and/or cultural groups distinct from their own. Cultural, linguistic, political, and social issues in learning have until recently been seen as distant to and have had little impact on the teaching and learning of STEM. But the problems of "others" that are "different" from "us" are now a reality (Babaci-Wilhite 2016; Markic and Abels 2016). If STEM education is to become an equitable practice, there is a continuing need for research that takes seriously an understanding of the complexity of the teaching and learning in multilingual situations and the possible benefits these may have.

Research into the teaching and learning of STEM in multilingual situations is normally closely linked to the phenomena of worldwide refugee migration that many recipient societies see as a problem (Australia and Germany as two examples among many) rather than as an opportunity as Atweh and Clarkson (2001) have argued. But this is only one context that gives rise to multilingual classrooms. Other contexts include Papua New Guinea where one of the official languages, English in a land of 820 languages, is deemed to be the language of teaching, although few students understand English when entering school, and many beginning teachers are still learning the language (Clarkson 2016). In the USA the long-term multiple-generation Latino populace often has to learn in English. Hence the learning/teaching is more complicated than supposed in much of the research literature when a monolingual context is normally assumed. There is a clear need for research that understands the complexity of the teaching and learning of STEM in multilingual

P. Clarkson (✉) • L. Carter
Australian Catholic University, Melbourne, Australia
e-mail: philip.clarkson@acu.edu.au

© Springer International Publishing AG 2017   233
K. Hahl et al. (eds.), *Cognitive and Affective Aspects in Science Education Research*, Contributions from Science Education Research 3,
DOI 10.1007/978-3-319-58685-4_18

situations and an exploration of the possibilities this may have for more equitable societies. A new positioning for STEM teaching/learning is needed.

In this chapter we review some assumptions often made in research. We then explore some different language contexts that give rise to diversity. We first look at the diversity that exists within the teaching language and how this can impact on different students. We then turn our attention to language diversity that arises because of the different languages that may be present within a classroom. Finally we turn our attention to researching STEM in the midst of such diversity.

Some Assumptions and Inherent Complexity

The narrow focus of much STEM research stems from related assumptions of a "monolingual" context, students belonging to the dominant culture, and that they possess the social habitus of the middle class (Clarkson et al. 2001). Thus it is often assumed that mathematics and science can be taught in the absence of a common language because STEM subjects are "universal" and independent of language. For those who make this assumption, a STEM classroom is not the best place to learn the language and the norms of the school. It is taken for granted that students have already a mastery of the language of instruction and its subtleties, and this is somehow automatically linked by the students to the discourses of different subjects taught in the school. It is also a common assumption that the students know the "norms" of the school. But such is just not the case for many students, particularly those from migrant communities. Compounding this context, it is particularly difficult for children from a non-Western background, migrating to a Western or Westernized country, to learn Western science and mathematics when these are understood as part of Western culture. Further, these curricula are embedded in the wider school curriculum, and are intended for monolingual, middle-class students, belonging to the dominant social group. Often neglected is the reciprocity of this dynamic: Learning is influenced by language and culture is accepted, but language and culture also influence what is taught and what is researched, and indeed the research methods used are not always understood (Chellougui et al. 2015).

In fact we know STEM teaching/learning is a process where cognitive, affective, emotional, social, cultural, and linguistic factors are deeply intertwined (Bishop 1988; Lave 1988; Lee 2005). Further, the multiple links among these factors make the teaching of STEM a complex task even in a monolingual context, let alone in multilingual/multicultural contexts. In a classroom, neither the teacher nor the researcher may now assume that they are part of, or with, a homogenous group. Indeed there should be recognition by teacher and researcher that there is a great heterogeneity among the several multilingual and/or multicultural situations that can, and probably are, present in any one classroom. The complexity of the research contexts requires the use of multilayered theoretical perspectives and a deep sensitivity for different cultures present.

One advance that goes beyond the above stereotypical assumptions interestingly focuses on achievement and assessment. For a long period, most of the research concerning ethnic, cultural, or linguistic minorities and their learning of STEM subjects focused on the achievement of those groups. It is only recently that researchers' interests have turned to the understanding of how and why it is for many of these students that they normally obtain low achievement scores and why it is that there are very interesting exceptions for a particular small group of such students who gain above average scores in STEM (Clarkson 2007; Cummins 2000). This new direction for research has not been at the expense of a focus on achievement per se. The societal need for high achievement in STEM is normally present when there is an emphasis on schooling, and hence achievement outcomes cannot be neglected by education research. The new direction is more of opening up, another parallel line of investigation, with the belief that both are interrelated. However, there is also an understanding with this new direction that "achievement" should no longer be looked upon as the sole arbiter of whether students are "succeeding" or a particular program is "performing well."

So gradually the notion of "achievement" as the ultimate measure of quality in all things is coming under challenge, although whether this change can be brought about in the understanding of society in general is more problematic. One interesting example comes from work with small groups in classrooms. In the search for an understanding of the STEM learning of individuals belonging to groups that are culturally different to the dominant one, the idea of "participation" seems to be crucial. Participation refers to both participation in verbal conversation and in the broader discourse that takes place in the small group, within the classroom, as well as participation in the wider school culture (Clarkson 1992). All seem to be crucial. Participation is an essential process for inclusion. It has to be mediated at least in part by the teacher and has to take into account both the students' backgrounds and foregrounds. The formal STEM education of an individual requires his/her participation in an institutional network of practice where empowerment, recognition, and dialogue are tools to face conflict in a positive way. Conflict should be understood not only as cognitive conflict but also as cultural, social, and linguistic conflict. Within this broader sense, it must be seen also as a tool for learning. Indeed it may turn out to be the critical strategy for learning. Once this type of thinking is entered into, achievement seems to be a very gross measurement for a conglomerate of interconnected processes that function when a student is learning.

Diverse Contexts

As noted above, it is a given that linguistic diversity exists in most classrooms, in most schools, throughout the world. But there are a variety of linguistic contexts of multilingual STEM classrooms that are generated by different societal determinants. Such situations include classrooms where the language of instruction is different from the first language of the students: For example, the teaching of recently arrived immigrant students. At least in some places, for example, the southern states of USA, there are complete Latino classes of students who speak the same language, although it is a non-English language, but in a school system where English is designated as the language of instruction (Cuevas et al. 1995).

However the situation can be more complicated than this. In some European countries, the new influx of migrants mean that schools are admitting students who come from a number of different language groups, and they sit with students who speak the language of instruction as their first language: For example, in Germany, many classes have a majority of Turkish-speaking students in various regions but who share classes with mono German-speaking students (Markic 2013). In the Catalan region of Spain, it is even more complicated. Catalan is the language of instruction, although Spanish is also an important language in use. Clearly local students speak both. But there are now migrant students attending these schools who speak neither language as their home language (Espinet et al. 2015). In some other countries such as Australia, there is yet again a different variation. There is a continuing flow of new migrants from different language groups being added to older migrant families who speak other languages. For example, many schools that still have first- or second-generation migrant families who came from southern European countries and still speak Greek, Italian, Croatian, etc. in their homes are being joined by students from Vietnam, India, and Cambodia, but the teaching language for all is English (Wotley 2001). Recently the authors were working in a school that has 50% Chaldean students (very recent arrivals), about 30% Vietnamese (both first- and second-generation migrants), and an assortment of other groups including some Somali students who attended school for the first time at this school although they were aged 10 or 11.

A further scenario is when the teaching of STEM may be in a language which is not the first language of the teacher or students. In Papua New Guinea, this happens often where the teaching language is English, but students and teachers may well speak multiple non-English languages in their homes (Clarkson 2016; Muke and Clarkson 2011). And yet another situation was found in Malay schools until recently. The policy from 2003 was for mathematics and science to be taught in English, but all other subjects were still taught in the language common to both teacher and students, Bahasa Malay (Clarkson and Indris 2006; Heng and Tan 2006). Interestingly that policy was changed abruptly in 2011 to revert to the use of Bahasa Malay for all subjects including science and mathematics (Lim and Presmeg 2011).

How communication and learning take place when the languages spoken are not shared, how the fluency of the language of instruction is related to the mastery of the broader notion of scientific discourse, and how using a particular language is linked to different ways of learning are all questions that need further exploration. But the quite different possible contexts – only some of which are listed above – in which such questions arise, need to be taken seriously in our research.

Diversity Within the Same Language

The impacts of other social factors such as social class also impact classrooms. Even within monolingual classrooms, this creates diversity. The communication patterns of the middle class are dominant within most classrooms, in Australia at least. During the authors' many decades of working in schools, it is noticeable that most teachers are drawn from this sector of society. Even if the teacher is from a lower class, they enter the teaching profession as a way of progressing "up" away from the lower class, and hence utilization of the middle-class patterns of communication for such teachers is accepted as the norm. But for students from the lower class, this can be a barrier for their ease of communication. Hence knowing when to ask questions, of whom it is appropriate to ask questions, and how to respond to questions in todays' STEM classrooms are important for quality learning. So even though the language of teaching may well be an official language of the society and may be shared as the home language of both teacher and most students, this classroom context can be described as multilingual and can impact on students' performance and their sense of belonging in the STEM classroom.

For example, in the Australian context, students from the western suburbs of Melbourne, by and large a lower socioeconomic status area, tend to speak fast, run words together, truncate words, and use an inordinate amount of slang terms. Although all of these characteristics may well be noted by non-Australian speakers of English as partly what denotes "Australian English," the extent of all these characteristics is much more noticeable in schools in the west of Melbourne. The authors have noted young beginning teachers completing their school practice as part of their preservice degree, trying to understand the "English" these western suburban students are talking. Clearly this "English" is not spoken in the homes the young teachers grew up in.

Another "within a language" diversity is the specific STEM language embedded in the official teaching language. This crucial variation is probably the most studied aspect of this complicated question (Ryan and Childs 2013) and at times seems to be the only consideration given to issues of language and STEM. As implied in the argument presented here, that is far too limiting a context to understand the messy language complexity of most STEM classrooms. But nevertheless "STEM languages" are a critical aspect of this complexity.

Within an official teaching language, there is the obvious science vocabulary, more or less exclusive to one or more of the subjects that make up STEM. Clearly students need to master these. You will not get far in chemistry without understanding "pH," "acids and bases," and so on. It is more complicated with other words. For example, a common logical connective "or" changes meaning slightly when shifting from ordinary, everyday English to mathematical and scientific language, and hence vocabulary whose meaning is context specific, can make life complicated for students. There are also shifts between written and verbal language. Common verbal Australian English heard from students such as "gotta" or "yer" is rarely written down. Some researchers and/or teachers might say such language is not proper

English, and yet it is the verbal language of many students. On the other hand in STEM written vocabulary, words like "explain" or "therefore" can be found, but they are much less frequent in STEM verbal language.

It is just not vocabulary that students have to come to understand in the context of STEM learning. To some extent, grammar and other aspects of the language also change. In English there is more use of logical connectives in STEM than in other subjects (Gardner 1975; Wellington and Osborne 2001). Often diagrams are incorporated into text, not just as beautifying elements, but as important fundamental aspects of the text without which students will not understand what is being communicated (Clarkson 1981, 1994). The incorporation of many symbols and the truncating of sentences are also elements of the written STEM language quite different to everyday language and can be confusing (Thomas 1986). Again in English, "words" that sound the same in verbal language but can have different spellings (sine, sign, sin) can also be a cause of difficulty. Although some or all of these aspects of language seem to occur in languages other than English, it would be interesting to know whether they cause similar problems for understanding science or whether other characteristics of specific languages cause or mitigate confusion and any promotion of or lack of understanding.

STEM Languages Across Languages

More research concerning the interactions between STEM languages embedded in different languages spoken by students in the same classroom is needed. All languages have embedded within them specific language that deals with this area of the culture. Some STEM languages are more elaborated than others depending on how a culture has met the needs it has faced. Hence, Tok Pisin, a pervasive lingua franca in Papua New Guinea, has measurement language that can be used for wholes and halves but has not been elaborated for further fractions.

Across languages, there may be analogous vocabulary, but symbolism may well differ. For example, "." (a "full stop" in ordinary "Australian English" but a "period" in "US English") is a decimal point used to separate the whole number part of a number from the decimal part; hence 100.98 would be written. In Australia the decimal point is normally placed on the line of writing, but can be used above the line. In Greece and Malta, this decimal marker must be on the line. In Hong Kong and Singapore, it is used above the line. In Chile and Italy, "." is used to separate thousands and hundreds, but in Australia a gap is left to show this, and in a number of countries such as Greece and Holland a "," is used. Actually in Australia "." can also be used to notate a product. So for migrant students, a statement such as "3.4 + 4 = …" may well have different meanings for students who might be sitting in the same classroom in Australia (some of these uses of symbols may have changed in the intervening years since Thomas 1986).

Teachers'/Students' Research Questions

In developing important research questions, one avenue that should perhaps be taken far more seriously is what teachers are saying. The changing and complex role teachers are asked to live out in STEM classrooms has already been noted. What teachers have to contend with in their day-to-day teaching experiences may not readily match the theoretical thinking and rhetoric expounded on at various research conferences and in research journals. This could lead to a gap between accepted theoretical knowledge and teacher knowledge. Such a gap can give rise to potential dilemmas, but if problematized these in turn can lead to insightful questions. To this end, we need good practical descriptions of teaching within multicultural classrooms which may be best generated by teachers. This would give researchers the classroom context as seen by teachers in order to inform the research questions developed perhaps by teachers in consultation with researchers. In other words, the culture of the practice of teaching should be a rich resource for research questions and may well lead to possible ways forward in our theorization as well as in our attempts to help generate more insightful practice. It is probable in this dialogue that the researchers' perspective with its wide-ranging resources and knowledge of theory may well give a general frame for such teacher-generated questions. Hence a dialogue between the two is needed, as both teachers and researchers stand in the overlap of their domains.

Students do not always behave as we anticipate. This is particularly so when it comes to language. They use their language abilities as they wish in order to communicate among other things their understandings, confusions, delights, and annoyances, at a specific point in time with each other and the teacher. They do this in various ways including gestures and using their language(s). Students live their lives without recourse to knowledge of the way researchers have interpreted other students' behaviors in the past. Barwell et al. (2015) have critiqued the way researchers have formalized switching codes and how this can inform future research. They suggest, although many insights have been gained, that one needs to remember that this, as all research, is an interpretation of reality. Barwell (2016) goes further with such a critique and wonders whether if we could really view the context as a student sees it, and with all the nuances and possibilities they see, we might well choose to ask different research questions than those we ask at present. Clearly we cannot see the world as a student sees it. But attempting to may well foreground different aspects of the language context of learning in multilingual situations that never occur to us and hence in turn suggest research questions that would be well worth pursuing.

Framing Research More Broadly

The recognition of complexity, sketched out above, when producing insightful research questions may well prove to be important. However there are other approaches that might also prove to be useful, such as the mapping out of different types of broad contexts within which research questions focused on multilingual STEM classrooms could sit. Two such sets of contexts are noted here. The first is the complexity of language linked to STEM education. This gives rise to at least four practical issues:

- Different "levels" of language (families of languages, distance between languages)
- Different language contexts (indigenous, multilingual, immigrants)
- Contexts within language (speaking, listening, writing, reading) as well as the immediate context (conversational compared with academic)
- Content realities (cultural, social, political)

There are also at least four theoretical issues that seem to be relevant and important, although there are clearly more:

- The structural relation between language and STEM
- The registers and discourses relating to STEM
- The interactions in STEM classrooms
- The different theoretical tools and approaches (e.g., linguistic approach, Vygotsky's social/cultural approach, education didactic approach)

The interplay between such broad descriptors as these may well be a framework for generating useful research questions.

Conclusion and Implications

This chapter has underlined the fact that many classrooms in which STEM is taught are multilingual. With the recognition that STEM, and more clearly what and how STEM is taught, is influenced by culture, language, and the social milieu of the classroom, school, and the wider society, deeper and complex issues for research immediately become the foreground. There are implications with such recognition for some traditional markers of what makes a successful student and/or program. For example, assessment may no longer be considered the only marker of success. However, an analysis of these issues shows that there are differing contexts that may be important in such research. The complexity that comes with the recognition that multilingualism is the norm for most students engaging with STEM suggests that this must become a normal aspect of our research and indeed becomes a fundamental that must be foregrounded for preparing teachers to teach STEM.

In this chapter different contexts and situations that arise with language have been briefly explored, but the same can be also undertaken for culture and other influences. There was no attempt here at deeply analyzing the implications of such complexity. The crucial aim of this chapter is to draw attention to such complexity and argue that it is undoubtedly important. Hence our research needs to accommodate this complexity, as suggested above, even though this undoubtedly will make out research more complex and messy. Likewise, we need to ensure that teachers become aware of the complexity that language issues give rise to in teaching STEM (Martin et al. 2016) and engage with strategies designed to cater for such diversity (Markic 2016).

References

Atweh, B., & Clarkson, P. (2001). Internationalization and globalization of mathematics education: Towards an agenda for research action. In B. Atweh, H. Forgasz, & B. Nebres (Eds.), *Socio-cultural aspects of mathematics education* (pp. 77–94). New York: Lawrence Erlbaum.

Babaci-Wilhite, Z. (Ed.). (2016). *Human rights in language and STEM education*. Rotterdam: Sense Publishers.

Barwell, R. (2016). Mathematics education, language and superdiversity. In A. Halai & P. Clarkson (Eds.), *Teaching multilingual students mathematics* (pp. 25–39). Rotterdam: Sense Publications.

Barwell, R., Clarkson, P., Halai, A., Kazima, M., Moschkovich, J., Planas, N., Phakeng, M., Valero, P., & Villavicencio, M. (2015). Introduction: An ICMI Study on language diversity in mathematics education. In R. Barwell, P. Clarkson, A. Halai, M. Kazima, J. Moschkovich, N. Planas, M. Phakeng, P. Valero, & M. Villavicencio (Eds.), *Multilingual contexts for teaching and learning mathematics* (pp. 1–22). Dordrecht: Springer.

Bishop, A. (1988). *Mathematical enculturation*. Dordrecht: Kluwer.

Chellougui, F., Thu, H., & Winsløw, C. (2015). Language diversity in research on language diversity in mathematics education. In R. Barwell, P. Clarkson, A. Halai, M. Kazima, J. Moschkovich, N. Planas, M. Phakeng, P. Valero, & M. Villavicencio (Eds.), *Multilingual contexts for teaching and learning mathematics* (pp. 263–277). Dordrecht: Springer.

Clarkson, P. C. (1981). A study in visual ability with Papua New Guinea students. In J. Baxter & A. Larkin (Eds.), *Research in mathematics education in Australia* (pp. 30–48). Adelaide: Mathematics Education Research Group of Australasia.

Clarkson, P. (1992). Evaluations: Some other perspectives. In T. Romberg (Ed.), *Mathematics assessment and evaluation* (pp. 285–300). New York: SUNY Press.

Clarkson, P. C. (1994). To see or not to see. *Australian Mathematics Teacher, 50*(4), 40–42.

Clarkson, P. (2007). Australian Vietnamese students learning mathematics. *Educational Studies in Mathematics, 64*, 191–215.

Clarkson, P. (2016). The intertwining of politics and mathematics teaching in Papua New Guinea. In A. Halai & P. Clarkson (Eds.), *Teaching multilingual students mathematics* (pp. 43–56). Rotterdam: Sense Publications.

Clarkson, P., & Indris, N. (2006). Reverting to English to teach mathematics. *Journal of Science and Mathematics Education in South East Asia, 29*(2), 69–96.

Clarkson, P., Bishop, A., Seah, W. T., & FitzSimons, G. (2001). An undervalued legacy of mathematics education. In G. FitzSimons, J. O'Donoghue, & D. Coben (Eds.), *Adult and lifelong education in mathematics* (pp. 47–70). Melbourne: Language Australia.

Cuevas, G., Silver, E., & Lane, S. (1995). *QUASAR students' use of Spanish/English in responding to mathematical tasks*. Paper presented at American Educational Research Association conference, San Francisco, CA.

Cummins, J. (2000). *Language, power and pedagogy: Bilingual children in the crossfire*. Clevedon: Multilingual Matters.

Espinet, M., Junyent, M., Amat, A., & Casteltort, A. (2015). Moving schools towards ESD in Catalonia, Spain: The tensions of a change. In R. Jucker & R. Mathar (Eds.), *Schooling for sustainable development in Europe* (pp. 177–200). New York: Springer.

Gardner, P. L. (1975). Logical connectives in science: A preliminary report. *Research in Science Education, 5*, 161–176.

Heng, C. S., & Tan, H. (2006). English for mathematics and science: Current Malaysian language-in-education policies and practices. *Language and Education, 20*(4), 306–321.

Lave, J. (1988). *Cognition in practice*. New York: Cambridge University Press.

Lee, O. (2005). STEM education with English language learners. *Review of Educational Research, 75*(4), 491–530.

Lim, C., & Presmeg, N. (2011). Teaching mathematics in two languages: A teaching dilemma of Malaysian Chinese primary schools. *International Journal of Science and Mathematics Education, 9*, 137–161.

Markic, S. (2013). *Comics in language-sensitive STEM lessons*. Paper presented at 2013 ESERA conference, Nicosia, Cyprus. Available at: www.esera.org/media/eBook_2013/strand%2012/ Silvija_Markic_19Jan2014.pdf

Markic, S. (2016). Learning language and intercultural understanding in science classes. In J. Lavonen, K. Juuti, J. Lampiselkä, A. Uitto, & K. Hahl (Eds.), *Science education research: Engaging learners for a sustainable future* (Electronic proceedings of the ESERA 2015 conference) (Strand 12; pp. 1864–1870). Helsinki: University of Helsinki. Available at: https://www. esera.org/conference-proceedings/19-esera-2015/270-science-education-research-engaging-learners-for-a-sustainable-future-proceedings-of-esera-2015

Markic, S., & Abels, S. (Eds.). (2016). *Science education towards inclusion*. New York: Nova Publishing.

Martin, S., Chu, H., & Park, J. (2016). Challenges for inquiry and language based science? Teaching to culturally and linguistically diverse students in Korea. In J. Lavonen, K. Juuti, J. Lampiselkä, A. Uitto, & K. Hahl (Eds.), *Science education research: Engaging learners for a sustainable future* (Electronic proceedings of the ESERA 2015 conference) (Strand 12; pp. 1910–1921). Helsinki: University of Helsinki. Available at: https://www.esera.org/con-ference-proceedings/19-esera-2015/270-science-education-research-engaging-learners-for-a-sustainablefuture-proceedings-of-esera-2015

Muke, C., & Clarkson, P. C. (2011). Teaching mathematics in the Papua New Guinea Highlands: A complex multilingual context. In J. Clark, B. Kissane, J. Mousley, T. Spencer, & S. Thornton (Eds.), *Mathematics: Traditions and [new] practices* (pp. 540–547). Adelaide: Australian Mathematics Teachers Association and Mathematics Education Research Group of Australasia.

Ryan, M., & Childs, P. (2013). *Irish STEM teachers' views on linguistic*. Paper presented at 2013 ESERA conference, Nicosia, Cyprus. Available at: www.esera.org/media/eBook_2013/ strand%2012/Real_one_ESERA_2013_PAPER_on_questionnaires_for_Conference_proceed-ings.pdf

Thomas, J. (1986). *Number does not equal maths*. Melbourne: Victorian Ministry of Education.

Wellington, J., & Osborne, J. (2001). *Language and literacy in science education*. Buckingham: Open University Press.

Wotley, S. (2001). *Immigration and mathematics education over five decades*. Unpublished Doctorate of Philosophy thesis. Monash University, Clayton, Australia.

Part V
Professional Development

Models and Modelling in Pre-service Teacher Education: Why We Need Both

Digna Couso and Anna Garrido-Espeja

Introduction: Scientific Practices for Teacher Education

Viewing science learning as participation in the practice of science is a framework gaining momentum in both the science education research literature and recent policy documents (NRC 2007). Within this framework, both science and school science are viewed as discursive, cognitive and social activities consisting of developing explanations, carrying out investigations and evaluating and arguing with evidence (Osborne 2014). This signals the key role of the scientific practices of modelling, inquiry and argumentation in school science.

The introduction of scientific practices in the science classroom can be justified in terms of both epistemic adequacy and learning potential. On the one hand, there is the growing recognition that epistemic knowledge is part of scientific knowledge (Osborne 2014) and needs to be actively considered in teaching and learning. On the other hand, there are the sociocultural and historical perspectives of learning, in which learning is seen as participation in the social activities and pursuits of communities (Lave and Wenger 1991). The two ideas combined call for new perspectives for teaching and learning able to overcome 'the traditional methods that ignore both the epistemic frameworks used when developing and evaluating scientific knowledge, and the social processes and contexts that shape how knowledge is created, communicated, represented, and argued' (Grandy and Duschl 2007: 144). Teaching science within the scientific practice framework is proposed as an alternative in which students participate in 'school science' activities that are socially embedded (both discursive and cognitive in nature) and which are coherent with those of real science (analogous to but not the same as those done by scientists). In our view, promoting students' participation in scientific practices is allowing

D. Couso (✉) • A. Garrido-Espeja
Universitat Autònoma de Barcelona, Barcelona, Spain
e-mail: digna.couso@uab.cat

© Springer International Publishing AG 2017   245
K. Hahl et al. (eds.), *Cognitive and Affective Aspects in Science Education
Research*, Contributions from Science Education Research 3,
DOI 10.1007/978-3-319-58685-4_19

students to experience what makes science different from other ways of knowing in a genuine manner that facilitates their learning.

These school scientific practices, however, are not taking place in schools. As some authors have pointed out, in most science classrooms a focus on the products rather than the processes of science is prevalent (Duschl and Grandy 2008). When referring to the elementary school, we usually face a dichotomy between a traditional transmissive teaching approach, mostly conceptual and textbook-centred, and a supposedly innovative one, founded on inquiry-based teaching that follows a stereotypical 'scientific method' (Windschitl et al. 2008). In particular, the practice of modelling is rarely incorporated into the educational experiences of elementary school pupils (Schwarz et al. 2009), in part because models, as abstract entities, are often considered inadequate for young children.

Introducing the framework of scientific practices, and particularly modelling, in primary school is necessarily a matter of teacher education that should start in pre-service training. For teachers to be able to involve their pupils in scientific practices, they should first be able to engage themselves in such practices actively and adequately (Davis 2003), experimenting at first-hand the details and characteristics of an adequate way of learning science. However, this new framework poses great challenges for pre-service teachers, and it demands well-designed teacher education courses (Reiser 2013). Research shows that teachers will need specific support both with the practices (modelling) and with the scientific ideas addressed by those practices (scientific models) (NRC 2007). With the aim of contributing to this line of research, in this study, we design and investigate a teacher education course for pre-service primary school teachers (PTs) who are asked to construct scientific models by participating in modelling practices.

Our Vision of Models and Modelling

In agreement with last century developments in philosophy of Science, most scholars agree that models are 'essential to the production, dissemination and acceptance of scientific knowledge' (Gilbert 2004: 116). In consequence, science education ascribes a very important role to models and modelling practices.

Modelling is not only a privileged scientific practice to be learned (a new content) but also can refer to a way of learning it: a didactic approach that views learning of a scientific practice as a matter of participating in it. We agree with Schwarz and colleagues that modelling should be understood as a social and personal engagement in 'sensemaking around developing ideas' (Schwarz et al. 2009: 637), rather than the common approach to modelling as the sharing or usage of finalised scientific ideas. Within this view, the aim of 'epistemologically adequate' school science is to put students in the situation of building themselves 'adequate enough' explanations of how the world works. How to develop these explanations by pupils (and therefore first by pre-service primary school teachers) has been discussed widely in the literature.

According to different authors, school modelling encompasses the practices of co- and self-construction and evaluation of models, following iterative cycles of generation, evaluation and modification (GEM) of models (Clement 2008; Khan 2007). In this vein, the well-known proposal of Schwarz et al. (2009) on modelling encompasses the processes of construction, use, evaluation and revision of models. Our own approach elaborates on these ideas but considers that the generation or construction of models is actually the global process of modelling rather than a particular stage in the modelling practice. From this perspective, the construction of increasingly sophisticated models is seen as the overarching activity in which pupils are engaged when using their mental models to predict or explain phenomena, when explicitly expressing their mental models, when evaluating their mental models against available evidence and when revising their mental models accordingly.

Our view of modelling also stems from the assumption that participation in scientific practices is not only to learn to engage in these practices (the procedural dimension) or to learn about these practices (the epistemic dimension) but also to learn the conceptual knowledge in which to frame them. In recent literature developments, this conceptual knowledge is seen rather as a small number of big (Harlen 2010) or core ideas (NRC 2012) than as a set of many concepts and theories. In our view, these ideas are structured in school-based scientific models that have the potential to explain a lot of different phenomena (Izquierdo-Aymerich and Adúriz-Bravo 2003). Without entering into the current discussions within model-based views of science education about the nature and actual definition of models (Oh and Oh 2011), we conceive school-based scientific models as school-adequate versions of scientific models which are conceptual in nature and have representational, explanatory and predictive power (Hernández et al. 2015; Izquierdo-Aymerich and Adúriz-Bravo 2003). Examples of such models are the particle model of matter or the model of Newtonian interactions.

Acquisition of these school-based scientific models by students is not an easy task. Students come to the classroom with their own ideas about phenomena, which are structured in their own mental models (DiSessa 1988; Norman 1983). These mental models can be more or less in agreement with the targeted school-based scientific models to be learned, and as such, they can be interpreted as different versions or levels of the school-based scientific models (Gutierrez and Pinto 2010), which are expected to evolve and increase their complexity alongside instruction. The analysis of this evolution, which is an aim of this research, can help us identify the common intellectual path or empirical learning progression of students' ideas (Corcoran et al. 2009; Duschl et al. 2011).

Research Aims and Context

To introduce these views of modelling and models in elementary school, PTs should start by experiencing this teaching and learning scenario themselves, adequately engaging in modelling practices and constructing good-enough versions of key

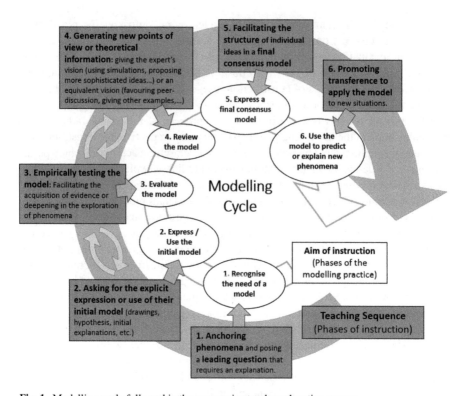

Fig. 1 Modelling cycle followed in the pre-service teacher education course

school-based scientific models. In order to evidence how these two learning objectives are being accomplished in a teacher education programme, the aim of this research is to analyse the modelling practices in which PTs engage and the versions of the model that can be inferred from their discussions and productions in an instructional context that promotes both.

The teaching scenarios that provide opportunities for PTs to engage fruitfully in modelling practices to learn the aforementioned school-based scientific models are those that promote interaction within a classroom culture that motivates students to 'figure things out' (Reiser 2013). A plausible context is that of small-group laboratory-based discussions where the need for an explanation arises from work on phenomena that can be interpreted with key scientific models.

To organise this teaching, we have defined an instructional modelling cycle consisting of six phases (Fig. 1). This instructional cycle is based on different proposals for model-based instruction available in the literature (Baek et al. 2011; Clement 2008; Hernández et al. 2015; Schwarz et al. 2009; Windschitl et al. 2008), which divide the teaching sequence into phases according to the modelling practice with which the students are expected to engage (learning objectives, inner cycle). In addition to this, we include in our modelling cycle details of the instruction or teaching activity that will promote these modelling practices (instructional phases, outer cycle).

The context of this research is an existing teacher education course called Didactics of Science, which is part of the third year of the primary school teacher education degree in Catalonia, Spain. The course is compulsory and divided into 12 lecture and seminar sessions of between 2 and 4 h during one trimester. The course was originally designed collaboratively by a group of science education researchers based on the literature on model-based instruction for primary school teacher education (Mikeska et al. 2009) and subsequently modified specifically to follow the modelling cycle included in Fig. 1.

To exemplify how this is done in the teacher education programme, we outline here part of a teaching and learning sequence to promote PTs' construction of the particle model of matter. In an initial task, PTs were asked to predict what would be the final total volume when we mix 50 ml of water and 50 ml of alcohol and to draw how they imagined the 'inside' of water and alcohol in order to justify their prediction. The aim was for PTs to express their initial mental model (phase 2 of the modelling cycle). Then, PTs carried out the actual experiment of mixing water and alcohol in order to evaluate their initial model (phase 3 of the modelling cycle); finally, they were asked to discuss the results and improve their initial drawings by revising their models (phase 4 of the modelling cycle).

Methods

To analyse how PTs' modelling practices and their versions of the school-based scientific model evolved over the aforementioned teaching and learning sequences, we video and audio recorded all the course sessions of a sample of six PT working groups from the cohort of 2014–2015, a total of 80 PTs being enrolled in the course. Each small group was formed by four to six pre-service teachers who worked together in the university lab. We also collected each participant's written productions and tasks.

Data shown in this chapter belong to two of the small groups participating in the already described teaching sequence about the particle model of matter. The first group was formed by six PTs (identified as S1, S2, S3, S4, S5 and S6) and the second group by five PTs (identified as S7, S8, S9, S10 and S11).

In a first stage of data analysis, we selected video episodes of PTs' discussions and lab work activity that were rich in terms of the modelling practice taking place and/or the level/version of the school-based scientific model being built. Later, these video episodes were transcribed and coded using qualitative analysis software (Atlas.ti). To code the data, we followed discourse analysis techniques, selecting as the unit of analysis any piece of discourse (utterance, intervention or a short part of a discussion) where a particular modelling practice was taking place, and/or a new version of the PTs' model could be inferred (Lemke 2012).

Table 1 Categories used to codify the modelling practices of PTs

Modelling practice	Definition of category (actions and discourse to…)	Examples (transcript extracts)
Use the model (USE)	Use their version of the scientific model to describe, explain or predict phenomena	'Water [goes] down because it is more dense, and alcohol [goes] up because it is less dense' (S7, Final written production)
Express the model (EXP)	Explicitly express, either orally or in writing/drawing, aspects of their version of the model in an abstract or general form	'I would put some A particles, some B and in the middle some AB' (S7, discussion in Episode 1)
Evaluate the model (EVA)	Test their model, analysing the level of adjustment of their ideas with reality	'Okay, but then why do they occupy less space?' (S7, discussion in Episode 1)
Revise the model (REV)	Make more sophisticated and improve specific aspects of their model to increase its descriptive, predictive and explanatory power	'But I think that we should make clear that A is water and B is alcohol so we can show that it is not mixed water at the bottom and it is not mixed alcohol at the top' (S7, discussion in Episode 1)

Following our theoretical framework for modelling practices, PTs' discourse when using, expressing, evaluating or revising their models was categorised accordingly. The definition of the categories used in our analysis is included in Table 1.

A parallel analysis was carried out to identify the versions of the school-based scientific model (PTs' mental models) that could be inferred from the PTs' discussions and productions. These versions were organised in levels as a four-step progression, from initial or naïve understandings (level 1) to a more adequate or sophisticated versions of the target model (level 4), to identify the most common learning pathway. These levels were initially predefined taking into account the literature on students' ideas and learning progressions of the studied models, in this case about the particle model of matter (Smith et al. 2006; Talanquer 2009). The initial levels were empirically and iteratively modified and refined, identifying in our data the ideas that act as stepping-stones in the PTs' construction of the key target models. The different levels were labelled with a letter referring to the specific target model (e.g. M for the particle model of matter) and a number for the level of the version of the model (e.g. M3 for level 3). Table 2 shows the levels and their progressive evolution (empirical learning pathway), which were identified and used to interpret PTs' versions of the model in each episode analysed.

To share the analysis carried out both in terms of modelling practices and levels of the model, Table 3 shows an example of the transcript and analysis of a section of a video episode from the second group analysed (PTs S7–S11). In this episode, PTs are involved in the task previously described, explaining through the particle model of matter the reduction of volume when mixing alcohol and water. The categorisation includes the modelling practice taking place (USE, EXP, EVA or REV) and the version of the target model that PTs' hold (level of the empirical learning pathway: M1–M4).

Table 2 Empirical learning pathway obtained from the iterative analysis of the literature and data. The levels of the particle model of matter in the learning pathway are used to classify PTs' versions of the model displayed in their discussions or productions

Level of the model	Definition of category	Examples (transcript extracts)
M4	Specific disposition/interaction between particles causes the properties of the material. Concrete microscopic configuration (disposition, velocity of vibration…) causes concrete macroscopic properties (density, temperature…).	"The mix of water and alcohol is not 100 ml, but a little bit less. This is due to the different densities of alcohol and water that make that alcohol (which is less dense) occupy the spaces that water molecules leave empty. The interaction of these two liquids causes this result." (S3, Final production).
M3	The internal disposition of particles somehow affects the properties of the material. Microscopic configuration is related to macroscopic properties.	"The particles of alcohol occupy the spaces between particles of water." (S5, Final production)
M2	Particles have the same properties as the material. Microscopic and macroscopic properties are the same.	"The new particles created occupy less space." (S9, discussion in Episode 1)
M1	Only macroscopic properties are taken into account.	"Alcohol goes up and water goes down and there is a part in the middle that mixes, because the water accepts alcohol." (S8, discussion in Episode 1)

To analyse the influence that the modelling activity had on each participant's learning, we also analysed their individual written and graphical productions following a typical qualitative analysis framework (Miles and Huberman 1984) to identify which versions of the model were expressed. Table 4 exemplifies this analysis, showing one example of each level of the model inferred in PTs' productions (M2, M3 and M4). To be able to identify possible modelling patterns and the evolution of the levels of PTs' versions of the model during the activity, we integrated the previous analysis into a more visual representation tool we have called the models and modelling evolution graph (MMEG). The graphs have been elaborated for each identified episode. These are particularly rich moments of PTs' discussions characterised by (1) PTs' focus on the task and (2) their ideas being expressed and moved forward.

Figures 2 and 3 in the results section are examples of MMEG. In these graphs, each unit of analysis (fragment of PTs' discussion) is represented as a box with a code for the participating PTs (S7, S8, etc.). The boxes are situated vertically regarding the modelling practice in which the PTs are engaged and shaded according to the version of the model expressed in or inferred from their intervention (following

Table 3 Analysis of one section of transcript of the video episode regarding the modelling practice and version of the model of the second group of PTs (S7–S11)

Teaching scenario: Demand from the teaching sequence to express the model (drawing)		
Transcription of PTs' group work dialogue	Modelling practice	Version of the model
S8: "Alcohol goes up and water goes down and there is a part in the middle that mixes, because the water accepts alcohol. In the middle they have the same density, right?"	USE	M1
S9: "But we have seen that when alcohol is mixed with water, the total volume is less than the total amount that we put. One of the two, alcohol or water, must have gone inside alcohol or water. There should be empty gaps here, because if not, where this has gone?" S8: "Water also accepts alcohol, I think. It's like the 'holder'." S11: "But in fact we have lost... a little bit [of volume]." S9: "Yes, 2 ml." S7: "That's why I said that if it is left longer, the volume would drop even more... If they [water and alcohol] would end up mixing [completely] or not."	EVA	M3
S9: "But the mixture has already been done, and therefore I do not believe that it will go further down. One or two ml. There comes a point that it no longer... [mixes]." S7: "Mmm... [nodding] no more. [...] What I mean is that... if it happens the same as with water and salt, it's not that one is placed inside the other, it's just that some of the particles join. But it is not that something goes within the other, right? Neither alcohol enters in water nor water enters in alcohol, but instead... " S9: "The new particles created occupy less space...?" [Trying to finish the sentence] S7: "Yes... I do not know the process followed by the particles, but ..." S8: "Ok, I think the same. Do we answer that? Because we should go a little bit faster..."	REV	M2

the shade used in Table 2). Horizontally, the graphs show time evolution within an episode. This sort of representation allows for an analysis of modelling patterns and the evolution tendency regarding the level of PTs' (mental) models over the group discussions.

Results and Discussion

The results discussed in this chapter refer to two episodes during teacher education where two groups of PTs engaged in modelling practices for constructing the particle model of matter. The two episodes were selected because they represent a

Table 4 Examples of PTs' final individual written and graphical productions to express their model (particle model) when interpreting the reduction of volume in a water and alcohol mix. The version of the model identified in the analysis is included

PTs	Drawing	Written explanation	Version
S10	Aigua= A Water AB Alcohol= B Alcohol Novas particulas= AB New particles	We think that **the particles that have been created** from mixing water and alcohol **occupy less space that particles of each liquid separately.** Therefore, little bit of liquid was missed.	M2
S5	] alcohol] aigua → + densa	Water is **more dense than alcohol** (stays at the bottom of the tube). The **particles of alcohol occupy the spaces between particles of water.**	M3
S3	aigua aigua + alcohol	The mix of water and alcohol is not 100mL., but a little bit less. This is due to the **different densities of alcohol and water** that make that **alcohol** (which is less dense) **occupy the spaces that water molecules leave empty.** The interaction of these two liquids **causes** this result.	M4

discursive sequence of laboratory-based group talk where (1) PTs engaged in a diversity of modelling practices (use, express, evaluate and revise) and (2) different versions of PTs' (mental) models emerged.

Figure 2 represents the discussion between the first group of PTs (S1–S6), and Fig. 3 represents the same discussion in the second group (S7–S11). The specific task proposed required that PTs evaluate their initial models in the light of new evidence (phase 3 of the modelling cycle, Fig. 1) and revise their models to produce a final representation (phase 4 of the modelling cycle, Fig. 1). Figures 2 and 3 also include (in the last part of the graph) the models expressed by PTs in their final individual written productions (as detailed in Table 4).

Regarding the Modelling Practices of Pre-service Teachers

The results in both episodes show that engaging PTs in the analysed tasks allows a diverse range of modelling practices to emerge—that is, that PTs engage in use, expression, evaluation and revision of their models. This shows the adequacy of the aforementioned modelling instructional sequence of Fig. 1 and the context of laboratory-based small group discussions as fruitful scenarios for developing scientific practices in initial teacher education.

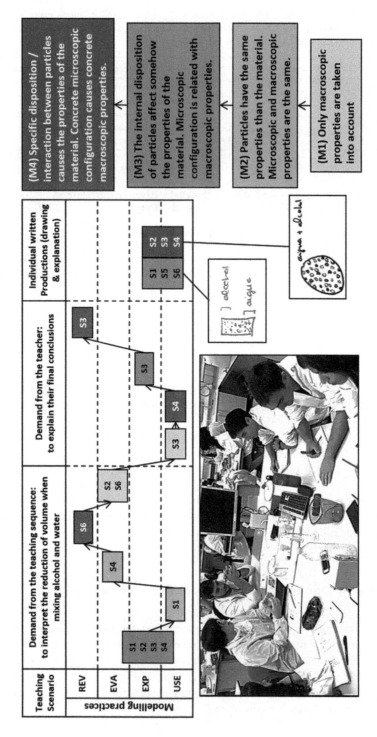

Fig. 2 First episode of MMEG: Representation of PTs' modelling practices and the versions of the particle model of matter when discussing the reduction of volume in a mixture of alcohol and water (Group 1: S7–S11)

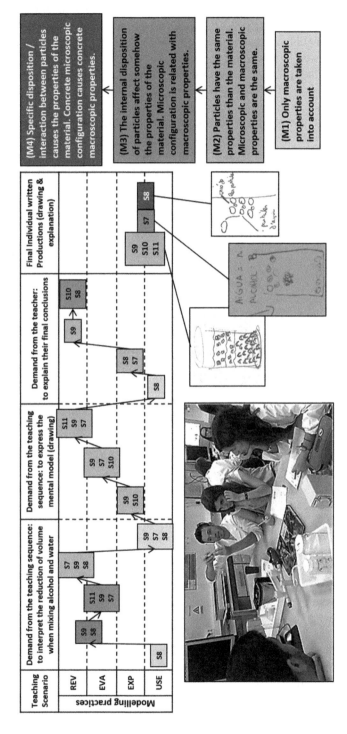

Fig. 3 Second episode of MMEG: Representation of PTs' modelling practices and the versions of the particle model of matter when discussing the reduction of volume in a mixture of alcohol and water (Group 2: S1–S6)

Interestingly, across these tasks PTs participate in all modelling practices, from the more simple (expression and use of models) to the more sophisticated (evaluation and revision of models). This occurs in spite of the fact that the tasks are designed as specific phases of the teaching and learning sequence and thus intend to promote particular modelling practices within the modelling cycle (e.g. the evaluation and revision of models). This has also been observed by us in other tasks across all phases of the modelling cycle, where all modelling practices were also present (Garrido 2016). These results are in line with results obtained by previous authors who already identified that students do not engage in each modelling practice only at the phase of the instructional cycle in which that practice is promoted (Louca and Zacharia 2015). Unlike these authors, however, we do not think this fact implies that some phases of the instructional cycle are not necessary but that students need to continually engage in diverse modelling practices to make sense of phenomena and construct their models.

Our results also diverge from the literature in the identification of an important role for the particular practice of expressing the model. In the literature, this practice has been either neglected or considered part of the practice of model construction (Achér and Reiser 2010; Clement 2008; Khan 2007; Louca et al. 2011; Schwarz et al. 2009). For these authors, the construction (or generation) of the model is seen as a specific modelling practice that occurs while students build a physical model, a drawing, a simulation, etc., generally at an initial phase (the beginning of the instruction or the beginning of each iteration of the modelling process). Apart from engaging in the elaboration of these artefacts as initial expressions of the model, students from these studies are not reported to engage in other expressions of their models along the process. In our case, on the contrary, data show that PTs express their models, in more or less abstract forms, across all phases of the modelling cycle and even when this is not explicitly demanded from the instructional task. We therefore consider that the expression of the model is a crucial modelling practice for the overall construction of the school-based scientific models in which PTs need to engage and do engage autonomously.

The results also show that PTs continually engage in the different modelling practices without following any specific order but in a quite dynamic and complex way. This is in agreement with the lack of a pre-established order in PTs' engagement in modelling found by other authors (Louca and Zacharia 2015). However, different constructs of the literature—such as the well-known GEM iterative structure of generation, evaluation and modification of models (Clement 2008; Khan 2007) or the empirically based sequence of construction, use, evaluation and revision (CUER) of models (Campbell and Oh 2015; Louca et al. 2011; Schwarz et al. 2009)—seem to describe the modelling process as a more ordered and structured one than we and other authors have found. A possible explanation for this apparent discrepancy could be the use of different granularities or scales of analysis. In mesoanalysis at episode level, we find a global tendency in modelling practices that resonate with both the GEM and the CUER structures. This tendency shows discussion time being devoted increasingly to the evaluation and revision of models and decreasingly to the use of the model along the modelling cycle (see also the results of Garrido

2016). Despite agreement with this global tendency, the microanalysis at utterance level conducted here enables us to describe the actual modelling process within each episode as less structured and considerably more complex than previously found.

The above-described tendency in modelling—that of episodes starting with PTs involved in the simpler modelling practices of using/expressing the model and progressing towards more complex ones (evaluation and revision)—is not the only modelling pattern identified in our study. As we have reported elsewhere (Garrido 2016), an iterative pattern of evaluation-revision of the model emerges through PTs' discussions, in particular for some student groups. This pattern is slightly identifiable in the two examples of Figs. 2 and 3, at the beginning of both episodes. Interestingly, these modelling patterns, despite reflecting increased sophistication in the modelling practice, do not necessarily relate to evolution towards the construction of higher versions of the school scientific model. In other words, a rich modelling practice does not always produce equally rich conceptual learning. As can be seen in Fig. 3, most PTs do not immediately improve the initial version of their model (some of them even step back to less sophisticated versions of the model), although they engage in successive patterns of rich modelling practice: use-express-evaluate-revise. From our viewpoint, this signals the different natures of modelling and models in a teacher education course and why we need both: while engaging in modelling practices is needed to construct models and happens quite autonomously among PTs whenever the teaching and learning scenario is designed to promote it, for these models to be adequate (a version of the model closer to the targeted one), additional teaching strategies are needed.

Regarding the Models Constructed by Pre-service Teachers

Figures 2 and 3 show that for both episodes and regarding two different discussion groups, there is change and evolution in PTs' versions of the models towards the school-based scientific target ones. This evolution, however, is disorganised: it shifts back and forth from more simple to more complex ideas during PTs' activities and discussions. This back and forth movement—and even the clear setbacks that some PTs experience during the discussions—seems to have some learning potential in the desired evolution from PTs' initial models towards more sophisticated ones. An example of this can be appreciated in Fig. 3, where the explicit discussion of non-adequate models seems crucial in allowing a step forward towards a more sophisticated one. In this example, the discussion around M1 and M2 by S7–S11 (while they had already expressed higher versions of the model) produces a return to previous versions of the model, which in the end allows for the emergence of better versions of the model: M3 (for S10 and S7) and even M4 (for S8).

However, the emergence of more sophisticated versions of the model at a certain moment in the small-group discussions does not necessarily imply that this specific version of the model is mastered by all PTs at that moment. An example is S8 in

Fig. 3, who expresses sophisticated versions of the model (M3) that led him/her to an upper level in his/her final productions (M4) but did not imply this sort of evolution for her/his peers, who remain in M2 or M3 in their final productions. In fact, our data show that during the teaching activities and for different student groups and models, many emergent high-level ideas are not productively discussed and even get lost for those who expressed them. For example, in Fig. 3, S7, S8, S9 and S11 expressed a high version of the model (M3) at the beginning of the episode, but the group did not take advantage of this, and some of them (S9 and S11) even used a less sophisticated model (M2) in their final written productions.

All of these findings signal the difficulty of self-guided scenarios benefiting from the rich modelling activity that is taking place, again suggesting the need for the teacher's specific guidance and help in some crucial moments of the discussion. As seen in the last part of the episode in Figs. 2 and 3, the teacher intervention triggers the emergence of more sophisticated models in both groups. This is especially important if we want PTs to construct key school-based scientific models that are generally complex and challenging but also central to science education (Izquierdo-Aymerich and Adúriz-Bravo 2003; NRC 2012). The role of the teacher in these scenarios proves crucial and needs to be further explored.

Conclusions and Implications

As a main conclusion, our research shows that the model-based instructional sequence followed, based on a modelling cycle (Fig. 1) in an educational context of small-group laboratory-based discussions, is a fruitful scenario for a rich modelling activity and a positive evolution of PTs' models to emerge in pre-service teacher education. We see this experience of learning both in terms of modelling practices and scientific models as a prerequisite for PTs to be able to use a model/modelling-based approach to science education. However, this experience is surely not enough: explicit reflection on this process would be necessary for PTs to acknowledge and master the used instructional approach in the future, as other authors have interestingly discussed (Schwarz and White 2005; Schwarz et al. 2009).

Despite the disorganised participation of PTs in the modelling practices and the role of expression of the model, which to some extent challenges previous literature findings, we have found some interesting modelling patterns coinciding with previous research (such as the use-express-evaluate-revise) that are not directly or immediately related to an improvement of the version of the mental models of PTs. From our viewpoint, these results indicate that the modelling process is much more dynamic and complex than we initially thought (and was suggested in the literature): although some interesting patterns arose, PTs need to engage in all the practices throughout time to be able to construct a key school-based scientific model.

Regarding models, the evolution is not easy for PTs who, instead of moving forward level by level, actually move back and forth on multiple occasions. However, there is a global tendency of evolution and increased richness in terms of the versions of PTs' models towards the targeted ones. As a consequence, we can conclude that a lot of discussion and activity time is needed to reach higher levels of the model, in general in a non-permanent fashion and with strong individual differences.

Our research also suggests the key influence that some teaching scenarios may have on the modelling practice and the evolution of PTs' versions of the model, namely, the teacher's role in group discussions (e.g. when asking participants to explain their final conclusions), which seems to be crucial to let PTs move to more complex versions of their model. Nevertheless, more research is needed to determine what type of influence the different teaching scenarios (whether driven by the task, the teacher educator or the students themselves) have on the modelling practices and on the evolution of the PTs' models. We will continue this type of analysis in further work, aiming to determine what types of teaching scenarios, mechanisms or strategies have an influence on promoting modelling practices and the evolution of models.

Finally, we would like the reader to know that we have used the knowledge gained in this study to redesign some aspects of the PT course, which has been analysed in depth and published elsewhere (Garrido 2016).

Acknowledgement This research was partially funded by the Spanish Government (EDU2015-66643-C2-1-P).

References

Acher, A., & Reiser, B. (2010). *Middle school students and teachers making sense of the modeling practices in their classrooms*. Paper presented at the annual conference of the National Association for Research in Science Teaching (NARST), Philadelphia, PA.

Baek, H., Schwarz, C., Chen, J., Hokayem, H., & Zhan, L. (2011). Engaging elementary students in scientific modeling: The MoDeLS fifth-grade approach and findings. In M. S. Khine & I. M. Saleh (Eds.), *Models and modeling* (pp. 195–218). Dordrecht: Springer.

Campbell, T., & Oh, P. S. (2015). Engaging students in modeling as an epistemic practice of science: An introduction to the special issue of the Journal of science education and Technology. *Journal of Science Education and Technology, 24*(2–3), 125–131.

Clement, J. J. (2008). Student/teacher co-construction of visualizable models in large group discussion. In J. J. Clement & M. A. Rea-Ramirez (Eds.), *Model based learning and instruction in science* (pp. 11–22). Dordrecht: Springer.

Corcoran, T., Mosher, F. A., & Rogat, A. (2009). *Learning progressions in science: An evidence-based approach to reform. CPRE research reports*. Philadelphia: Consortium for Policy Research in Education.

Davis, K. S. (2003). 'Change is hard': What science teachers are telling us about reform and teacher learning of innovative practices. *Science Education, 87*(1), 3–30.

DiSessa, A. A. (1988). Knowledge in pieces. In G. Forman & P. B. Pufall (Eds.), *Constructivism in the computer age* (pp. 49–70). Hillsdale: Lawrence Erlbaum Publishers.

Duschl, R. A., & Grandy, R. E. (2008). *Teaching scientific inquiry: Recommendations for research and implementation*. Rotterdam: Sense Publishers.

Duschl, R. A., Maeng, S., & Sezen, A. (2011). Learning progressions and teaching sequences: A review and analysis. *Studies in Science Education, 47*(2), 123–182.

Garrido, A. (2016). *Modelització i models en la formació inicial de mestres de primària des de la perspectiva de la pràctica científica* [Models and modelling practices in primary-school initial education from the Scientific Practices perspective]. Doctoral thesis. Universitat Autònoma de Barcelona, Barcelona, Spain.

Gilbert, J. K. (2004). Models and modelling: Routes to more authentic science education. *International Journal of Science and Mathematics Education, 2*(2), 115–130.

Grandy, R., & Duschl, R. A. (2007). Reconsidering the character and role of inquiry in school science: Analysis of a conference. *Science & Education, 16*(2), 141–166.

Gutierrez, R., & Pinto, R. (2010). From mental models to scientific models: Similarities in structures and its importance in scientific knowledge construction. In *Proceedings of the GIREP-ICPE-MPTL 2010 international conference* (pp. 80–81). Reims: University of Reims Champagne-Ardenne.

Harlen, W. (2010). *Principles and big ideas of science education*. Hatfield: ASE.

Hernández, M. I., Couso, D., & Pintó, R. (2015). Analyzing students' learning progressions throughout a teaching sequence on acoustic properties of materials with a model-based inquiry approach. *Journal of Science Education and Technology, 24*(2–3), 356–377.

Izquierdo-Aymerich, M., & Adúriz-Bravo, A. (2003). Epistemological foundations of school science. *Science & Education, 12*, 27–43.

Khan, S. (2007). Model-based inquiries in chemistry. *Science Education, 91*(1), 877–905.

Lave, J., & Wenger, E. (1991). *Situated learning: Legitimate peripheral participation*. Cambridge: Cambridge University Press.

Lemke, J. L. (2012). Analyzing verbal data: Principles, methods and problems. In B. J. Fraser, K. Tobin, & C. J. McRobbie (Eds.), *Second international handbook of science education* (pp. 1471–1484). Dordrecht: Springer.

Louca, L. T., & Zacharia, Z. C. (2015). Examining learning through modeling in K-6 science education. *Journal of Science Education and Technology, 24*(2–3), 192–215.

Louca, L. T., Zacharia, Z. C., & Constantinou, C. P. (2011). In quest of productive modeling-based learning discourse in elementary school science. *Journal of Research in Science Teaching, 48*(8), 919–951.

Mikeska, J. N., Anderson, C. W., & Schwarz, C. V. (2009). Principled reasoning about problems of practice. *Science Education, 93*(4), 678–686.

Miles, M. B., & Huberman, A. M. (1984). *Qualitative data analysis: A sourcebook of new methods*. Los Angeles: SAGE publishing.

Norman, D. A. (1983). Some observations on mental models. In D. Gentner & A. L. Stevens (Eds.), *Mental models* (pp. 7–14). Hillsdale: Lawence Erlbaum Associates Inc..

National Research Council. (NRC). (2007). *Taking science to school: Learning and teaching science in grades K-8*. Committee on Science Learning, Kindergarten Through Eighth Grade. R. A. Duschl, H. A. Schweingruber, & A. W. Shouse (Eds.). Board on Science Education, Center for Education. Division of Behavioral and Social Sciences and Education. Washington, DC: The National Academies Press.

National Research Council. NRC. (2012). *A framework for K-12 science education. Practices, crosscutting concepts and core ideas*. Committee on a conceptual framework for new K-12 science education standards. Board on science education, division of behavioral and social sciences and education. Washington, DC: The National Academies Press.

Oh, P. S., & Oh, S. J. (2011). What teachers of science need to know about models: An overview. *International Journal of Science Education, 33*(8), 1109–1130.

Osborne, J. (2014). Teaching scientific practices: Meeting the challenge of change. *Journal of Science Teacher Education, 25*(2), 177–196.

Reiser, B. J. (2013). *What professional development strategies are needed for successful implementation of the next generation science standards?* Paper presented at invitational research symposium on science assessment. K-12 center at ETS, Washington, DC.

Schwarz, C., & White, Y. (2005). Metamodeling knowledge: Developing students' understanding of scientific modeling. *Cognition and Instruction, 23*(2), 165–205.

Schwarz, C. V., Reiser, B. J., Davis, E. A., Kenyon, L., Achér, A., Fortus, D., et al. (2009). Developing a learning progression for scientific modeling: Making scientific modeling accessible and meaningful for learners. *Journal of Research in Science Teaching, 46*(6), 632–654.

Smith, C. L., Wiser, M., Anderson, C. W., & Krajcik, J. (2006). Implications of research on children's learning for standards and assessment: A proposed learning progression for matter and the atomic-molecular theory. *Measurement: Interdisciplinary Research & Perspective, 4*(1–2), 1–98.

Talanquer, V. (2009). On cognitive constraints and learning progressions: The case of 'structure of matter'. *International Journal of Science Education, 31*(15), 2123–2136.

Windschitl, M., Thompson, J., & Braaten, M. (2008). Beyond the scientific method: Model-based inquiry as a new paradigm of preference for school science investigations. *Science Education, 92*(5), 941–967.

Designing a Course for Enhancing Prospective Teachers' Inquiry Competence

Marios Papaevripidou, Maria Irakleous, and Zacharias C. Zacharia

Introduction

Inquiry, which refers to "the diverse ways in which scientists study the natural world and propose explanations based on the evidence derived from their work" (NRC 1996: 23), is at the core of scientific endeavor. Numerous research reports (e.g., Abd-El-Khalick et al. 2004, Bartos and Lederman 2014, Capps et al. 2012) indicated that learners can similarly benefit from this approach through their engagement in learning activities centered on inquiry, and the resulting outcome is the development of inquiry learning. Inquiry learning is considered as an approach to learning that entails "a process of exploring the natural or material world, and that leads to asking questions, making discoveries, and rigorously testing those discoveries in the search for new understanding" (NRC 2000: 2). The key for a successful design and implementation of science instructional settings through which learners will be scaffolded in developing inquiry skills is the *teacher*, given that teachers are considered to be the "linchpin" in any effort to change science education across nations (NRC 2012).

Consequently, reform documents in science education have underlined the increasing importance of preparing teachers, who will play key roles in guiding students through cognitive activities centered on inquiry (NRC 2000). Davis et al. (2006) indicated that to design and enact science instruction centered on inquiry, teachers must have strong understandings of inquiry and abilities to teach inquiry. Similarly, the National Research Council stressed that "for students to understand inquiry and use it to learn science, their teachers need to be well-versed in inquiry and inquiry based methods" (2000: 87).

M. Papaevripidou (✉) • M. Irakleous • Z.C. Zacharia
University of Cyprus, Nicosia, Cyprus
e-mail: mpapa@ucy.ac.cy

© Springer International Publishing AG 2017   263
K. Hahl et al. (eds.), *Cognitive and Affective Aspects in Science Education
Research*, Contributions from Science Education Research 3,
DOI 10.1007/978-3-319-58685-4_20

Despite this persistent call, evidence from the literature revealed that a vast majority of teachers have unsophisticated understandings of inquiry and do not routinely adopt inquiry-based instruction within their practices due to a number of systemic and other barriers (Crawford 2000, 2007; Davis et al. 2006; Kazempour and Amirshokoohi 2014; Saad and BouJaoude 2012). Consequently, the key to overcome this gap is to invest on teachers' professional development (PD) both at pre- and in-service level. A critical challenge that emerges is to identify the key features that PD programs should entail in order to succeed in changing teachers' epistemic knowledge of the nature of scientific inquiry, helping teachers appreciate the impact of inquiry-based learning to students' scientific literacy, and assisting them in understanding how to design inquiry-oriented instruction in their classrooms (Capps et al. 2012), and consequently influencing the development of their pedagogical content knowledge for scientific inquiry (Davis and Kracjick 2005).

Additionally, it is equally important to identify the role of teachers within such a program in order to maximize their professional expertise on teaching science through inquiry. Prior research (e.g., Clarke and Hollingsworth 2002, Kazempour and Amirshokoohi 2014) indicates that positioning teachers in the role of *active learners* rather than as information gatherers and letting them experience themselves the same learning journeys that their students are expected to follow could be beneficial for their professional development; this role might result in teachers' construction of meaningful knowledge about inquiry and skills for inquiry teaching (Loucks-Horsley et al. 1998). A second role that is important for teachers to encounter during their participation in a PD program is the role of *thinkers* of both the learning experiences gained through the inquiry hands-on activities and the underlying design principles of the curriculum materials they engaged with as learners. Theoretical readings, class discussions, and other reflective activities may facilitate this role of teachers, as they allow themselves to reflect on their developing understandings, enhance their knowledge about certain aspects of inquiry-based learning, and can shed light on prior established misconceptions about inquiry and science in general (Akerson et al. 2007). Lastly, given that reflective practice, which refers to the capacity to reflect on action that leads in engagement in a process of continuous learning (Schön 1983), can be a beneficial form of teachers' professional development (Ferraro 2000). Hence, a third role that is considered essential for teachers to follow during a PD program is that of *reflective practitioner*. This role is facilitated through allowing teachers to implement curriculum materials they developed or received within the context of a PD program into their own practice, make necessary adjustments to their teaching according to situations occurred at a particular time, collect evidence to evaluate and reflect on the effectiveness of their teaching, and bring reports of their field experiences to the course and analyze teaching strategies with their mentors and colleagues.

Purpose and Research Question

We present the structure of a PD program through which we aimed at impacting teachers' development of inquiry competence, namely, *inquiry skills*, *views and definitions of inquiry*, and *pedagogical content knowledge (PCK) for teaching science as inquiry*. Our approach draws on the constructs of constructivist learning (Driver et al. 1994) and situated cognition (Brown and Campione 1990). It also builds upon nine critical features[1] of effective inquiry PD suggested by Capps et al. (2012) and follows the recommendations for positioning teachers as learners (Phase 1), thinkers (Phase 2), and reflective practitioners (Phase 3) within the context of a PD program. The development of the curriculum materials incorporated within the course was grounded on the inquiry-based learning framework suggested by Pedaste et al. (2015).

The research question that we aimed to address was: How did teachers' (i) development of inquiry skills, (ii) views and definitions of inquiry, and (iii) PCK for teaching science as inquiry change along the course? Specifically, what learning outcomes did teachers gain during participating in each of the three consecutive phases of the PD program?

Methodology

The participants were 72 preservice elementary teachers who attended a science method course in Cyprus, within which the PD program was implemented. During the previous semester, all teachers attended a content course that made use of the Physics by Inquiry curriculum (McDermott 1996), whereas none of them taught science during their school practicum.

The PD course, taught by two university instructors and three graduate assistants, was organized into twelve 1.5-hour sessions and split in three phases. During Phase 1, a curriculum titled "Boiling and Peeling Eggs" was implemented, through which the teachers (groups of four) engaged in multiple inquiry cycles to answer "How to make perfect hard boiled eggs that are easy to peel?" Specifically, the teachers as *learners* defined the problem that merited solution; identified variables that might affect the boiling and peeling of eggs; formulated investigative questions and hypotheses; designed and performed experiments; collected, analyzed, and interpreted data; drew conclusions; and presented their findings in posters. During Phase 2, the teachers as *thinkers* were asked to study the curriculum they previously worked with to identify the phases of inquiry and their interconnections, in order to inductively formulate the underpinnings of the inquiry-based framework that guided the design of the curriculum. Next, the inquiry-based framework was intro-

---

[1] All nine critical features are presented in the Discussion section in relation to how they were addressed in the design and implementation of the PD program of the present study.

duced, and the teachers compared their perceived frameworks with the original one. Finally, during Phase 3, the teachers were assigned the role of *reflective practitioners* and were asked to design lesson plans and curriculum materials on a particular topic that they would use to engage an elementary student in inquiry-based activities. Throughout the meeting with their student, the teachers maintained reflective journals to record their student's inquiry-based learning progress, and all phases of inquiry were reported on a poster that was presented during a *science fair* organized in collaboration with the teachers and a local school. At the end of the course, the teachers made presentations of their science fair projects, shared their reflections and lessons learned with their peers, and received feedback from the instructors and peers.

We collected multiple forms of data: (a) *teachers' written definitions of inquiry*, as documented in questionnaires administered during the first, the seventh, and the last course meeting; (b) *reflective diaries*, in which teachers were asked to document their evolving understanding of inquiry-based learning (used as means for capturing their PCK for scientific inquiry); (c) *pre- and post-assessment of teachers' inquiry skills*; (d) *science fair project work*; and (e) *end-of-course individual interviews*.

An open coding scheme refined through the use of the constant comparative method (Glaser and Strauss 1967) was followed for answering study's research question. Specifically, teachers' responses on the various data collection instruments were classified along a three-level inquiry advancement scheme, namely, *novice inquiry*, *basic inquiry*, and *advanced inquiry*. Novice inquiry pertains to teachers' responses that revealed the presence of naïve ideas and misconceptions about inquiry. The second category (basic inquiry) reflected the presence of a limited number of ideas that point to informed understandings about inquiry combined with instances of naïve ideas, whereas the third category (advanced inquiry) evinced the presence of ideas consisted with informed understandings about inquiry.

Findings

The findings are presented in Table 1 and are discussed in the subsequent three subsections in relation to teachers' inquiry competence development along the three phases of the PD program. Representative examples are also included within each subsection as evidence of how we reached these results.

Inquiry Skills

The findings revealed that in the beginning of the course, the level of teachers' acquisition inquiry skills was at a moderate level (79%, 82% – basic inquiry – see Table 1). With regard to teachers' identification of experimental flaw skill, the majority of teachers' responses indicated that they failed to identify all experimental

Table 1 Percentage of teachers' inquiry competence classification across three levels of inquiry (naïve, basic, advanced) during each phase of the PD program

Phases of the PD															
	Phase 1: Teachers as learners			Phase 2: Teachers as thinkers						Phase 3: Teachers as reflective practitioners			Final assessment (4 weeks after the end of the course)		
	Pre			Post/pre			Post/pre			Post					
	N[a]	B[b]	A[c]	N[a]	B[b]	A[c]	N[a]	B[b]	A[c]	N[a]	B[b]	A[c]	N[a]	B[b]	A[c]
	%	%	%	%	%	%	%	%	%	%	%	%	%	%	%
Percentage of teachers' inquiry competence classification across three levels of inquiry															
Assessment of															
1. Inquiry skills															
1.1. Application of the control of variables skill-data interpretation	5	79	16	0	8	92	x[d]	X	x	0	4	96	0	3	97
1.2. Identification of experimental flaws – revision of experimental design	9	82	9	0	10	90	x	X	x	0	6	94	0	3	97
2. Definition of scientific inquiry	87	13	0	12	58	30	2	26	72	0	13	87	0	4	96
3. PCK for teaching science as inquiry															
3.1. Understanding of the instructional strategies and tools for supporting inquiry	96	4	0	33	67	0	31	56	13	11	12	87	5	7	88
3.2. Knowledge of children's understandings and misunderstandings associated with inquiry	98	2	0	91	9	0	88	12	0	0	31	69	1	8	91
3.3. Knowledge of appropriate curriculum for inquiry	75	25	0	35	62	3	11	25	74	1	11	88	0	4	96
3.4. Knowledge of assessment techniques for inquiry	84	16	0	15	79	6	15	78	7	2	8	90	0	4	96

[a]Novice inquiry: presence of naïve ideas and misconceptions
[b]Basic inquiry: presence of a limited number of ideas that point to informed understandings about inquiry combined with some instances of naïve ideas
[c]Advanced inquiry: presence of ideas consistent with informed understandings about inquiry
[d]No administration of assessment tasks

flaws in a given experimental design. We present below a task that was administered to evaluate this specific inquiry skill followed by a representative quote from a teacher's response to document this finding:

> Marina conducted an experiment to test if the material a hammer nail is made of affects its rusting time when placed inside a liquid. She used three test tubes, three different hammer nails and two types of liquids. In the first tube she put an iron hammer nail and water. In the second tube she put a cuprum hammer nail and vinegar. In the third tube, she put a steel hammer nail, vinegar and water. Then, she left them in the kitchen for a week. At the end of the week, she observed that only the iron nail rusted. Therefore, she concluded that water affects the rusting of a metal pin in a better way than the vinegar. Do you agree with Marina's conclusion? Explain the reasoning behind your response. (Adapted from Constantinou et al. 2004)

A representative response that documents the majority of teachers' failure to identify all experimental flaws and thus their classification in the basic inquiry level is as follows:

> I don't agree with Marina's conclusion, because she should have put the same type of liquid in each tube in order to find out if the type of the material of a hammer nail affects its rusting when placed inside a liquid. (Teacher #14)

The abovementioned response indicates that this particular teacher identified only the type of liquid as the variable that should have been kept constant in the given experimental design and failed to identify other variables (e.g., the volume of the liquid in each tube, the size and material of each tube, etc.) that should have been kept constant. In addition, the teacher did not notice that the conclusion derived from the experimental design is irrelevant to the investigative question being researched (i.e., the investigative question pertains to the type of material of the hammer nail, whereas the conclusion focuses on the type of the liquid).

At the end of Phase 1, teachers made a significant shift in terms of the development of their inquiry skills (90% and 92% in advanced inquiry level), which was slightly increased by the end of the course (97% in advanced inquiry level, see Table 1). Almost all teachers were able to identify all experimental flaws in the given experimental design and proposed revisions of the experimental design in order to perform a controlled experiment to answer the investigative question under study. Teachers' slight inquiry skills improvement by end of the course might be attributed to the teaching experience they gained during working with their students for the science fair project, since they had to help their students develop inquiry skills themselves through the curriculum materials and the assessment tasks they developed.

Definition of Scientific Inquiry

At the beginning of the course, all teachers held uninformed views of inquiry and teaching science as inquiry (87% – novice inquiry – see Table 1). A representative quote with regard to the definition of inquiry-based learning, provided by a teacher at the beginning of the course and categorized in the cluster of naïve inquiry, is as follows:

> Inquiry is a learning situation during which students and teacher interact, discuss, and experiment with an appropriate problem and at the end they reach a mutual response. (Teacher #43)

Their definitions of inquiry were continually changed and improved throughout the course, since they progressed from 0% of advanced inquiry at the beginning of the course to 30% at the end of Phase 1, 72% at end of Phase 2, 87% at the end of Phase 3, and 96% at the final assessment which was performed 4 weeks after the end of the course. The following is a representative example of a comprehensive definition of inquiry (clustered as advanced inquiry) provided by teacher #43 at the end of the course:

> Inquiry is a process, similar to the one scientists follow in their daily work, though which a learner engages with a problem and performs several actions for solving the problem. Inquiry involves defining the problem of interest, making some research on getting insight on the concepts that relate to the problem, formulating a question and generating a hypothesis based on the question, designing a controlled experiment to answer the question, collecting and interpreting data, and drawing conclusions in relation to the initial question. The process is not a linear one, since one can follow different paths depending on the type of problem, the conceptualization of the problem, etc., and you can always go back to further investigate your question or formulate and test new research questions. (Teacher #43)

PCK for Teaching Science as Inquiry

Teachers' PCK for teaching science as inquiry was found to be significantly enhanced only after the end of Phase 3, since at the end of Phases 1 and 2, the majority of teachers' PCK was clustered as either naïve or basic inquiry. For instance, with regard to the aspect "Knowledge of assessment techniques for inquiry" prior to the course, a teacher provided the following response:

> During the first lesson with electric circuits, I would ask students to form groups of four and then I would give them a wire, a light bulb and a battery and I would challenge them to find a way to make the bulb to lit. Hence, I would be able to observe their reactions, if they are able to collaborate with each other, and with appropriate guidance I would keep notes if they can learn something new by themselves. (Teacher #66)

At the end of the course, teachers' *knowledge of assessment techniques for inquiry* was significantly increased (96% – advanced inquiry – see Table 1). An indicative quote from a response by teacher #66 is provided below:

> I would ask students to describe what they should do if they wanted to learn whether the sun is essential for plants to growth. In scaffolding their work, I would present 6 different pictures that varied in the type of the plant, the size of the pot, the presence/absence of sun, and the amount of water that is added in each pot, and I would ask them to choose which two they should choose in answering the posed question. (Teacher #66)

Similarly, teachers' *knowledge of appropriate curriculum for inquiry* was significantly improved. The following extracts from a teacher's lesson plans provided at the beginning and end of the course, through a task that sought to evaluate teachers' knowledge of appropriate curriculum for inquiry, are particularly revealing:

> The objective of an inquiry-based lesson is to give students the opportunity to familiarize themselves with magnets, and especially with their magnetic poles. Initially, the teacher problematizes his students, and then students experiment and test their hypotheses. The teacher does not provide ready-made responses, but evaluates students through appropriate questions. (Teacher #29, before the course, cluster of inquiry: *basic*)

> The teacher introduces students to a problem that relates to why some objects sink and some others float in water. She prompts students to pose their initial ideas (these might relate to the identification of variables that might affect the sinking/floating of objects), and helps students to formulate hypotheses that would later test through experiments. Before formulating hypotheses, the students formulate investigative questions in the form "Does variable A affect variable B?", and for each question they formulate a hypothesis. Next, the students are asked to choose a question and design a controlled experiment (only one variable is altered while the rest are maintained constant) for answering it. During their experiment, they collect data, organize them in a table, and when they have collected enough data, they proceed in interpreting their data in relation to their initial hypothesis and investigative question. The students follow the same procedure for answering all investigative questions, and the support from the teacher faints out, as she observes that the students are able to transfer the experimental design strategy for investigating the effect of new variables in the sinking/floating of objects. (Teacher #29, at the end of the course, cluster of inquiry: *advanced*)

Teachers' knowledge of *children's understandings and misunderstandings associated with inquiry* has improved by the end of the course. During Phases 1 and 2, the majority of teachers were classified in the naïve inquiry level (see Table 1), and it was at the end of Phase 3 and 4 weeks after the course that they made a significant progress to the advanced inquiry level (69% and 91% in advanced inquiry, respectively, see Table 1). For instance, in a task that teachers were prompted to refer to the inquiry skills a student should master in order to engage in inquiry, a teacher in the beginning of the course stated the following:

> It is essential that students should be able to collaborate with each other and follow specific instructions. Also, it is important that students are not used of receiving ready-made knowledge, but be able to formulate conclusions themselves. (Teacher # 11, cluster of inquiry: *naïve*)

Based on the abovementioned response, it is obvious that this particular teacher failed to reflect and name some of the inquiry skills that a child should have already developed in order to meaningfully engage in inquiry activities. After teachers' participation in the three consecutive phases of the PD program and specifically after working with an elementary school student for the purposes of the science fair project, the majority of teachers appeared to be able to make statements on the skills that are fostered within an inquiry-oriented instruction. The following quote from a participant's response documents this assertion:

A student should have mastered several inquiry skills in order to enrol in inquiry activities. These skills are as follows: (i) identification of variables skill; (ii) formulation of investigative questions skill; (iii) control of variables skill; (iv) data interpretation skill; (v) hypothesis generation skill; (vi) hypothesis testing skill. (Teacher # 3, cluster of inquiry: *advanced*)

As far as teachers' *understanding of the instructional strategies and tools for supporting inquiry* is concerned, a similar pattern of improvement was revealed. Specifically, to evaluate this aspect of PCK for inquiry, we administered to the teachers a set of scenarios that illustrated how different teachers approached the teaching of the same topic with their students. The teachers were prompted to choose which of the scenarios involved instructional strategies and tools for supporting students' engagement in inquiry. One of the scenarios was as follows:

Mr. Lowe is a 3rd grade teacher. One of his eventual objectives is for students to learn (at a simple level) about the relationship between form and function. He begins a specific lesson on fish by showing an overhead transparency of a fish, naming several parts, and labelling them as shown. (Adapted from Schuster et al. 2007)

Prior to the course, the majority of teachers' responses were clustered as naïve, since they considered this lesson as inquiry-related and provided arguments like:

This is a good lesson, because the teacher aims to introduce the terms in a systematic way that the children will need while studying the fish.

Or,

I consider this a good lesson, because learning about fish function should start by introducing the names of the fish parts to students, and then proceed on studying how these affect the function of the fish.

At the end of the course, teachers' evaluations of the same lesson scenario appeared to have changed since they considered it as not an inquiry-oriented one. To document their evaluations, they provided responses like the one below:

This lesson is not appropriate, because it follows a content delivery approach (e.g. the teacher provides the names of parts of the fish to the children) and there is no evidence to show that the teacher aims to prompt students to develop questions and hypotheses of how and why each part of the fish affects its function.

This finding can also be attributed to the rich teaching and learning experience they received during their efforts to engage their students with inquiry-based activities and scaffold the development of their inquiry skills and understandings about critical aspects of inquiry (Phase 3 of the PD program).

Discussion

The purpose of this study was to investigate the effect of a PD program on teachers' development of inquiry competence. The findings demonstrate significant shifts of teachers from naïve to advanced inquiry in all three aspects of their inquiry competence (inquiry skills, definitions of inquiry, and PCK for teaching science as inquiry).

These promising findings can be attributed to two important aspects of the PD program that was designed and followed for the purposes of the present study. The first relates to the *features of the course*, such as, the format and structure, the curriculum materials, and the teaching approach. The second one is associated with the three distinct *participatory roles* that teachers were assigned to during their engagement in the three consecutive phases of the PD program. We briefly elaborate on each of them below.

Features of the Course

All *nine critical features of effective inquiry* derived from Capps et al. (2012) were addressed in the design and were successfully implemented during the course. As far as the *structural features* of the course are concerned, the *total time* of the course (12 weeks) compared with the duration of the reviewed studies by Capps et al. (from 1 to 6 weeks) provides a significant time difference that allowed both instructors and participants to work out several important learning and teaching activities without being constrained by the time factor. Consequently, PD programs should provide teachers with adequate time frames to deconstruct their understandings about learning and teaching through inquiry (Capps et al. 2012) and eventually to modify their teaching practices (Supovitz and Turner 2000).

Also, the *extended support* provided to teachers at various instances during each phase of the course might also account for the significant inquiry gains that were evidenced in their reports and presented in the Findings section. For instance, during Phase 3 (teachers as reflective practitioners), the teachers received feedback on their science fair project proposals by the instructors of the course. They also met with the instructors once a week on a volunteer basis to pose questions, discuss problems encountered during the meetings with their students, and get support on their future steps. The support received was also extended and enhanced via online communication; a social network page was created to offer teachers the opportunity to exchange ideas with their peers, to share learning experiences and discuss the lessons learned from the meetings with their students, and also to receive feedback on their lesson plans and curriculum materials from the science teachers of the local school that their students came from. Hence, it appears that extended support is vital during teachers' professional development. This is in agreement with the literature of the domain, which postulates that the provision of support influence teachers willingness to change their teaching practices (Simon et al. 2011).

The third structural feature of the course, namely *authentic experiences*, is also considered as an important factor for teachers' inquiry learning achievements. For instance, during Phase 1 (teachers as learners), the teachers were engaged with a curriculum developed for the purposes of this course titled "Boiling and Peeling Eggs," and they were prompted to answer "How to make perfect hard boiled eggs that are easy to peel?" Specifically, the teachers (working in groups of four) defined the problem that merited solution; identified variables that might affect the boiling and peeling of eggs; formulated investigative questions and hypotheses; designed and performed valid experiments to answer their questions and test their hypotheses; collected, analyzed, and interpreted data derived from their experiments; drew conclusions from the data; and presented their findings in posters to communicate with the rest of their peers. They neither received lecturing on what inquiry is and how it is performed nor were given ready-made experiments to follow in answering their questions. Instead, they worked in the science lab for an extended amount of time aiming to produce reliable knowledge on the topic of boiling and peeling eggs that could not be found in books, the Internet, etc. Accordingly, teachers who receive authentic inquiry experiences – similar to those they will implement at a later stage in their classroom – are expected to be able to better translate their learning experiences to their students, better communicate and relate concepts to their students, and have a higher impact on enhancing students' interest and achievement in science (Dubner et al. 2001).

As far as the *core features* of the course are concerned, we took into account the five features introduced by Capps et al. 2012. Firstly, with regard to the feature of *coherence*, a serious attempt was made to follow the inquiry paradigm while designing the course, given that inquiry-based learning is manifested in the national curriculum of Cyprus and the science textbooks' units are considered to have been developed on the tenets of the inquiry-based approach. Thus, the compatibility and coherence of the aims and content of the course with the national curriculum (Ministry of Education and Culture 2016) was expected to facilitate and support teachers' teaching practice when entering the school for the purposes of their school practicum the following academic year. This conjecture is in line with what Grant et al. (as cited in Garet et al. 2001: 927) claimed; namely, if the sources used for teachers' training "…provide a coherent set of goals, they can facilitate teachers' efforts to improve teaching practice, but if they conflict they may create tensions that impede teacher efforts to develop their teaching in a consistent direction."

Secondly, the *developed lessons* feature might account for teachers' significant development of their PCK for teaching science as inquiry (Akerson et al. 2009; Basista and Mathews 2002). Specifically, during Phase 3 (teachers as reflective practitioners), the teachers were asked to develop lesson plans and curriculum materials that they would use in engaging a student in inquiry-based activities for the purposes of the science fair project. In developing their lesson plans, the teachers formulated learning objectives and designed activities that were aligned with the principles of inquiry-based learning (e.g., students would learn how to formulate investigative questions, test hypotheses, develop and apply the control of variables skill, design and perform controlled experiments, make inferences from the data collected, use evidence to develop explanations, etc.).

Thirdly, the *modeled inquiry* feature enabled teachers to experience firsthand how inquiry-based instruction looks like in practice and thus to appear more ready and confident in their own field of practice for scaffolding their students' learning pathways while involved in inquiry-based activities (Putnam and Borko 1997; Radford 1998). Specifically, the participating teachers (working in groups of four) were assigned to the role of learners during Phase 1 of the course and followed the specially designed curriculum to complete activities and evaluation tasks in an attempt to learn firsthand how inquiry-based learning looks like in the curriculum. The teachers discussed the progress of their work with the course instructors during "checkout points" placed in specific stages of the curriculum. The instructors aimed to engage teachers in semi-Socratic dialogues during the checkout points, instead of merely answering questions or providing the correct answers to the activities of the curriculum.

Fourthly, the *reflect* feature enabled teachers to become thinkers of their evolved conceptualizations of various aspects related to inquiry along the course and thus to develop sophisticated understandings of inquiry and inquiry-based learning (Clift et al. 1990). This was accomplished in Phase 1 during which the teachers were asked to keep reflective diaries to record their evolved understandings of inquiry, the questions and problems that emerged during working with the curriculum to answer the investigative questions they formulated, and their impressions from the course. In addition, when teachers were involved in the teachers as thinkers phase (Phase 2), they were asked to reflect on the curriculum in which they were engaged in the previous stage as learners from the lens of its pedagogical rationale and discuss how inquiry skills and knowledge were fostered within specific learning activities. Through reflection – which is considered of pivotal importance for the success of teachers' professional development courses – teachers are empowered to apply changes in both the content and the pedagogy of their practices (Fenstermacher 1994).

The fifth core feature of the course, namely, *transference* (which might be associated with the development of teachers' PCK for teaching science as inquiry), was integrated in the course when teachers adapted the format and structure of the curriculum they were engaged with (Phase 1), in order to design their own curriculum to be used during the engagement of an elementary school student in inquiry-based activities for the purposes of the science fair. During the design of their curriculum materials, they received feedback from the instructors on certain aspects of their work, which was proven beneficiary in transferring the PD materials and experiences in their own field of practice.

Lastly, the course not only focused in engaging teachers in inquiry-based activities but also on helping them develop specific content knowledge, including understanding of certain aspects of the nature of science, the nature of scientific inquiry, and the science concepts that related to the context of the curriculum (e.g., boiling, heat and temperature, egg protein denaturation, etc.). Developing teachers' content knowledge was an important aspect of the study's PD course. This is in accordance with Capps et al. (2012) work, which claims that if teachers' development of adequate content knowledge is neglected within their training, "they will likely be uncomfortable with the material they teach and have difficulties when they attempt to teach the material" (Capps et al. 2012: 302). Additionally, the course gave emphasis on promoting teachers' development of inquiry skills, such as the control of variables, the design of controlled experiments, the data interpretation, the identification of experimental flaws in given experimental designs, etc. Based on the reported findings that relate to teachers' development of inquiry skills and informed understandings of inquiry, the core feature that relates to *content knowledge* is another source to take into account when interpreting the findings of the present study.

Participatory Roles of Teachers

Teachers' learning gains in terms of inquiry skills, definitions of inquiry, and PCK for teaching science as inquiry development can also be attributed to the three participatory roles that they were assigned to during each of the three consecutive phases of the PD program. Firstly, during Phase 1, the teachers as *active learners* experienced themselves how inquiry-based instruction looked like. Their engagement with the specially designed curriculum "Boiling and Peeling Eggs" enabled them to walk through the same learning journeys that their students were expected to follow, and based on the analysis of their responses in the pre- and post-assessment tasks that sought to evaluate the level of their inquiry skills, it appeared that the learning experiences received during Phase 1 enabled the significant development of inquiry skills (see Table 1 for more details). This finding is in line with Loucks-Horsley et al. (1998) claim that engaging teachers as learners in the context of PD programs impact on the construction of meaningful knowledge about inquiry and skills for inquiry teaching.

Secondly, the designed activities that teachers engaged with as *thinkers* during Phase 2 (e.g., identification of the phases and subphases of the inquiry learning framework that the curriculum they worked with during Phase 1 was designed on, reflection on the learning objectives that were fostered through certain activities of the curriculum of Phase 1, etc.) seemed to have helped them to improve their understandings of what inquiry is (see Definition of inquiry, Phase 2, post findings in Table 1) and their knowledge of appropriate curriculum for inquiry (see Knowledge of appropriate curriculum, Phase 2, post findings in Table 1).

Lastly, the findings at the end of Phase 3, during which teachers were positioned as *reflective practitioners* and were asked to design and implement curriculum materials for the purposes of the science fair project and to collect evidence to evaluate and reflect on the effectiveness of their teaching, demonstrate significant development in all three aspects of their inquiry competence. Hence, as Freese (1999) put it, these learning gains that resulted because of teachers' role of reflective practitioners are expected to affect positively their inquiry practices both during their preservice and in-service teacher placement.

Lessons Learned

In this study we aimed at developing a PD program that could positively impact teachers' development of inquiry competence. It appears that our approach, particularly the *features of the course* and the three distinct *participatory roles* that teachers were assigned to during their engagement in the three consecutive phases of the PD program, was particularly effective. The latter has a number of implications on how PD programs on inquiry should be enacted. For example, it is apparent that teachers would benefit from each of the three aforementioned roles in a way that would enable them to capture the inquiry competence in its entirety, because each role has something unique to offer that the other two roles do not entail. Of course, further research with larger samples is needed for reaching more concrete and generalizable conclusions.

Acknowledgments This study was conducted in the context of the European project "Ark of Inquiry: Inquiry Awards for Youth over Europe," funded by the European Union (EU) under the Science in Society (SiS) theme of the 7th Framework Programme (Grant Agreement 612252). This document does not represent the opinion of the EU, and the EU is not responsible for any use that might be made of its content.

References

Abd-El-Khalick, F., Baujaoude, S., Duschl, R., Lederman, N. G., Mamlok-Naaman, R., Hofstein, A., et al. (2004). Inquiry in science education: International perspectives. *Science Education, 88*(3), 397–419.

Akerson, V. L., Hanson, D. L., & Cullen, T. A. (2007). The influence of guided inquiry and explicit instruction on K-6 teachers' views of nature of science. *Journal of Science Teacher Education, 18*, 751–772.

Akerson, V., Townsend, J., Donnelly, L., Hanson, D., Tira, P., & White, O. (2009). Scientific modeling for inquiring teachers network (SMIT'N): The influence on elementary teachers' views of nature of science, inquiry, and modeling. *Journal of Science Teacher Education, 20*(1), 21–40.

Bartos, S. A., & Lederman, N. G. (2014). Teachers knowledge structures for nature of science and scientific inquiry: Conceptions and classroom practice. *Journal of Research in Science Teaching, 51*(9), 1150–1184.

Basista, B., & Mathews, S. (2002). Integrated science and mathematics professional development programs. *School Science and Mathematics, 102*(7), 359–370.

Brown, A. L., & Campione, J. C. (1990). Communities of learning and thinking, or a context by any other name. In D. Kuhn (Ed.), *Developmental perspectives on teaching and learning thinking skills* (Vol. 21, pp. 108–126). New York: Karger.

Capps, D., Crawford, B., & Constas, M. (2012). A review of empirical literature on inquiry professional development. *Journal of Science Teacher Education, 23*, 291–318.

Clarke, D., & Hollingsworth, H. (2002). Elaborating a model of teacher professional growth. *Teaching and Teacher Education, 18*, 947–967.

Clift, R. T., Houston, W. R., & Pugach, M. C. (1990). *Encouraging reflective practice in education an analysis of issues and programs.* New York: Teachers College Press, Teachers College, Columbia University.

Constantinou, C., Kalifommatou, N., Kyriazi, E., Constantinide, K., Nicolaou, C., Papadouris, N., et al. (2004). *The science fair as a means for developing investigative skills: Teacher's guide.* Nicosia: Ministry of Education.

Crawford, B. A. (2000). Embracing the essence of inquiry: New roles for science teachers. *Journal of Research in Science Teaching, 37*(9), 916–937.

Crawford, B. A. (2007). Learning to teach science as inquiry in the rough and tumble of practice. *Journal of Research in Science Teaching, 44*(4), 613–642.

Davis, E. A., & Krajcik, J. (2005). Designing educative curriculum materials to promote teacher learning. *Educational Researcher, 34*(3), 3–14.

Davis, E. A., Petish, D., & Smithey, J. (2006). Challenges new science teacher's face. *Review of Educational Research, 76*(4), 607–651.

Driver, R., Asoko, H., Leach, J., Mortimer, E., & Scott, P. (1994). Constructing scientific knowledge in the classroom. *Educational Researcher, 23*(7), 5–12.

Dubner, J., Silverstein, S. C., Carey, N., Frechtling, J., Busch-Johnsen, T., Han, J., … Zounar, E. (2001). Evaluating science research experience for teachers programs and their effects on student interest and academic performance: A preliminary report of an ongoing collaborative study by eight programs. In MRS Proceedings. Cambridge: Cambridge University Press.

Fenstermacher, G. D. (1994). The place of practical argument in the education of teachers. In V. Richardson (Ed.), *Teacher change and the staff development process: A case in reading instruction* (pp. 23–43). New York: Teachers College Press.

Ferraro, J. M. (2000). Reflective practice and professional development. ERIC Digest. Available at http://www.ericdigests.org/2001-3/reflective.htm

Freese, A. R. (1999). The role of reflection on preservice teachers' development in the context of a professional development school. *Teaching and Teacher Education, 15*(8), 895–909.

Garet, M. S., Porter, A. C., Desimone, L., Birman, B. F., & Yoon, K. S. (2001). What makes professional development effective? Results from a national sample of teachers. *American Educational Research Journal, 38*(4), 915–945.

Glaser, B., & Strauss, A. (1967). *The discovery of grounded theory*. New York: Aldine de Gruyter.

Kazempour, M., & Amirshokoohi, A. (2014). Transitioning to inquiry-based teaching: Exploring science Teachers' professional development experiences. *International Journal of Environmental Sciences, 6*(3), 285–309.

Loucks-Horsley, S., Hewson, P. W., Love, N., & Stiles, K. E. (1998). *Designing professional development for teachers of science and mathematics*. Thousand Oaks: Corwin Press, Inc..

McDermott, L. C. (1996). *Physics by inquiry* (Vol. 1, 2). New York: John Wiley and Sons, Inc..

Ministry of Education and Culture, Cyprus. (2016). *Science curriculum for elementary education*. Cyprus: Ministry of Education and Culture, Cyprus.

NRC. (1996). *National science education standards*. Washington, DC: National Academy Press.

NRC. (2000). *Inquiry and the national science education standards*. Washington, DC: National Academic Press.

NRC. (2012). *A framework for K–12 science education: Practices, crosscutting concepts, and core ideas*. Washington, DC: National Academies Press.

Pedaste, M., de Vries, B., Burget, M., Bardone, E., Brikker, M., Jaakkola, T., et al. (2015). Ark of inquiry: Responsible research and innovation through computer-based inquiry learning. In T. Kojiri, T. Supnithi, Y. Wang, Y.-T. Wu, H. Ogata, W. Chen, S. C. Kong, & F. Oiu (Eds.), *Workshop Proceedings of the 23rd International conference on computers in education ICCE 2015* (pp. 187–192). Hangzhou: Asia-Pacific Society for Computers in Education.

Putnam, R. T., & Borko, H. (1997). Teacher learning: Implications of new views of cognition. In B. J. Biddle, T. L. Good, & I. F. Goodson (Eds.), *International handbook of teachers & teaching* (pp. 1223–1296). Dordrecht: Kluwer.

Radford, D. L. (1998). Transferring theory into practice: A model for professional development for science education reform. *Journal of Research in Science Teaching, 35*(1), 73–88.

Saad, R., & BouJaoude, S. (2012). The relationship between teachers' knowledge and beliefs about science and inquiry and their classroom practices. *Eurasia Journal of Mathematics, Science & Technology Education, 8*(2), 113–128.

Schön, D. (1983). *The reflective practitioner. How professionals think in action*. London: Temple Smith.

Schuster, D., Cobern, W., Applegate, B., Schwartz, R., Vellom, P., & Undreiu, A. (2007). Assessing pedagogical content knowledge of inquiry science teaching. In *Proceedings of the National STEM Assessment Conference on Assessment of Student Achievement*. Washington, DC: National Science Foundation and Drury University.

Simon, S., Campbell, S., Johnson, S., & Stylianidou, F. (2011). Characteristics of effective professional development for early career science teachers. *Research in Science & Technological
Education, 29*(1), 5–23.
Supovitz, J. A., & Turner, H. M. (2000). The effects of professional development on science teaching practices and classroom culture. *Journal of Research in Science Teaching, 37*(9), 963–980.

Investigating Science Teachers' Transformations of Inquiry Aspects When Implementing Research-Based Teaching-Learning Sequences

Alessandro Zappia, Giuliana Capasso, Silvia Galano, Irene Marzoli, Luigi Smaldone, and Italo Testa

Introduction and Aims

The aims of science teaching are gradually shifting from understanding of science contents to understanding of science as an interpretative body of knowledge. Thus, teachers should stress the importance of common methodologies besides the contents of specific scientific disciplines, to let students be reflective about the procedures they adopt while practicing science. A possible approach to fulfill this aim is Inquiry-Based Science Education (IBSE), which is acknowledged as central in many curriculum reform documents since the mid-1990s (NRC 1996; NGSS 2013).

Inquiry approaches allow students to emulate the way in which professional scientists carry out their work, by playing the role of researchers and investigating real problems grounded in real contexts. The focus of such an approach is to increase students' ability to design experiments and to support/rebut the validity of a particular thesis or adopted procedure, stressing the importance of teamwork and discussion among peers. However, at secondary school level, the implementation of such approach is limited, especially in countries where the curriculum is still content-oriented, often due to a lack of appropriate training courses that could support teachers (Blanchard et al. 2009; Ortlieb and Lu 2011).

A. Zappia (✉)
University of Camerino, Camerino, Italy

University Federico II, Naples, Italy
e-mail: alessandro.zappia@unicam.it

G. Capasso • L. Smaldone • I. Testa
University Federico II, Naples, Italy

S. Galano • I. Marzoli
University of Camerino, Camerino, Italy

© Springer International Publishing AG 2017
K. Hahl et al. (eds.), *Cognitive and Affective Aspects in Science Education Research*, Contributions from Science Education Research 3,
DOI 10.1007/978-3-319-58685-4_21

On this subject, literature has shown that such courses may benefit from the collaborative reflection on classroom practice and on how it might be improved (Anderson 2002; Crawford 2007). Previous studies investigated how teachers' beliefs and practices evolve in response to professional development programs (Carleton et al. 2008; Lakshmanan et al. 2011; Marshall and Smart 2013) and the impact on students' learning of such approaches (Blanchard et al. 2010). However, very little is known about the way specific aspects of inquiry teaching are implemented and "transformed" in classroom practice (Pintò et al. 2003). The identification of such transforming trends may be useful to inform training courses with possible factors that may favor/hinder adoption of IBSE. This paper investigates this issue through the following research question: *what are the aspects of inquiry teaching that teachers mostly accept or transform?*

Theoretical Framework

Teachers' Transformations

By the term "transformation", we refer to teachers' selection and re-organization of the features of a didactic innovation (Pintò et al. 2003). Such transformations may concern original designers' objectives (Van Den Akker 1998) or the use of specific teaching approaches, technologies, and languages (Sassi et al. 2005). Transformations naturally occur during class-work, as well as when teachers interact with their peers in training courses when interpreting the description/instructions of innovations. Our use of the term *transformation* is therefore different to that used by Marshall and Smart (2013: 132) since they focus on how "the beliefs and practices of teachers regarding inquiry-based instruction evolve" with time. Here, we focus on how teachers adopt and modify in their practice specific aspects of inquiry when implementing a research-based teaching-learning sequence (TLS; Méheut and Psillos 2004).

According to the above definition, inquiry-based approaches are particularly prone to transformations and modifications since teachers may:

- Find significant challenges in implementing them (Luft and Pizzini 1998)
- Hold contradictory beliefs with respect to inquiry pedagogy (Windschitl 2003)
- Have never experienced themselves this approach (Kleine et al. 2002)

As a consequence, IBSE approaches have often not been naturally adopted by teachers but rather transferred from a research or policy level at classroom level. Since IBSE is not merely a teaching method but involves also aspects of how Nature of Science and Scientific Inquiry are conceptualized (Lederman 2006; Schwartz and Crawford 2006), discrepancies between what is expected from teachers and what is really implemented by them (Capps et al. 2016) should hence be investigated in the wider frame of knowledge transfer.

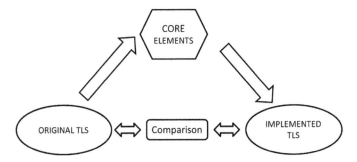

Fig. 1 The model used to analyze the classroom implementation of the proposed TLSs

Model of Inquiry Transfer and Adoption

To study teachers' transformations as defined above, we adopted a modified ARI (Adaption and Re-Invention) Model (Rogers 1983), which was originally introduced to describe industrial exchanges between different national contexts. The ARI model can be useful to describe how a TLS, developed for a certain educational context, is implemented in a different one (Testa et al. 2012).

The framework works as follows: first, "core" and "non-core" elements of a given TLS are identified. "Core" elements are those essential features of inquiry teaching which, according to designer's didactic aims, should not be changed while implementing a TLS, since they characterize it in a unique way. "Non-core" elements are complimentary features that mainly concern classroom management and activities timing and can be changed to better fit the TLS in school practice. In other words, core elements emphasize the extent to which the teacher implements specific inquiry-related aspects of a given TLS. Non-core elements identify whether the teacher organizes the classroom context as a place where students actively participate, work together, and fruitfully discuss the science content among themselves.

Then, after classroom implementations, the TLS, as enacted in teachers' practice, is compared with the original one to identify elements that have been adopted/transformed.

A graphical representation of the model implementation is reported in Fig. 1.

Methods

Training Workshops

Before implementing the activities, two groups of teachers were involved in a 30-h professional development (PD) course on IBSE, in academic years 2013/2014 and 2014/2015. Follow-up support during the implementation of the activities was provided for a total of 60 h in PD duration, which can be considered as satisfactory for

Table 1 Description of the teaching-learning sequences used in the study

Title	Context	What students do
Plants in space	Students, as researchers of a department of bio-astronomy, should develop suitable plans for a life-sustaining unit to use on possible future space flights	Investigations about dependence of photosynthesis on electromagnetic radiation wavelength
Mars-ology	Students, as researchers at the Institute of Planetary Research, are asked by the NASA to propose a research study to be carried out with a space probe on Mars	Investigations about dependence between viscosity of lava and shape of a volcano
Out of sight, out of mind	Students, as members of a city committee, are asked to study the risks of pollution due to landfills	Investigations about diffusion of polluting agents in soil
Green light	Students, as consultants of the Energy Efficient Lighting Committee, should produce a document about the main advantages of using compact fluorescent lamps	Investigations about energy dissipation of a compact fluorescent lamp and an ordinary filament lamp
Green heating	Students, as researchers of an advertising company, are asked to produce a document about advantages of solar thermal collectors for domestic use	Investigations about the role of materials in energy transfers between radiation and matter
ET Phone Earth	Students, as TV journalists, should prepare a television broadcast to discuss the possibility of extraterrestrial life	Argument about evidence in favor and against the existence of extraterrestrial life
Collision course	Students, as scientists of the "stellar center," are asked to produce a written report for NASA about possible risks for Earth due to collisions with asteroids	Investigations about momentum and energy of colliding objects

sustaining an inquiry practice (Lakin and Wallace 2015). During course workshops, teachers were introduced to IBSE through existing research-based modules (SHU 2009), properly adapted to the Italian educational context (Table 1). The teachers were engaged in the same activities of their students to increase awareness of possible difficulties their students would encounter when doing such activities (Stofflett and Stoddart 1994).

Overall, TLSs were grounded in real contexts and shared common meta-cognitive objectives related to the nature of scientific inquiry. More specifically, the aims of the TLS designers combined inquiry conceptions from NRC (1996) and NGSS (2013):

(i) To generate research questions and to make predictions, estimations, and/or hypotheses about the observed phenomenology
(ii) To justify the proposed experimental/simulation procedures
(iii) To use scientific ideas and models to explain phenomena
(iv) To draw meaningful conclusions from collected evidence

Corresponding core elements of the TLSs were:

- (CE1) using a meaningful context/problem to introduce the activities
- (CE2) generating research questions and hypothesis
- (CE3) collecting and analyzing data
- (CE4) discussing and communicating results using collected evidence
- (CE5) creating a community of practice

Non-core elements of the TLSs were:

- (NCE1) supporting students in using the materials about the addressed science content
- (NCE2) setting up of a fruitful and collaborative classroom culture
- (NCE3) managing time and school constraints
- (NCE4) designing of a suitable homework

After the PD workshops, all teachers accepted to implement at least one TLS in their classroom (for a minimum of 5 h). During TLSs in classroom delivery, videos and audios were recorded, and field notes were taken. Teachers were also invited not to use usual textbooks to avoid possible methodological inconsistencies with what was proposed in the TLSs.

Sample

Twenty volunteer secondary school science teachers participated in the PD in 2 years. The age of students involved was about 14–15 years. For the purpose of this study, the teachers had enough freedom in choosing the TLS, the timing, and number of teaching hours to match their didactic needs and to smooth the inclusion of the proposed activities into the school syllabus. Among the teachers involved in the PD workshops, only 13 were observed while implementing the TLS. The reason of such a choice was due to external duties and school constraints. Details are reported in Table 2.

Data Collection and Analysis

During classroom delivery, the activities were video recorded (overall about 60 h for all teachers). Audio recordings and field notes were also used as support to the analysis. Unlike the previous studies (Lakshmanan et al. 2011; Marshall and Smart 2013), we decided not to resort to observation protocols and self-reports since we were interested in teachers' actions that could reveal transformation of the inquiry elements of the TLSs. A five-step scheme was adopted to analyze teachers' transformations of inquiry aspects:

Table 2 Sample teachers' details

Teacher's short name	Subject taught	Type of school	Teaching experience (years)
T1	Math and Physics	Scientific Lyceum	20
T2	Biology and Earth Science	Scientific Lyceum	30
T3	Physics	Vocational School	15
T4	Chemistry	Vocational School	20
T5	Biology and Earth Science	Vocational School	20
T6	Chemistry	Vocational School	20
T7	Biology and Earth Science	Scientific Lyceum	20
T8	Biology and Earth Science	Scientific Lyceum	20
T9	Biology and Earth Science	Scientific Lyceum	25
T10	Math and Physics	Scientific Lyceum	25
T11	Biology and Earth Science	Scientific Lyceum	15
T12	Physics	Vocational School	10
T13	Physics	Vocational School	5

1. First, a coding category system was inductively developed using the constant comparative method (Strauss and Corbin 1998). A unit of analysis consisted of a period of 10 min in order to observe significant teachers' actions and students' reactions to teacher's triggers. About 30 frames (300 min, 5 h) for each teacher were analyzed. After 2 rounds of refinement, 19 categories were adopted. The emerging categories exemplify typical actions that the teacher and students carried out during the activities and correspond to specific inquiry *aspects*.

2. In the second step, using the ARI model, the categories were assigned to core and non-core elements of the implemented TLS. The assignment was made by comparing the coded action with the corresponding aims of the TLS.

Tables 3 and 4 display the adopted categories and the correspondence to the inquiry element.

3. In the third step, for each teacher, two raters scored independently the categories using a three-level classification. The aim was to evaluate the extent to which a specific inquiry aspect, as instantiated by the emerging category, was accepted or transformed with respect to TLS designers' intentions:

- *Low* if the aspect was completely modified
- *Medium* if the aspect was partially adopted or modified
- *High* if the aspect was fully adopted and not modified

Table 3 Categories emerged from the analysis corresponding to core elements of the TLSs

Indicator – the teacher	Short name	Core element
Encouraged to use elements of abstraction (e.g., symbolicrepresentations, diagrams, schemes)	Abstraction	CE4
Let students analyze data by themselves	Analysis	CE3
Let students focus on data collection	Collection	CE3
Involved students in active communication of their ideas	Communication	CE4
Introduced the context of the proposed activities	Context	CE1
Asked to make predictions, estimations, and/or hypotheses	Hypothesis	CE2
Guided students to develop their own investigations	Investigation-II	CE2
Guided students to generate their own research questions	Investigation-I	CE2
Required the students to link phenomena with scientific knowledge	Knowledge	CE4
Engaged students as members of a learning community	Members	CE5
Focused on students' argumentations according to research question	Question	CE4
Allowed students to be reflective about their researchprocedures	Reflection	CE4

Table 4 Categories emerged from the analysis corresponding to non-core elements of the TLSs

Indicator – the teacher	Short name	Non-core element
Stressed the importance of listening what others had to say	Discussion	NCE2
Required a final work consistent with what was requiredin the TLS	Homework	NCE4
Was able to manage students' questions and comments	Management	NCE2
Informed activities on the provided materials	Materials	NCE1
Acted as a resource person, working to support student investigations	Support	NCE1
Gave students enough time to discuss the materials	Time	NCE3
Focused on the required topic	Topic	NCE1

4. Then, a raw numerical score was assigned to each level (low = 1, medium = 2, high = 3). A score of 3 for a given category means that the corresponding inquiry aspect was fully adopted by the teacher, with no transformation. A score of 2 means a partial transformation, while a score of 1 means a complete transformation of the aspect. A final negotiation between the raters led to a consensual agreement with an inter-rater reliability Kappa value of 0.76. The final score for each aspect was calculated by averaging scores on all time frames where the representative category emerged. Examples of how teachers' actions in the time frames were coded and scored are reported in Zappia (in preparation).

5. Finally, to investigate whether a specific aspect was transformed across the sample, we introduced a numerical parameter, Δ. This parameter indicates the difference between the number of teachers who did not transform that aspect (score = 3)

Table 5 Scoring of core and non-core subscales

Δ values	The inquiry aspect was
$-9 \leq \Delta \leq -3$	Transformed
$-1 \leq \Delta \leq +1$	Partially transformed
$+3 \leq \Delta \leq +9$	Adopted

Table 6 Scoring of teachers' transformations subscales

Average score	The teacher has made
Score < 2 (majority of lows)	Heavy transformations
$2 \leq$ Score ≤ 2.5 (majority of mediums)	Some transformations
Score > 2.5 (majority of highs)	Almost no transformations

and teachers who partially or fully transformed it (score ≤ 2). Hence, depending on the calculated Δ value, each aspect was labeled as *heavily transformed, somewhat transformed,* or *adopted.* Minimum and maximum observed values were $\Delta = +7$ and $\Delta = -5$, which means that, in the best case (+7), the aspect was not transformed by ten teachers and transformed by three teachers, while, in the worst case (−5), the aspect was not transformed by four teachers and transformed by nine teachers. To give an idea of the degree of transformation of a given aspect, we assumed that if the majority of teachers (more than 7, $\Delta \leq -3$) transformed the aspect, it was transformed by our sample. If half of the teachers transformed the aspect, the aspect was partially transformed by our sample. If the number of teachers who transformed the aspect was between 5 and 4, the aspect was considered adopted by our sample (Table 5). Correspondingly, to assess each teacher's degree of transformation of core and non-core aspects, an overall score was obtained by averaging the scores in each aspect, using a three-level variable (Table 6).

Results

Figure 2 shows the Δ values for core and non-core aspects. Among the core aspects, *Collection* ($\Delta = 5$) was the only one adopted by the majority of teachers. Four aspects (*Hypothesis, Question, Investigation-II,* and *Members,* $\Delta = -1$) were partially adopted. The remaining seven aspects (*Knowledge, Investigation-I, Abstraction, Analyses, Context, Reflection, Communication*) were transformed by our sample ($\Delta = -5$ and -3). Among the non-core aspects, three were essentially adopted by the sample (*Support, Homework, Time,* $\Delta = +3$), three were partially adopted (*Discussion, Management, Materials,* $\Delta = +1$), and one (*Topic*) was transformed ($\Delta = -5$).

Average scores for core and non-core aspects for each teacher are reported in Table 7. Data show that, for all teachers, scores related to core aspects are generally lower than those of the non-core aspects.

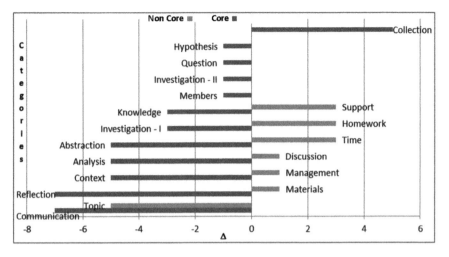

Fig. 2 Δ values for core and non-core aspects (see text for the definition of the delta parameter)

Table 7 Teachers' average scores for core and non-core aspects

Teacher	TLS Implemented	Average score	
		Core aspects	Non-core aspects
T1	Green light + green heating[a]	1.27	1.75
T2	Plants in space	1.27	1.25
T3	Green light + green heating[a]	2.54	2.75
T4	Mars-ology	2.09	2.38
T5	Plants in space	1.73	2.13
T6	Plants in space	2.64	2.88
T7	ET Phone Earth	1.45	2.25
T8	Out of sight, out of mind	2.55	2.50
T9	Collision course	2.73	2.38
T10	Plants in space	3.00	3.00
T11	Out of sight, out of mind	1.45	1.62
T12	Collision course	1.55	1.88
T13	Mars-ology	3.00	3.00

[a]T1 and T3 implemented two TLSs to reach the same number of implementation hours of the other teachers

Teachers T1, T2, T11, and T12 made *heavy transformations* in both core (average score = 1.37) and non-core (average score = 1.62) aspects, while teachers T5 and T7 did *heavy transformations* in core aspects (average score = 1.59) and *some transformations* in non-core ones (average score = 2.19). Teachers T3, T6, T10, and T13 made *almost no transformations* in both core and non-core aspects (average scores = 2.80 and 2.91, respectively); teacher T4 made *some transformations* in both core (score = 2.09) and non-core (score = 2.38) aspects. Finally, teachers T8 and T9 made *almost no transformations* in core aspects (average score = 2.64) and *some transformations* in non-core ones (average score = 2.44).

Four out of the six Biology and Earth Science teachers (T2, T5, T7, T11) made *heavy transformations* in both core (average score = 1.48) and non-core (average score = 1.88) aspects. The three Physics teachers (T3, T12, T13) made *some transformation* in core aspects (average score = 2.36). The two Math and Physics teachers show completely different performances: T1 made *heavy transformations* in both core and non-core aspects (average score = 1.27 for core aspects and 1.75 for non-core aspects), while T10 made *almost no transformation* (average score = 3 for both core and non-core aspects). Chemistry teachers show different performances, too: T4 made *some transformations* both in core (average score = 2.09) and non-core (average score = 2.38) aspects, while T6 made *almost no transformation* in both categories (average score = 2.64 for core aspects and 2.88 for non-core aspects).

Discussions and Conclusions

Overall, the reported evidence suggests that teachers in our sample were more resonant in non-core than in core aspects of the implemented TLS. A plausible reason for this result may be that the non-core aspects likely required teachers to make changes to their usual way of teaching that could be effectively implemented even after a short PD course. On the contrary, the core aspects were much more different with respect to the usual school practice and likely needed more follow-up sessions of discussion in the training course time to improve embedment in classroom practice (Marshall and Smart 2013).

At a closer look, the data show that, for the core aspects, only one (*Collection*) was completely adopted by the sample. This aspect refers to CE3 (collecting and analyzing data), an important feature of the proposed TLSs. Such result suggests that teachers, on average, stressed the importance of quantitative measurements during the implementation; however, it could be also that teachers in our sample likely viewed inquiry merely as laboratory (Lotter et al. 2007).

The most striking result is that the majority of the core aspects have been only partially adopted (*Hypothesis, Question, Investigation-II,* and *Members* with $\Delta = -1$) or completely transformed by the teachers (*Knowledge, Investigation-I, Abstraction, Analyses, Context, Reflection, Communication* with average $\Delta = -5$). In particular, among the transformed aspects, those with the smallest value of Δ ($= -7$) are *Reflection* and *Communication*. *Reflection* is a quite complex aspect of scientific inquiry to be implemented in classroom practice, since it includes both the skill of relating the aims of the investigation to experimental evidences and, at the same time, giving enough room to the students to analyze data by themselves and reflect on the procedures they adopted during the activities (CE4). Hence, this aspect measures the extent to which students have actually taken control of their own learning and self-assessment (Bybee 2006). Similarly, *Communication* refers to the capability of the teacher to let students communicate the results of their investigations in an appropriate way, their interpretation of evidences, and their conclusions referring to initial research question (CE4). Our result confirms that students, when implementing inquiry tasks, need support on reflective aspects and procedural knowledge (Arnold et al. 2014). A possible explanation for partial/non-adoption of such aspects by the teachers of our sample could be that investigations/research questions were often not proposed by students. In some cases, teachers even introduced the concepts related to the module before doing activities, giving students relevant hints on how to work and what to do.

The majority of non-core aspects were partially or fully adopted by the sample. In particular, the most adopted non-core aspects were *Time, Homework*, and *Support* ($\Delta = 3$). These aspects concern (i) a good management of the timings of the class activities (NCE3), (ii) the development of a final work consistent with what was required in the TLS (NCE4), and (iii) the role of teacher as resource person to help students use the TLS materials (NCE1). This was an expected result since the peer discussion was also implemented during the PD workshops when teachers acted as students, stressing the importance of the delivery of the final work, which was different for each TLS. Moreover, our results confirm those reported by Forbes and Davis (2007), which found that inquiry implementation in practice requires the use of a variety of resources to fit new materials into existing curricula.

Aspects partially adopted (*Materials, Management,* and *Discussion*, with $\Delta = 1$) were more related to the actual use of the provided materials and to the ability of teachers to create a favorable learning environment, by managing students' questions and comments, guiding the classroom discussions, and giving all students opportunities to express their ideas (NCE2). Such evidence suggests that effective implementation of inquiry requires significant changes in classroom organizational practices (Harris and Rooks 2010).

One non-core aspect was heavily transformed by the teachers, *Topic* ($\Delta = -5$). Such aspect concerns the capability of using the inquiry activities to focus on a specific scientific content (NCE1). Our findings suggest that most of the teachers exploited the activities mainly to introduce topics with which they were more familiar, although not targeted by the TLS.

The analysis of the transformations made by the individual teachers shows that T1, T2, T11, and T12 obtained low scores (less than two-thirds) in both core and non-core aspects. For the core aspects, in particular, T1 delivered the TLS as teacher-centered lessons, for instance, enacting the experimental activities as desk demonstrations, thus transforming the entire approach of the TLSs (CE2 and CE3). T2, T11, and T12 adopted a student-centered attitude, but, to get the "right" result within time constraints (Blanchard et al. 2009), poor data analysis was suggested or there was little effort to let students reflect on what had been done, again overturning TLS aims related to CE4. In the case of T1 and T2, also the fact that their students had much more familiarity with a teacher-centered lesson may have influenced the delivery of the TLSs. For non-core aspects, in particular, T1 and T2 did not focus on the required topic, since they exploited the TLS aiming mainly at students' assessment of syllabus contents (NCE1). Although relying on provided materials, T11 and T12 had difficulty in managing students' questions and comments (NCE2), so they focused for most of the time on topics not strictly related to what students were asked to investigate.

T5 and T7 made *heavy transformations* in core aspects (average scores, respectively, 1.73 and 1.45). In particular, they gave scarce importance to group work and often did not engage the rest of the class in general discussions and in communication and sharing of ideas (CE5). Moreover, they overemphasized experimental activities, with the consequence that students could not reflect on adopted procedure and on the validity of collected data (CE4).They also made *some transformations* in non-core aspects (average score = 2.13 and 2.25, respectively).

Only one teacher (T4) made *some transformations* in both categories while implementing the proposed TLS (average score = 2.09 for core aspects and = 2.38 for non-core aspects). For core aspects, in particular, T4 gave the students too much guidance in the formulation of the research questions and in the design of experiments to support the validity of their thesis (CE2). For non-core aspects, T4 did not focus on argumentation and general discussions, letting students implement activities as a sequence of predetermined steps (NCE2).

T8 and T9 adopted all core aspects, since they let students design investigations and experiments, to collect and interpret data (CE2 and CE3). These teachers, however, made *some transformations* in non-core aspects (average score = 2.50 and 2.38, respectively), since they did not give students enough time to discuss the materials provided, their hypothesis, and possible investigations (NCE3).

Four teachers (T3, T6, T10, T13) obtained high scores (greater than 2.5/3) in both core and non-core categories. These teachers after the course seemed more proficient in a wide range of inquiry practical skills, especially in managing students' questions and suggestions. They used a motivating context to introduce the activities (CE1), enabled students to generate and develop their investigations (CE2), and let them analyze data by themselves (CE3). Two teachers (T10 and T13), in particular, obtained the maximum score (3 for both core and non-core aspects). We note that, although both teachers taught Physics as subject in their classrooms, they had a very different teaching experience (25 and 5 years, respectively). A plausible reason for this evidence can be that, although both teachers were never engaged in any specific scientific inquiry-training course, they were involved in previous training activities in Physics education. In particular, teacher T13 followed a 2-year course (1 year was devoted to practicum in a school setting), while teacher T10 had been involved in after-class activities at the Physics department for about 30 h.

In conclusion, the present study adds to the research field by answering the question about how teachers embed inquiry aspects in their practice. While previous studies broadly investigated how PD programs changed teachers' practice toward a more inquiry-oriented one, we analyzed how teachers adopted and transformed inquiry approaches to embed them in their practice. Specifically, we provide a list of teachers' actions, extracted from an extended audio/video analysis, which, while not exhaustive, can be used to describe in detail the gap between intended aims of an inquiry-based TLS and the actual implementation in a classroom. The ARI model used in this study may help researchers find out what are the most difficult aspects of inquiry for the teachers to implement. More specifically, our findings may inform

PD courses by showing to teachers possible transformations of specific teaching-learning aspects that could occur when implementing inquiry activities. In such a way, teachers may become aware at an earlier stage of the training if their intended practice is resonant to core aspects of inquiry, thus avoiding overreporting or over-rating of their own teaching practices (Capps et al. 2016; Lakin and Wallace 2015).

The separation between core and non-core aspects also allows to interpret and justify previous research results, as the apparently contradictory ones by Marshall and Smart (2013). They found that teachers after a PD course evolved in their beliefs and conceptions about inquiry, but such improvement was only partially reflected in their practice. Our theoretical framework explains that, likely, beliefs and conceptions concerned mainly non-core aspects of inquiry practice, while to fully adopt an inquiry approach, core aspects are essential. Our results may also spread light on the factors at the basis of resonant teaching behaviors, which can plausibly favor the adoption of inquiry approaches in school practice. To this regard, a first follow-up to our study concerns the extent to which teachers' beliefs and conceptions affect implementation of core and non-core aspects of inquiry.

Further steps of this research study involve the analysis of post-implementation interviews and teachers' portfolios to collect more evidence about factors underlying transformations/difficulties in adopting inquiry-based approaches.

Acknowledgment Research carried out in this paper was supported by the EC project, Chain Reaction: A Sustainable Approach to Inquiry-Based Science Education, FP7-SCIENCE-IN-SOCIETY-2012-1, contract n° 321278.

References

Anderson, R. D. (2002). Reforming science teaching: What research says about inquiry. *Journal of Science Teacher Education, 13*(1), 1–12.

Arnold, J. C., Kremer, K., & Mayer, J. (2014). Understanding students' experiments—What kind of support do they need in inquiry tasks? *International Journal of Science Education, 36*(16), 2719–2749.

Blanchard, M. R., Southerland, S. A., & Granger, E. M. (2009). No silver bullet for inquiry: Making sense of teacher change following an inquiry-based research experience for teachers. *Science Education, 93*(2), 322–360.

Blanchard, M. R., Southerland, S. A., Osborne, J. W., Sampson, V. D., Annetta, L. A., & Granger, E. M. (2010). Is inquiry possible in light of accountability? A quantitative comparison of the relative effectiveness of guided inquiry and verification laboratory instruction. *Science Education, 94*, 577–616.

Bybee, R. W. (2006). Scientific inquiry and science teaching. In L. B. Flick & N. G. Lederman (Eds.), *Scientific inquiry and nature of science* (pp. 1–15). Dordrecht: Springer.

Capps, D. K., Shemwell, J. T., & Young, A. M. (2016). Over reported and misunderstood? A study of teachers' reported enactment and knowledge of inquiry-based science teaching. *International Journal of Science Education, 38*(6), 934–959.

Carleton, L. E., Fitch, J. C., & Krockover, G. H. (2008). An in-service teacher education program's effect on teacher efficacy and attitudes. *The Educational Forum, 72*(1), 46.

Crawford, B. A. (2007). Learning to teach science as inquiry on the rough and tumble of practice. *Journal of Research in Science Teaching, 44*(4), 613–642.

Forbes, C. T., & Davis, E. A. (2007). *Beginning elementary teachers' learning through the use of science curriculum materials: A longitudinal study*. Paper presented at the Annual Meeting of the *National Association for Research in Science Teaching* (NARST), April 2007, New Orleans.

Harris, C. J., & Rooks, D. L. (2010). Managing inquiry-based science: Challenges in enacting complex science instruction in elementary and middle school classrooms. *Journal of Science Teacher Education, 21*, 227–240.

Kleine, K., Brown, B., Harte, B., Hilson, A., Malone, D., Moller, K., et al. (2002). Examining inquiry. *Principal Leadership, 3*(3), 36–39.

Lakin, J. M., & Wallace, C. S. (2015). Assessing dimensions of inquiry practice by middle school science teachers engaged in a professional development program. *Journal of Science Teacher Education, 26*, 139–162.

Lakshmanan, A., Heath, B. P., Perlmutter, A., & Elder, M. (2011). The impact of science content and professional learning communities on science teaching efficacy and standards-based instruction. *Journal of Research in Science Teaching, 48*, 534–551.

Lederman, N. G. (2006). Syntax of nature of science within inquiry and science instruction. In L. B. Flick & N. G. Lederman (Eds.), *Scientific inquiry and nature of science* (pp. 301–317). Dordrecht: Springer.

Lotter, C., Harwood, W. S., & Bonner, J. J. (2007). The influence of core teaching conceptions on teachers' use of inquiry teaching practices. *Journal of Research in Science Teaching, 44*(9), 1318–1347.

Luft, J. A., & Pizzini, E. L. (1998). The demonstration classroom in-service: Changes in the classroom. *Science Education, 82*(2), 147–162.

Marshall, J., & Smart, J. (2013). Teachers' transformation to inquiry-based instructional practice. *Creative Education, 4*(2), 132–142.

Méheut, M., & Psillos, D. (2004). Teaching-learning sequences: Aims and tools for science education research. *International Journal of Science Education, 26*(5), 515–535.

National Research Council (NRC). (1996). *National science education standards*. Washington, DC: National Academy Press.

Next Generation Science Standards (NGSS). (2013). *A framework for K-12 science education*. Washington, DC: National Academy Press.

Ortlieb, E. T., & Lu, L. (2011). Improving teacher education through inquiry-based learning. *International Education Studies, 4*(3), 41–46.

Pintò, R., Ametller, J., Couso, D., Sassi, E., Monroy, G., Testa, I., & Lombardi, S. (2003). Some problems encountered in the introduction of innovations in secondary school science education and suggestions for overcoming them. *Mediterranean Journal of Educational Studies, 8*(1), 113–134.

Rogers, E. M. (1983). *Diffusion of innovations*. New York: The Free Press.

Sassi, E., Monroy, G., & Testa, I. (2005). Teacher training about real-time approaches: Research-based guidelines and materials. *Science Education, 89*(1), 28–37.

Schwartz, R. S., & Crawford, B. A. (2006). Authentic scientific inquiry as context for teaching nature of science: Identifying critical element. In L. B. Flick & N. G. Lederman (Eds.), *Scientific inquiry and nature of science* (pp. 331–355). Dordrecht: Springer.

Sheffield Hallam University (SHU). (2009). *Earth and Universe Project Research Briefs*. www.chreact.eu. Retrieved 26 Dec 2016.

Stofflett, R., & Stoddart, T. (1994). The ability to understand and use conceptual change pedagogy as a function of prior content learning experience. *Journal of Research in Science Teaching, 31*(1), 31–51.

Strauss, A., & Corbin, J. (1998). *Basics of qualitative research: Techniques and procedures for developing grounded theory*. London: Sage.

Testa, I., Molohidis, A., Lombardi, S., Psillos, D., Monroy, G., & Hatzikraniotis, E. (2012). Transfer of a teaching-learning sequence from Greek to Italian school: Do similarities in educational systems really help? In C. Bruguière, A. Tiberghien, & P. Clément (Eds.), *E-Book Proceedings of the ESERA 2011 Conference: Science Learning and Citizenship* (pp. 76–82). Lyon: European Science Education Research Association. ISBN: 978-9963-700-44-8.

Van Den Akker, J. (1998). The science curriculum: Between ideals and outcomes. In B. J. Fraser & K. G. Tobin (Eds.), *International handbook in science education* (Vol. 2, pp. 421–447). Dordrecht: Kluwer.

Windschitl, M. (2003). Inquiry projects in science teacher education: What can investigative experiences reveal about teacher thinking and eventual classroom practice? *Science Education, 87*(1), 112–143.

Zappia, A. (in preparation). *How secondary school science teachers implement scientific inquiry in classroom practice*. PhD dissertation manuscript in preparation. Camerino: School of Advanced Studies.

Part VI
Expanding Science Teaching and Learning

Cross-Curricular Goals and Raising the Relevance of Science Education

Nadja Belova, Johanna Dittmar, Lena Hansson, Avi Hofstein, Jan Alexis Nielsen, Jesper Sjöström, and Ingo Eilks

Introduction

In a recent review, Stuckey et al. (2013) analysed the science education literature of the last 50 years concerning the different meanings of the terms 'relevance' and 'relevant' when used in connection to science education. They suggested a definition and a model to understand the different dimensions and characteristics of relevance in science education. The definition is based on the question of whether the gain of certain science knowledge and related skills has or will have potential to make a difference to the learner's life and future. Stuckey et al. (2013) also identified three dimensions of relevance of science education, namely, individual, societal and vocational relevance. All of the three dimensions incorporate present and future, as well as intrinsic and extrinsic aspects.

In their analysis of the understanding of relevance for the science curriculum, Stuckey et al. (2013) showed that the emphasis between the different dimensions in science education is not always well balanced. More specifically, the societal dimension and essential parts of the vocational dimension were found to be neglected in

N. Belova • J. Dittmar • I. Eilks (✉)
University of Bremen, Bremen, Germany
e-mail: ingo.eilks@uni-bremen.de

L. Hansson
Kristianstad University, Kristianstad, Sweden

A. Hofstein
The Weizmann Institute of Science, Rehovot, Israel

J.A. Nielsen
University of Copenhagen, Copenhagen, Denmark

J. Sjöström
Malmö University, Malmö, Sweden

© Springer International Publishing AG 2017
K. Hahl et al. (eds.), *Cognitive and Affective Aspects in Science Education Research*, Contributions from Science Education Research 3,
DOI 10.1007/978-3-319-58685-4_22

many science teaching curricula worldwide. In an attempt to build up a comprehensive understanding of what makes science education relevant, Eilks and Hofstein (2015) edited a first book discussing numerous issues of relevance of science education in the case of chemistry education. This work examined 20 different foci in the curriculum and practices in chemistry education that were considered valuable contributions to raise the relevance of science education in general and chemistry education in particular.

This chapter discusses further perspectives for making science education relevant by considering certain cross-curricular goals from a chemistry education perspective. Cross-curricular goals are suggested to be generally accepted educational demands across all school subjects and all educational levels. They are not achievable by merely adding another unit to the curriculum. This chapter discusses challenges by four such cross-curricular goals for raising the relevance of science education and students' perception thereof, namely, (I) education for sustainability, (II) critical media literacy, (III) innovation competence and (IV) vocational orientation and employability. These four exemplary cross-curricular goals were selected since the discussion in this chapter emerged from a symposium at the ESERA 2015 conference covering corresponding issues. Directions for research and curriculum development will be also suggested that emerge from more thoroughly taking into account the perspective of cross-curricular goals in the science education.

Education for Sustainability

Education for Sustainability (EfS) builds on the notion that students have to be prepared to shape their society in a sustainable fashion. Sustainability issues have been on the political agenda for a long time, and recently the UN decided on '17 goals to transform our world' (see www.un.org/sustainabledevelopment/). In line with this and other policy documents, EfS offers a framework for an education aiming at the younger generation becoming responsible global citizens and professionals (Thomas 2009). All learning domains, and thus also science education, are asked to contribute to this demand, as it was explicitly discussed for the case of chemistry education by Burmeister et al. (2012). This makes EfS a cross-curricular goal which can contribute to enhancing the relevance of science education, especially its societal and individual relevance (Stuckey et al. 2013).

The modern society can be described as a globalized risk society which is characterized by the increasing complexity and unpredictable consequences of techno-scientific innovations and production. Examples of global environmental issues include climate change, ocean acidification, ozone depletion, decreasing freshwater quality and availability and distribution of resources (Rockström et al. 2009). Another example is the 'chemicalization' of our society, our bodies and nature. One could even talk about *chemical oppression*, where people are exposed to different risk-related chemicals, such as additives and contaminants, generally without being aware of the fact. Two examples which are sometimes reported on in the public debate include phthalates and perfluorinated chemicals. Our chemicalized society is part of the 'risk society'. Hodges (2015) talks about the 'chemical life'. Citizens must be able to manage health and environmental risks, of which scientific knowledge is only one of several necessary skills (Elmose and Roth 2005). Schooling has to focus on such 'epoch-typical' issues, which would make education relevant for the individual learner as well as for the society as a whole (Sjöström and Eilks 2017).

Science education, like all other domains of education, has to react to local and global challenges (Sjöström et al. 2016). Science education for sustainability needs to be interdisciplinary, holistic and value-driven, promoting critical thinking. It needs to prepare for critical, democratic, participatory decision-making (Sjöström et al. 2015). Decision-making on such complex issues involves taking many different aspects into consideration – not just scientific but also economic and ethical perspectives and also values and worldviews (Hansson et al. 2011). For this aim science education should contribute educating critical-democratic citizens who are ready for socio-political action (Bencze and Alsop 2014; Hodson 2011).

The literature suggests different ways to educate students for citizenship. Different kinds of action-oriented science education have been suggested (Bencze and Alsop 2014). For example, Santos (2009) discussed the implications of critical pedagogy (a Freirean perspective). In such education the aim is to teach with the goal of change and transformation in the society. It wishes to give students various skills of action in order to break different kinds of oppression, such as the chemical oppression described above. In education, chemical applications are often treated in a non-problematized manner (Sjöström 2013). When focused mainly on benefits but not on discussing risks, education acts to strengthen chemical oppression, rather than to break it. Critical pedagogy instead opens up for breaking oppression through education. By putting emphasis in school on chemical oppression, society will produce educated citizens who are able to assess and value situations, e.g. information about pollutants and chemical risks, and also to take educated action.

EfS includes working with socio-scientific issues (SSIs) in the science classroom. Starting from chemistry-related issues, Eilks and his coworkers have developed a framework for socio-critical and problem-oriented science teaching (Marks and Eilks 2009; see also Sjöström et al. 2015). The corresponding lesson plans for the whole range of secondary science education start with current, authentic and controversial problems being debated in public. Such topics include debates about alternative fuels, climate change, diets and risky chemicals in consumer products. All of the lesson plans integrate the learning of scientific content knowledge and experiments with reflections on the authentic handling of scientific information and decision-making in society based on authentic issues and media. This kind of lesson plan can serve as a starting point for teachers who are to teach for sustainability in science class, e.g. along the evaluation of different plastics in their ecological, economic and societal impacts as suggested in Burmeister and Eilks (2012). Such a teaching could give students a genuine chance to control their own lives and act as responsible citizens in society.

Critical Media Literacy

Achieving critical (scientific) media literacy is another cross-curricular goal. Mass media are one of the main channels which citizens use to access information in their everyday lives. Due to the growing role of the mass media in the first half of the twentieth century, media literacy was a topic first addressed internationally in 1946 by UNESCO as an element of fundamental education for everyone (Holmes et al. 1947). The growing role of the media in our lives has also influenced formal education during the past few decades. Media literacy has therefore been suggested as a topic to be embedded in school curricula all over the world (Belova et al. 2015).

Different definitions and concepts of media literacy can be traced in the literature. Two main dimensions of media literacy were outlined by UNESCO (2006) as 'reading' and 'writing' media. On the one hand, students need to understand the different forms of communication used in the media. On the other hand, they have to be able to create their own media products (such as blogs, videos or advertisements). The four concrete goals of media literacy are accessing, analysing, evaluating and creating media (Hobbs 2003). In line with the growing importance of media literacy in general, science in the media currently has also become a field in science education research. However, media research in science education still remains focused on specific media types, mostly traditional print-based media such as newspapers or magazines (McClune and Jarman 2012). We believe that science education in our modern digital age should expand its perspectives to other types of media.

Generally speaking, the objectives for strengthening learning about science in the media include both linking classrooms to the outside world and making science lessons more relevant in the eyes of the students. The corresponding skills encompass being able to critically evaluate media offers and gaining competence in dealing with understanding socio-scientific issues debated in both the media and the public arena; these goals strongly relate to the individual as well as societal dimension of relevant science education (Stuckey et al. 2013). In summary, dealing with science-related media and being able to discuss such items in a profound as well as critical way is viewed as a crucial requirement of becoming a modern citizen (Elliott 2006; McClune and Jarman 2012). Moreover, research has revealed that the media strongly affect pupils' perceptions of both science and its nature (Dhingra 2003).

As one potential step towards a broader view of media implementation in the science classroom, Belova and Eilks (2014) recently evaluated and justified the use of science-related advertising by conducting teacher interviews on the use of advertising in the science classroom and implementing an advertising-based lesson plan on natural cosmetics. Advertising often contains scientific information or is related to it, but students lack the necessary knowledge to recognize, understand and evaluate it in many cases. Advertising research suggests that activities intensively dealing with advertising and increasing student awareness of the corresponding knowledge can lead to a more critical view of advertising (Rozendaal et al. 2012). Anyhow, most available teaching ideas so far rarely focus on critical media literacy goals and the critical evaluation of advertising. Science education can bridge the gap between science content and advertising. However, such connections are seldom brought into being in the science classroom, although corresponding objectives can be derived from UNESCO's concept of 'advertising literacy' as a part of general media literacy including its scientific component (UNESCO 2011). The teaching ideas that have been suggested in the literature mostly focus on motivational purposes or on contextualizing content ('learning with advertising'). A deeper evaluation as well as creation of media ('learning *about* advertising') only appears very rarely in science education but is possible as the case by Belova and Eilks (2014) shows.

In a lesson plan on natural cosmetics by Belova and Eilks (2015), the students start learning chemistry by discussing the credibility of advertising claims such as 'Natural cosmetics are chemistry-free'. The students receive authentic slogans, which must be rated regarding their attractiveness, scientific background and credibility. Then the students learn about the composition and components of a generic skin cream and also about the use of controversial ingredients in cosmetics. The pupils prepare their own skin cream and have to decide whether or not to use various ingredients, e.g. parabens. For the final activity, the students receive a summary of positive and negative information about ingredients in cosmetics. They have to select what they want to use for creating an advertisement for their product. Finally, the students present their advertising and reflect the use of science-related information and claims in advertising in general.

Advertising used as both a learning tool and as a teaching topic has the potential to enhance the relevance of science education. Science education with and about advertising can contribute to the development of critical media literacy and critical consumerism. Dealing with broad-based advertising methods and with nontraditional media forms like the Internet or social media may open up new opportunities for better social contextualization of science learning. Therefore, science teaching should more often focus on science in the media. It should take more thoroughly into consideration the whole spectrum of today's media landscape beyond traditional print-based news media. Case studies on using Internet forums and blogs give indications that these media have similar potential as advertising has, even though these studies have not yet been fully evaluated and published. They all allow educators to open new and relevant perspectives in science learning and also contribute to a broader view of media literacy.

Innovation Competence

There is a growing consensus among policy-makers that relevant education for the twenty-first century also involves fostering students' innovation competence (EU-Commission 2010). While the term 'innovation' often strongly connotes economic and pecuniary enrichment (e.g. Nielsen and Holmegaard 2014; 2015), educational scholars have argued that innovation competence can be given an interpretation that is meaningful in a school context. This is meaningful in terms of existing science education and aligns with state-of-the-art learning theory. Roughly stated, in this interpretation innovation competence is the ability to address *authentic* issues from a *field of practice* from outside of school and work to suggest *improvements* to these issues in a way that can be *valuable* to the field of practice (see also Hobel and Christensen 2012). So science teaching *for* innovation competence would be a subtype of what educators know as inductive project-oriented learning (Prince and Felder 2006) and authentic science education (Bencze and Hodson 1999).

An example of a science teaching activity that aims at developing students' innovation competence is students in upper secondary school working together with a marine biologist from the local municipality. They cooperate in order to (I) identify and describe the main factors behind sea water quality, (II) generate solutions to improve sea water quality and (III) jointly reflect on the realizability and value potential of their suggested solution. In detail, the students are 'given' the problem by the marine biologist as an authentic problem form of his or her daily work life. The students could then start to investigate or explore the problem in order to identify the main factors – e.g. as an inquiry process. Based on which factors the students choose to mainly work with (there could be several, of course), the students then start to generate ideas for how these factors could best be alleviated – both from the perspective of biology and by drawing on knowledge and methods from other disciplines (e.g. chemistry, mathematics, social sciences). The key is that the students should be supported both in generating ideas and later prioritizing ideas based on their requisite realizability or value creation potential.

Many of the dynamics in such teaching will resonate with the dynamics that are ongoing when students address socio-scientific issues (Nielsen 2010, 2013). The key is helping students to draw on their science knowledge and skills in order to suggest how to handle an authentic issue. This includes self-reflection on the realizability and value potential of their suggestions. This means that science teaching for innovation competence is radical in the sense that the students are not engaged with science just because the teacher asks them to (teaching for innovation cannot be scholastic). Rather their work is prompted by a real field of practice. When working in this manner, reality communicates things to the students.

In a recent empirical study, Nielsen (2015) identified a way to operationalize innovation competence as a learning goal. This was done by identifying five dimensions of innovation competence, namely, creativity, collaboration, navigation, implementation and communication. Together, these aspects comprise innovation competence. This can be understood as the ability to 'individually or together with others, and on the basis of relevant knowledge, (a) generate ideas or solutions to an issue from an existing practice; (b) to asses these ideas in terms of their utility, realizability, and value-creation potential; (c) to implement selected ideas, possibly in sketch-form; and (d) to communicate ideas to different stakeholders' (Nielsen and Holmegaard 2015: 322). This framework can guide teachers' formative and summative assessment of the progressional development of students' innovation competence. In order to validate the framework, a study about of the possibility of using the framework as an assessment guide in examination settings in five different disciplines in upper secondary school was conducted, including biology and mathematics. It was found that in all five disciplines, three to four independent teacher-assessors were able to reliably assess students' innovation competence in the dimensions mentioned above. Cronbach's alpha values ranging from 0.71 to 0.95 were found.

Studies into the possibility of assessing innovation competence are an important first step in making science education more relevant. However, more curriculum development and research is needed in order to locate and understand the effects which the new learning goals have on teaching practices and students.

Career Orientation and Employability

Relevance in science education also includes vocational relevance (Stuckey et al. 2013) that includes career orientation as well as providing students with skills for further education and later employment. It has been suggested that high school science education needs to provide students with both orientation around and preparation for science-related careers. In order to become a science literate employee in science-related fields, students need to understand basic science. However, they also need to understand its related societal and technological applications and implications. In addition, students should develop skills that are needed for future employment (CORD 1999).

Over the years, the predominant model for teaching sciences was the 'structure of the discipline' approach that was developed since the late 1960s and early 1970s accordingly. This approach is still widely used in many countries (Gabel 1999). The main objectives of such science curricula were to train and prepare future scientists (chemists, physicists and biologists). These programmes are centred around a list of chemical key concepts with only very minor (if any) societal, environmental and/or technological manifestations. This approach was found to be adequate and motivating to only a small portion of the students. Primarily it attracted students who hoped or were already certain that they would embark on professional careers in the sciences and related fields. Unfortunately, until today, many high school science teachers still see their role almost exclusively in preparing their students for future careers in science-related professions (Gabel 1999; Eilks and Hofstein 2015).

Context-based and socio-scientific issue-driven approaches have the potential to enhance students' understanding with regard to the central role of the sciences in life and in many professions (Holman 1987). It gives them a view that science can have an impact on their future lives and potential careers. Suitable professional contexts also have the potential to expose the learners to much wider issues related to the concepts taught in school. One example that was applied in the context of teaching chemistry in Israel was the industrial case study approach. One of the cases is based on the Dead Sea industries, namely, 'bromine and its compounds' (Hofstein and Kesner 2006).

In the teaching and learning module by Hofstein and Kesner (2006), the learning about the chemistry of bromine and its compounds is embedded into the learning about the bromine industry in Israel. The students learn the chemical and techno-logical issues behind the production of bromine compounds from salt from the Dead Sea. However, the students are also provided with information about jobs and careers in the corresponding industry, and they also reflect the economic, societal and ecological implications of operating chemical plants near the fragile ecosystem of the Dead Sea. Dealing with bromine plants and other similar industries provides learners with orientation on career opportunities in many professions directly or indirectly related to the production and use of chemicals. The industrial case studies combine subject-matter learning (about the chemistry of bromine and its com-pounds) with environmental and societal issues, technological applications and ideas regarding careers and employment in industry (Hofstein and Kesner 2006).

Connecting science learning with industrial case studies provides ample oppor-tunities to vary the classroom environment for more relevant science education. This allows the development of a wider range of skills and strongly corresponds with the vocational dimension of the model of relevance by Stuckey et al. (2013). In addition, the students receive many opportunities to discuss the complexity of oper-ating such industrial plants. It provides insights into the work opportunities for engi-neers, chemists, lab technicians, geologists, environmentalists, lawyers, marketing experts, etc. One of the key elements in the industrial case studies is to provide the students with opportunities to experience the industrial environment firsthand. In order to accomplish this goal, three strategies were developed. The first brings the students to the industrial plant. While visiting the industrial plant, students are pro-vided with opportunities to interview different employees and to learn about their jobs and educational backgrounds. The second strategy is to engage the students in industrial ideas by bringing them into the classroom via the Internet. The third approach is to bring different lecturers (from the various industries) into the class-room to describe the various facets of plant operation. All the approaches need to teach about industry and its related issues in a holistic manner for providing career orientation and raising the employability of future workers.

On the basis of several years of research regarding the teaching and learning by industrial case studies, we have come to the conclusion (Hofstein and Kesner 2006) that there are two major obstacles that inhibit the effective implementation of these teaching strategies. The first hindrance is teachers who genuinely believe that they need to employ clear conceptual approaches. The second obstacle is the issue of how to exactly assess students' achievement while learning by industrial case stud-ies. This is in fact a call for science educators to develop assessment tools and CPD initiatives to provide teachers with effective strategies. Although there are obstacles in implementation, it is suggested to better connect chemistry learning with issues and experiences from outside the classroom for enhancing the relevance of chemis-try education and as a result increasing students' interest and motivation.

Conclusions

Cross-curricular goals as discussed above represent a great challenge for all the school subjects, including science education. Without a doubt, the science education curricula in many countries do not consider such cross-curricular demands thoroughly, e.g. for the case of EfS (Burmeister et al. 2012; Sjöström et al. 2015), nor do many of them attempt to raise the societal and vocational relevance of science education sufficiently (Stuckey et al. 2013). The ideas and studies discussed above provide evidence-based guidance and first strategies to integrate contemporary cross-curricular goals with science education. From the foci and cases discussed in this paper, it is suggested to open the chemistry curriculum more towards new and societal relevant questions and goals. Issues of sustainable development, from business and industry, or the media, have potential to enrich the chemistry curriculum and also to enhance the relevance of chemistry education. Such approaches should also be connected to open the chemistry classroom for the inclusion of new pedagogies, implementation of new types of media or even leaving the classroom for nonformal learning by field trips to chemically relevant sites and businesses. However, more thorough research, curriculum development and continuous professional development of teachers still seem to be needed on these exemplary cross-curricular goals as well as on others, e.g. health education. Corresponding research and curriculum development need to encompass both cooperation among the science subjects and a harmonizing of the science curricula. They also need to connect science more thoroughly to the humanities and social sciences. Stronger cooperation between domain-specific educational research and development with more general fields like media education or vocational education also needs to be addressed. Before this occurs, however, several things must happen. First, we need further discussion of which cross-curricular goals are especially challenging for science education. We also need to identify what the specific contribution coming from science education in general and chemistry education in particular for achieving each of the goals is and should be. Finally, we need to address and define exactly how the learning process can be started and how learning progress can be outlined and evaluated.

References

Belova, N., & Eilks, I. (2014). Promoting societal-oriented communication and decision making skills by learning about advertising in science education. *Centre for Educational Policy Studies Journal, 4*, 32–49.

Belova, N., & Eilks, I. (2015). Learning with and about advertising in chemistry education with a lesson plan on natural cosmetics – A case study. *Chemistry Education Research and Practice, 16*, 578–588.

Belova, N., Chang Rundgren, S.-N., & Eilks, I. (2015). Advertising and science education: A multi-perspective review of the literature. *Studies in Science Education, 51*, 169–200.

Bencze, L., & Alsop, S. (Eds.). (2014). *Activist science and technology education*. Dordrecht: Springer.

Bencze, L., & Hodson, D. (1999). Changing practice by changing practice: Toward more authentic science and science curriculum development. *Journal of Research in Science Teaching, 36*, 521–539.

Burmeister, M., & Eilks, I. (2012). An example of learning about plastics and their evaluation as a contribution to Education for sustainable development in secondary school chemistry teaching. *Chemistry Education Research and Practice, 13*, 93–102.

Burmeister, M., Rauch, F., & Eilks, I. (2012). Education for sustainable development (ESD) and secondary chemistry education. *Chemistry Education Research and Practice, 13*, 59–68.

Center for Occupational Research and Development. (1999). *Teaching science contextually*. Waco: CORD.

Dhingra, K. (2003). Thinking about television science: How students understand the nature of science from different program genres. *Journal of Research in Science Teaching, 40*, 234–256.

Eilks, I., & Hofstein, A. (Eds.). (2015). *Relevant chemistry education – From theory to practice*. Rotterdam: Sense.

Elliott, P. (2006). Reviewing newspaper articles as a technique for enhancing the scientific literacy of student teachers. *International Journal of Science Education, 28*, 1245–1265.

Elmose, S., & Roth, W.-M. (2005). *Allgemeinbildung*: Readiness for living in a risk society. *Journal of Curriculum Studies, 37*, 11–34.

EU-Commision. (2010). *Europe 2020: A strategy for smart, sustainable and inclusive growth*. Brussels: EU-Commission.

Gabel, D. (1999). Improving learning of chemistry through research: A look at the future. *Journal of Chemical Education, 76*, 548–554.

Hansson, L., Redfors, A., & Rosberg, M. (2011). Students' socio-scientific reasoning in an astrobiological context during work with a digital learning environment. *Journal of Science Education & Technology, 20*, 388–402.

Hobbs, R. (2003). Understanding teachers' experiences with media literacy in the classroom. In B. Duncan & K. Tyner (Eds.), *Visions/revisions: Moving forward with media education* (pp. 100–108). Madison: NTC.

Hobel, P., & Christensen, T. S. (2012). Innovative evner og de gymnasiale uddannelser [Innovative abilities and the upper secondary education programmes]. In M. Paulsen & S. H. Klausen (Eds.), *Innovation og Læring* (pp. 49–73). Aalborg: Aalborg Universitetsforlag.

Hodges, N. (2015). The chemical life. *Health Communication, 30*, 627–634.

Hodson, D. (2011). *Looking to the future: Building a curriculum for social activism*. Rotterdam: Sense.

Hofstein, A., & Kesner, M. (2006). Industrial chemistry and school chemistry: Making chemistry studies more relevant. *International Journal of Science Education, 28*, 1017–1039.

Holman, J. (1987). *Science and technology in society. General guide for teachers*. Hatfield: ASE.

Holmes, H. W., et al. (Eds.) (1947). *Fundamental education, common ground for all peoples* (Report of a special committee to the Preparatory Commission of the UNESCO, Paris, 1946). New York: Macmillan.

Marks, R., & Eilks, I. (2009). Promoting scientific literacy using a socio-critical and problem-oriented approach to chemistry teaching: Concept, examples, experiences. *International Journal of Environmental and Science Education, 4*, 131–145.

McClune, B., & Jarman, R. (2012). Encouraging and equipping students to engage critically with science in the news: What can we learn from the literature? *Studies in Science Education, 48*, 1–49.

Nielsen, J. A. (2010). Functional roles of science in socio-scientific discussions. In I. Eilks & B. Ralle (Eds.), *Contemporary science education* (pp. 83–96). Aachen: Shaker.

Nielsen, J. A. (2013). Delusions about evidence: On why scientific evidence should not be the main concern in socioscientific decision-making. *Canadian Journal for Science, Mathematics, and Technology Education, 13*, 373–385.

Nielsen, J. A. (2015). Assessment of innovation competence: A thematic analysis of upper secondary school teachers' talk. *Journal of Educational Research, 108*, 318–330.

Nielsen, J. A., & Holmegaard, H. T. (2014). On the educational goals of innovation and employability. In I. Eilks & B. Ralle (Eds.), *Science education research and education for sustainable development* (pp. 169–180). Aachen: Shaker.

Nielsen, J. A., & Holmegaard, H. T. (2015). Innovation and employability: Moving beyond the buzzwords – A theoretical lens to improve chemistry education. In I. Eilks & A. Hofstein (Eds.), *Relevant chemistry education – From theory to practice* (pp. 317–334). Rotterdam: Sense.

Prince, M. J., & Felder, R. M. (2006). Inductive teaching and learning methods: Definitions, comparisons, and research bases. *Journal of Engineering Education, 95*, 123–138.

Rockström, J., Steffen, W., Noone, K., Persson, Å., Chapin, F. S. I., Lambin, E., et al. (2009). Planetary boundaries: Exploring the safe operating space for humanity. *Ecology and Society, 14*, 32–63.

Rozendaal, E., Buijzen, M., & Valkenburg, P. M. (2012). Think-aloud process superior to thought-listing in increasing children's critical processing of advertising. *Human Communication Research, 38*, 199–221.

Santos, W. L. P. (2009). Scientific literacy: A Freirean perspective as a radical view of humanistic science education. *Science Education, 93*, 361–382.

Sjöström, J. (2013). Towards *Bildung*-oriented chemistry education. *Science & Education, 22*, 1873–1890.

Sjöström, J., & Eilks, I. (2017). Reconsidering different visions of scientific literacy and science education based on the concept of *Bildung*. In J. Dori, Z. Mevarech, & D. Baker (Eds.), *Cognition, metacognition, and culture in STEM education*. Dordrecht: Springer.

Sjöström, J., Rauch, F., & Eilks, I. (2015). Chemistry education for sustainability. In I. Eilks & A. Hofstein (Eds.), *Relevant chemistry education – From theory to practice* (pp. 163–184). Rotterdam: Sense.

Sjöström, J., Eilks, I., & Zuin, V. (2016). Towards eco-reflexive science education. *Science & Education, 25*, 321–341.

Stuckey, M., Mamlok-Naaman, R., Hofstein, A., & Eilks, I. (2013). The meaning of 'relevance' in science education and its implications for the science curriculum. *Studies in Science Education, 49*, 1–34.

Thomas, I. (2009). Critical thinking, transformative learning, sustainable education, and problem-based learning in universities. *Journal of Transformative Education, 7*, 245–264.

UNESCO. (2006). *Media education. A kit for teachers, students, parents and professionals*. http://unesdoc.unesco.org/images/0014/001492/149278e.pdf. Accessed 01 Apr 2016.

UNESCO. (2011). *Media and information literacy. Curriculum for teachers*. http://unesdoc.unesco.org/images/0019/001929/192971e.pdf. Accessed 01 Apr 2016.

Integrating Mathematics into Science: Design, Development and Evaluation of a Curriculum Model

Gráinne Walshe, Jennifer Johnston, and George McClelland

Introduction

In today's fast-changing world, the ability to integrate STEM concepts is said to be a prerequisite for solving the complex and multidisciplinary problems society faces (Rennie et al. 2012). One of the biggest challenges for second-level education is that few guidelines or models exist for teachers regarding how to teach using STEM integration approaches in their classroom (Roehrig et al. 2012). Integration of science and mathematics, for example, has long been recommended as a way to make meaningful connections between these two subjects for students, but models for how to integrate them have been found to vary considerably (Czerniak and Johnson 2014; Pang and Good 2000). Second-level teachers do not often get the opportunity to experience integration nor do they have ready access to integrated instructional materials (Stinson et al. 2009). This research therefore was concerned with the design, development and evaluation of a curriculum model to assist teachers in supporting student transfer of mathematical knowledge and skills into their learning of lower second-level science in Ireland.

The theoretical premise of this research is that disciplines and subjects have epistemological and methodological boundaries. Students should become knowledgeable in the explanatory framework of the discipline of science while nonetheless being given the opportunity to draw on and apply interdisciplinary knowledge (Moore 2011). This premise formed the basis of the design of the curriculum model. The model was evaluated by teachers, principals, subject experts and other curriculum stakeholders across a number of microcycles of research during a doctoral

G. Walshe (✉) • G. McClelland
University of Limerick, Limerick, Ireland
e-mail: grainne.walshe@ul.ie

J. Johnston
University of Lincoln, Lincoln, UK

© Springer International Publishing AG 2017
K. Hahl et al. (eds.), *Cognitive and Affective Aspects in Science Education
Research*, Contributions from Science Education Research 3,
DOI 10.1007/978-3-319-58685-4_23

309

research project (Walshe 2015). The initial analysis of the data indicated that, with some minor modifications, the model could support teachers to integrate science and mathematics in terms of providing them with a valid mechanism for coordinating subject planning across the curriculum and for developing their own integrated activities. However, attitudinal, structural and affective issues that would impact on the likelihood of teachers implementing the model came to the fore across the various microcycles of research. This chapter aims to answer the following research question: 'what are the key elements emerging from participants' perceptions of the feasibility of crossing the sociocultural and disciplinary boundaries of school science and mathematics?'

Theoretical Framework

The renewed interest in STEM integration in recent years has sometimes led to conflation of desirable outcomes for students (development of their capacities to problem-solve and utilise critical thinking in interdisciplinary contexts) with dissolution of disciplinary boundaries as a principle for curriculum organisation (Young and Muller 2010). However, this can lead to students missing out on critical steps in their knowledge of the disciplines (Basista and Mathews 2002; National Research Council 2014). The theoretical framework developed in this project locates the rationale for interdisciplinary curriculum design within a knowledge-based rather than purely applications-led curriculum design (Young and Muller 2010). In this framework, interdisciplinarity is not posited as an *alternative* to disciplinarity in curriculum (Moore 2011) but rather builds upon it. In this way students are not denied access to the abstract theoretical disciplinary knowledge required in order to apply it in real-world and multidisciplinary contexts (National Research Council 2014). Moreover, the principle of curriculum coherence, which implies progressive learning of concepts and practices over time and across subjects, underpins the model (Geraedts et al. 2006).

Curriculum Model

This research focused on designing a curriculum model for integrating mathematics into science. Mathematics is integral to the modes of inquiry science employs in its unique concern with developing empirically compatible models that allow us to explain and predict phenomenon in the natural world (Irzik and Nola 2011; Osborne 2014). The identification of curricular overlaps and subsequent development of a progression of essential mathematical practices relevant to science in the context of the Irish lower second-level curriculum (age 12–15 years) is a central aspect of the model. Importantly, the integration is focused on the needs of the main subject, science in this case, but with explicit integration of mathematical concepts, skills and language, in order to support student transfer of knowledge (National Research Council 2014).

The curriculum model is composed of the following curricular artefacts:

- A syllabus map of the Irish lower second-level mathematics curriculum on to the lower second-level science curriculum, and an Integrated Science and Mathematics Teaching and Learning Sequence for Irish lower second level
- Three exemplary integrated lesson units that explicitly integrate mathematical concepts and skills from the mathematics curriculum into lower second-level science topics

The model is intended to support curriculum designers, schools and teachers in taking a systematic and coherent approach to integrating science and mathematics, but which does not deprive students of access to important subject-specific learning. The primary purpose of the curriculum model is with providing schools and teachers with a valid and practical mechanism for integrating mathematics into science.

Disciplinary Disconnect and Boundary Crossing

In order to contextualise the participants' reactions to the curriculum model, the literature on both science and mathematics integration, and on boundary crossing between different zones of professional practice, is delineated in this section. The integration literature indicates that teachers generally have positive perceptions of the value of implementing science and mathematics integration for their students, in terms of, for example, the impact it could have on student motivation to learn and on student understanding of the utility of these subjects (Czerniak and Johnson 2014; Lee et al. 2013). However, a number of barriers to integration have been identified. At the level of the school, these include lack of time for collaborative planning, lack of resources and lack of supportive school structures (Czerniak and Johnson 2014; Pang and Good 2000; Ní Ríordáin et al. 2016). Increasingly important are the systemic barriers imposed by the imposition of assessment measures that only focus on discipline-based learning (Berlin and White 2012; Lee et al. 2013), and which in the Irish context have resulted in an educational culture of teaching to the test, with the curriculum narrowed to the highly prescribed subject-specific syllabuses (Ní Ríordáin et al. 2016). At the level of the individual teacher, the literature emphasises that teacher subject content knowledge can be insufficient to integrate both subjects fully (National Research Council 2014; Pang and Good 2000; Stinson et al. 2009). Moreover, teachers often do not have opportunities to experience education for integration (Basista and Mathews 2002). Not surprisingly, teachers can have negative perceptions of the practicality of implementing science and mathematics integration in the light of these various barriers (Czerniak and Johnson 2014).

The literature on STEM integration tends to view the disconnect so often found between school subjects as resulting from obstacles such as these that must be overcome, with a strong emphasis on teacher deficits of knowledge and attitude. However, the literature on professional boundary crossing permits a more nuanced view of what is at stake for teachers in implementing integration. In addition to the disciplinary differences between school subjects, there are also sociocultural boundaries of professional practice between science and mathematics teachers (Hobbs 2013). Teachers' communities of practice at second level are defined by the subjects they teach and their relationship to them (Hobbs 2013). The literature on the concept of 'boundary crossing' between different communities of practice (Akkerman and Bakker 2011; Kent et al. 2007; Nicolini et al. 2012) provides a framework for understanding the mechanisms by which practitioners interact across subject boundaries. In their review of this literature, Akkerman and Bakker (2011) define boundaries as sociocultural differences that effectively establish and define expertise and rights of participation within different domains of professional practice. Boundaries between contexts constitute an ambiguous and ill-defined space, but they also represent opportunities for creativity and learning. Boundary crossing then entails the activities of individuals or groups to establish continuity in action or interaction across different practices (Akkerman and Bakker 2011). This is often mediated via boundary objects, artefacts that perform a bridging function between communities (Kent et al. 2007).

Teaching a subject with which you are unfamiliar is not dissimilar to the experience of teaching out of field. Teachers' subject identities are formed through participation in and recognition by the relevant subject subculture (Hobbs 2013). Thus, teaching out of field has the potential to disrupt a teacher's sense of competence, self-efficacy and well-being (Pillay et al. 2005 cited in Hobbs 2013). Teachers who participate in activities at the boundaries between subjects can have strong emotional reactions to the perceived dilution of their subject and hence their own expertise (Edwards 2011; Olson and Hansen 2012). However, from a boundary-crossing perspective the point is not to seek to dissolve subject boundaries; but rather a boundary is potentially a site of professional and social learning (Akkerman and Bakker 2011). Hobbs (2013) draws on the boundary-crossing literature to theorise that out-of-field teachers can establish the action or interaction across subject boundaries through the use of boundary objects as professional learning opportunities that can lead to identity expansion and their reconceptualisation of their practice (Hobbs 2013).

In summary, crossing boundaries of professional practice can be a very fraught activity for teachers, but boundary objects, such as the curriculum model in this project, may support them in negotiating school subject boundaries.

Methodology

This chapter outlines some of the findings from a doctoral research study (Walshe 2015). The overarching methodology of the doctoral study was Curriculum Development Research, a variant of Educational Design Research, characterised by

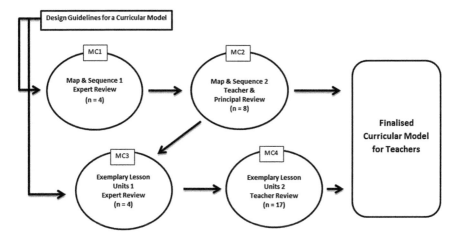

Fig. 1 The four microcycles (MC) of iterative design and formative evaluation of the curriculum model

iterative design and formative evaluation of curricular interventions in complex real-world settings (van den Akker 2013). Working with practitioners is an essential part of this methodology. Here, this included teachers and also principals in their capacity as primary organisers of a school's curriculum. Figure 1 illustrates the four microcycles of iterative design and formative evaluation of the curriculum model. Taken together, these microcycles were concerned with evaluating the model's validity, practicality and expected effectiveness (Plomp and Nieveen 2013) as a process for assisting schools and teachers to integrate mathematics into science.

This chapter presents data primarily drawn from the convenience samples of end users who participated in microcycle 2 and microcycle 4. In microcycle 2, four science and/or mathematics teachers and four school principals reviewed the map and sequence. Participant data was collected mainly from interviews. In microcycle 4, 17 science teachers (nine of whom also taught mathematics) reviewed the integrated lesson units. Participant data was collected from interviews (n = 9) and a survey (n = 16). Table 1 presents a summary of the data collection methods, research focus and approach to data analysis in microcycles 2 and 4.

The primary researcher coded and analysed the data, with checks performed by the other researchers and with codes reviewed by and discussed with a critical friend. The emergent themes and categories from the first cycle of analysis were developed deductively, utilising theoretical contributions from the literature on boundary crossing, in a second cycle of data analysis (Fereday and Muir-Cochrane 2008). The themes thus developed are outlined below.

Table 1 Summary of data collection methods and analysis

Data collection	Semi-structured interviews; survey with 28 Likert items and 3 open questions
Focus of the microcycles	Participants' perceptions of strengths, weaknesses and any recommendations for change of the particular curricular artefacts under review. More generally their perceptions of the feasibility and expected effectiveness of implementing the curriculum model in Irish second-level schools and classrooms, with a view to refining the model as appropriate.
Sampling procedure	Convenience sampling (volunteers)
Type of analysis	Thematic analysis (NVivo) of qualitative data. First cycle of analysis: evaluation coding strategy (Saldana 2013). Second cycle: developing emergent themes from the first cycle deductively (Fereday and Muir-Cochrane 2008). Responses to Likert items: means, medians, standard deviations and boxplots generated (SPSS). Quantitative and qualitative data cross-referenced to build a deeper picture of teachers' reactions to the exemplary integrated lessons.

Results

The focus in this chapter is on the two major themes that emerged from the second cycle of data analysis, 'disciplinary disconnect' and 'boundary crossing'. The 'disciplinary disconnect' between science and mathematics is comprised of subject subculture and teacher subject identity subthemes. The second theme indicated the potential of the curricular materials to promote interdisciplinary 'boundary crossing', through the conversations they encourage across subject lines – both external, when participants consulted their colleagues, and internal, when they began to indicate their reconceptualisation of their subject.

Disciplinary Disconnect

The factors identified as potentially militating against the implementation of the model by schools and teachers are a set of interdependent systemic, cultural and attitudinal factors.

Culture of Subject Separation

The data indicated that there are very insular or closed subject subcultures in some of the schools where participants worked. A comment from a science-only teacher illustrates this divide: 'there is no link between the maths and science department in the school and we do not plan together'. School factors such as the school size and the number of shared personnel across the science and mathematics departments impacted teachers' perceptions that integration of these subjects would be practical to implement in their school. As one teacher pointed out:

> Teachers and students work in very atomised/discrete units. This has a cultural and exam based root. It will require large effort to change this… Also this is accentuated by the large [number] of maths/science teachers in any school. (Science-and-mathematics teacher)

School leadership also emerged as an important factor; of the four principals consulted, two indicated that they did not see integration as a practical proposition (despite positive reactions to the curricular planning documents). One said 'the school would not put science and maths together usually… The Sequence would work if science and maths are [taught by] the same person… Another teacher wouldn't have the interest to find out the science links, etc.' This principal detached herself from any role in leading the implementation of integration in the school; locating this within the happenstance of individual teacher interest. These are likely to be the kind of incidental factors that will support or prevent teachers integrating subjects.

These factors are underscored by the subject-based examination culture in Irish schools, as mentioned by the teacher quoted above. One principal was emphatic that unless the curriculum model was based on the national subject syllabuses, teachers would not implement it, which five teachers also made clear. For some of the teachers, this was compounded by their concern that their lack of knowledge of either science or mathematics would make it difficult and time-consuming to plan for integration. Similarly, the two principals who were most pessimistic about the possibility of implementing integration in their schools had no background in either subject, suggesting lack of subject knowledge may also be an influencing factor on leadership capacity.

Teacher Subject Identities

All of the above will impact upon teacher subject identity, and there was evidence in the data that some teachers had a more insular subject identity. The relevance of teaching particular mathematical concepts and skills in science was questioned, if, as one teacher said, 'they're going to do it anyway in maths'. This is not just to do with lack of mathematical knowledge as even teachers who teach both subjects had strong perceptions of what constitutes science or mathematics. This may be the reason why one teacher perceived that the lessons were too mathematical:

> Your different ways of doing it I thought was interesting, too. But does this mean that you're doing maths in the science class? (Science-and-mathematics teacher)

This teacher is very familiar with the mathematical concepts, but it does not fit with his understanding of what science entails to explicitly teach mathematics in science.

Teachers also indicated that there are affective factors for them in incorporating mathematics into their science teaching, especially if they do not teach mathematics.

A science-only teacher explained that the day when she teaches her first year students (age 12–13 years) how to draw a solubility curve 'is the most stressful day of the year, because they haven't a clue... So then I don't want to be the first person doing this with them'. Several science-only teachers expressed similar anxiety about teaching unfamiliar mathematical concepts and skills.

However, some of the attitudes were more negative: a science-and-mathematics teacher took the view that teaching mathematics in science class 'would waste too much time on maths side instead of "science" aspect', indicating a dismissive attitude towards incorporating mathematics into science. Similarly changing the sequence of topics to accommodate learning in another subject may be problematic for teachers. Another science-and-mathematics teacher did not see why he would change the 'natural order' of his science topics to suit integration with mathematics. These kinds of comments suggest some tension between subject subcultures, and some defensive guarding of the subject boundaries, even by some of the teachers who have knowledge of both.

Boundary Crossing

The 'boundary-crossing' category delineates those themes in the data that demonstrate that participation in the review of the model (or one of its elements) provided participants with a locus for interdisciplinary dialogue in the sense that it provided them with an avenue of entry into the discourses at the boundaries between school science and mathematics.

Facilitating Planning, Coordination and Collaboration

Some participants took tentative steps to cross the subject boundaries of science and mathematics as a result of engaging in the review of the map and sequence and/or the integrated lessons. Some participants consulted formally with their colleagues in staff meetings about these documents, invited the researcher to give a workshop on the concepts they embodied or used the ideas to develop their own workshop and lesson plans, while two of the principals considered adapting them for implementation in their school. Participants were also prompted to consult their colleagues informally for further explanation and information about the other domain. Two of the subject matter experts consulted colleagues about the first- and second-level science and mathematics curricula, respectively. One science teacher asked her mathematics colleague for a 'mini-tutorial' as she put it to explain the mathematics concepts to her. They subsequently had an informal conversation about how these concepts could play out in either of their subjects. Several teachers indicated the casual collegiality that characterised their boundary crossing. An example of this is when a science teacher chatted to his colleague about the mathematical topics she was teaching to their shared students, when they met at the school photocopier.

Two science-and-mathematics teachers took on the role of brokers at the boundaries of the two subjects. The first of these volunteered to trial run the integrated lesson ideas at a very early stage of their development, contributing hugely to the design-thinking underpinning the model. The second decided, having being introduced to the integrated lesson units, to recruit four science teachers from her school to implement them in their classrooms as part of their formative evaluation. She also coordinated the group, dispensing materials and gathering data instruments from them to return to the researcher.

It is important to highlight that none of these efforts to cross the subject boundaries were required of the participants as part of their participation in the review process but rather were voluntary initiatives they themselves undertook.

Reconceptualisation of One's Subject

Most notably, many participants indicated small changes of perspective whereby they began to reconceptualise a subject because of their participation in the review process:

> Stuff like the leaf-and-stem graphs, I've never encountered them before. I looked it up and I was like…that'll be brilliant for doing the heart rates, just as an example…. It's just a lovely way of presenting stuff. (Science teacher)

This opening up of new perspectives on one's subject(s) was not limited to the science-only or mathematics-only teachers. Science-and-mathematics teachers also remarked on the ways in which the model provided them with new ideas. One teacher perceived that the lesson units could provide a useful model for teachers who teach both subjects:

> Even just to get people- even for me, just to realise there for the first time ever that the data-handling cycle is exactly the same thing as the scientific inquiry. I was going, "Oh, yeah. I never thought of that before,"…But we don't go looking for [the connections]. We're too busy teaching in our own little narrow tunnel, you know? (Science-and-mathematics teacher)

The quote indicates that having knowledge of both subjects does not imply that teachers automatically consider the ways in which they can integrate them together. For some teachers, the feedback interview clearly indicated the internal dialogue and process of reflection on their practice that had been initiated by their review of the model. As one science teacher put it:

> I like the idea of the integration of the two subjects but I feel the science course is long enough…without adding the maths parts that are not relevant… Maybe I just need to embrace it and decide to make it more relevant! (Science teacher)

This teacher has been prompted to consider changing her practice as a result of her participation in the evaluation of the curriculum model. She and other teachers thus demonstrated the beginnings of an internal dialogue regarding subject boundaries, suggesting a reconceptualisation of their perception of their subject, or subjects.

Discussion

The categories identified in the second layer of analysis of the data are a combination of interrelated systemic, structural and personalised factors. The category of 'disciplinary disconnect', for example, indicates that broader systemic issues to do with the organisation of the curriculum at the national level (very atomised subject curricula) affects how they are organised at the school level and at the teacher level, in terms of what is perceived as valid and feasible to implement. Similarly, the category of 'boundary crossing' indicates that, where schools and teachers have the support of a curriculum model to make connections across subject-specific curricula, teachers and principals can, in some instances, begin to change their conceptualisations of what is feasible and possible to do in their classrooms.

It is not surprising that teachers who are not well versed in the content and pedagogies of an unfamiliar subject would be reluctant to integrate that subject into their teaching (Czerniak and Johnson 2014). The findings indicate that this will be compounded by other factors to do with the particular school structures and (lack of) leadership (Rennie et al. 2012), in an educational culture with a strong focus on subject-specific examinations. There are also affective factors at play in incorporating an unfamiliar subject into one's own (Edwards 2011). It can be conjectured that lack of knowledge of how to draw a graph is not the cause, for example, of the stress one teacher mentioned about teaching graphs to new students, but rather that this mathematical practice falls into a liminal zone between the two subjects. Mathematical knowledge is clearly required for, but is somehow not part of, the particular teacher's conceptualisation of her subject; therefore, she has not acquired the confidence and self-efficacy required for teaching this boundary topic. It is here that teacher attitudes and feelings about their subjects are sharpest because their professional identities are bound up in their subject teaching (Hobbs 2013). As recounted in the introduction, the majority of participants had no major objections to the validity of the curriculum model or to the abstract principle of integration of mathematics into science as such. Nonetheless, the evidence of the findings described in this chapter indicates that for some it conflicted with their identities as teachers of science and/or mathematics.

Thus, while lack of subject knowledge will clearly be a constraint, an important finding from this research is that knowledge is a necessary but not a sufficient condition to prompt teachers to incorporate content from another subject into their teaching. Simply because a teacher is familiar with the contents and methods of both subjects does not automatically imply that they have a schema of how to integrate them and, more importantly, that they believe they should integrate them. Teachers have a crucial role as translators across disciplinary systems (Nikitina 2006).

However, crossing subject boundaries can be disruptive to a teacher's identity and self-efficacy, even for subject insiders.

On the other hand, through their participation in the review of the curriculum model, some participants were indeed prompted to take the first steps into interdisciplinary activities in a variety of ways. The formal and informal consultations they initiated with colleagues were indicative of this. The artefacts provided participants with a means and motive to have such interdisciplinary conversations. As the discussion above of the systemic and other factors that contribute to teachers not having such conversations indicates, this is not a minor outcome of the review process. The willingness of certain participants to support the development of the model, though, for example, volunteering to try out lessons and coordinating the activities of colleagues, further indicates the interest of teachers in being part of a boundary dialogue with their fellow professionals conducted via the design of the curriculum model. In this way, the model performed a bridging function as boundary object across the different communities of professionals involved (Akkerman and Bakker 2011).

Most importantly, the moments of reflection captured in the 'Boundary Crossing' theme signify that, for some teachers, participation in the review process led to professional learning at the boundaries of school science and mathematics. There was evidence of participants achieving new understandings through reflection and identity development, considered as essential precursors to any change in practice (Akkerman and Bakker 2011). This occurs because engaging with a boundary object such as the model stimulates reflection on what is being proposed, on how it fits with current practice and on how current practice could change to accommodate it, or not, as the case may be. The material artefacts themselves may be adapted by teachers and schools in a variety of ways. As boundary objects, they provide a motive and an occasion for communication and collaboration to occur (Nicolini et al. 2012), and even more importantly, they present an opportunity for participants to develop their ideas, perspectives and potentially expand their subject identities. Without suitable boundary objects to stimulate this process, change is possible but less likely to occur.

Conclusions

In summary, it may be said that curriculum integration is more than a matter of upskilling teachers or providing them with the right kinds of planning tools and lesson materials. This chapter has shown that curriculum models need to take account of the subject subculture, school structural and teacher subject identity issues that impact on the curricular choices that teachers make. While the model has the capacity to provide teachers and schools with a practical and coherent approach to integration of mathematics into science, it was through the review process that teachers and other participants in the research project embarked on social and professional learning at the boundary. The planning tools and exemplary lessons in and of themselves can always be improved; their main value is in providing a focus for 'photocopier conversations', that is, for opening up new perspectives on one's subject and channels of inter-subject dialogue that may ultimately lead to change in practice. This, more than the physical materials themselves, is the catalyst needed for teachers to take the step into unfamiliar territory and to make it their own.

References

Akkerman, S. F., & Bakker, A. (2011). Boundary crossing and boundary objects. *Review of Educational Research, 81*(2), 132–169.

Basista, B., & Mathews, S. (2002). Integrated science and mathematics professional development programs. *School Science and Mathematics, 102*, 359–370. doi:10.1111/j.1949-8594.2002. tb18219.x.

Berlin, D. F., & White, A. L. (2012). A longitudinal look at attitudes and perceptions related to the integration of mathematics, science, and technology education. *School Science and Mathematics, 112*, 20–30. doi:10.1111/j.1949-8594.2011.00111.x.

Czerniak, C. M., & Johnson, C. C. (2014). Interdisciplinary science teaching. In S. K. Abell & N. G. Lederman (Eds.), *Handbook of research on science education* (2nd ed., pp. 395–411). London: Routledge.

Edwards, A. (2011). Building common knowledge at the boundaries between professional practices: Relational agency and relational expertise in systems of distributed expertise. *International Journal of Educational Research, 50*(1), 33–39.

Fereday, J., & Muir-Cochrane, E. (2008). Demonstrating rigor using thematic analysis: A hybrid approach of inductive and deductive coding and theme development. *International Journal of Qualitative Methods, 5*(1), 80–92.

Geraedts, C., Boersma, K. T., & Eijkelhof, H. M. (2006). Towards coherent science and technology education. *Journal of Curriculum Studies, 38*(3), 307–325.

Hobbs, L. (2013). Teaching 'out-of-field' as a boundary-crossing event: Factors shaping teacher identity. *International Journal of Science and Mathematics Education, 11*(2), 271–297.

Irzik, G., & Nola, R. (2011). A family resemblance approach to the nature of science for science education. *Science & Education, 20*, 591–607. doi:10.1007/s11191-010-9293-4.

Kent, P., Noss, R., Guile, D., Hoyles, C., & Bakker, A. (2007). Characterizing the use of mathematical knowledge in boundary-crossing situations at work. *Mind, Culture, and Activity, 14*(1–2), 64–82.

Lee, M. M., Chauvot, J. B., Vowell, J., Culpepper, S. M., & Plankis, B. J. (2013). Stepping into iSMART: Understanding science–mathematics integration for middle school science and mathematics teachers. *School Science and Mathematics, 113*, 159–169. doi:10.1111/ssm.12015.

Moore, R. (2011). Making the break: Disciplines and Interdisciplinarity. In F. Christie & K. Maton (Eds.), *Disciplinarity: Functional linguistic and sociological perspectives* (pp. 87–105). London/New York: Continuum.

National Research Council. (2014). *STEM integration in K-12 education: Status, prospects, and an agenda for research*. Washington, DC: National Academies Press.

Nicolini, D., Mengis, J., & Swan, J. (2012). Understanding the role of objects in cross-disciplinary collaboration. *Organization Science, 23*(3), 612–629.

Nikitina, S. (2006). Three strategies for interdisciplinary teaching: Contextualizing, conceptualizing, and problem-centring. *Journal of Curriculum Studies, 38*, 251–271. doi:10.1080/00220270500422632.

Ní Ríordáin, M., Johnston, J., & Walshe, G. (2016). Making mathematics and science integration happen: Key aspects of practice. *International Journal of Mathematical Education in Science and Technology, 47*, 233–255. doi:10.1080/0020739X.2015.1078001.

Olson, J., & Hansen, K.-H. (2012). New directions in science education and the culture of the school: The CROSSNET project as a transnational framework for research. In K.-H. Hansen, W. Gräber, & M. Lang (Eds.), *Crossing boundaries in science teacher education* (pp. 9–30). Munster: Waxmann Verlag.

Osborne, J. (2014). Scientific practices and inquiry in the science classroom. In N. G. Lederman & S. K. Abell (Eds.), *Handbook of research on science education* (2nd ed., pp. 579–599). London: Routledge.

Pang, J., & Good, R. (2000). A review of the integration of science and mathematics: Implications for further research. *School Science and Mathematics, 100*, 73–82. doi:10.1111/j.1949-8594.2000. tb17239.x.

Plomp, T., & Nieveen, N. (Eds.). (2013). *Educational design research part A: An introduction* (2nd ed.). Enschede: SLO, Netherlands Institute for Curriculum Development.

Rennie, L., Venville, G., & Wallace, J. (Eds.). (2012). *Integrating science, technology, engineering, and mathematics: Issues, reflections, and ways forward*. New York: Routledge.

Roehrig, G. H., Moore, T. J., Wang, H.-H., & Park, M. S. (2012). Is adding the E enough? Investigating the impact of K-12 engineering standards on the implementation of STEM integration. *School Science and Mathematics, 112*, 31–44. doi:10.1111/j.1949-8594.2011.00112.x.

Saldana, J. (2013). *The coding manual for qualitative researchers*. Los Angeles: Sage.

Stinson, K., Harkness, S. S., Meyer, H., & Stallworth, J. (2009). Mathematics and science integration: Models and characterizations. *School Science and Mathematics, 109*, 153–161. doi:10.1111/j.1949-8594.2009.tb17951.x.

van den Akker, J. (2013). Curricular development research as a specimen of educational design research. In T. Plomp & N. Nieveen (Eds.), *Educational design research part A: An introduction* (2nd ed., pp. 5–71). Enschede: SLO, Netherlands Institute for Curriculum Development.

Walshe, G. (2015). *Integrating Mathematics into Science: Design, Development and Evaluation of a Curricular Model for Lower Second-Level Education* (Unpublished PhD thesis). University of Limerick, Limerick.

Young, M., & Muller, J. (2010). Three educational scenarios for the future: Lessons from the sociology of knowledge. *European Journal of Education, 45*, 11–27. doi:10.1111/j.1465-3435.2009.01413.x.

Fostering European Students' STEM Vocational Choices

Irina Kudenko, Cristina Simarro, and Roser Pintó

Introduction

Young people's competence in Science, Technology, Engineering and Mathematics (STEM) education and their interest in related careers have consistently been a major concern in Europe. The last assessment of STEM competencies (EC 2013) and employability skills (EC 2015a) confirmed the proliferation of negative trends and predicted the internal supply of STEM-related professionals in the next decade to fall short of the EU labour market needs. These political concerns fuel academic research on students' interest in STEM learning, career aspirations and choices, particularly on factors that shape students' views and influence their actions related to STEM education and careers (DeWitt et al. 2014).

Research identifies four basic groups of interrelated factors affecting student career choices in STEM (ECB-InGenious 2011a). First, good subject knowledge, competence in STEM disciplines and students' engagement in learning (factor group A) are commonly recognised as essential prerequisites to positive attitudes to STEM learning and careers. However, on their own these factors are often not enough to stimulate career aspirations of students (The Royal Society 2004), and researchers point to the importance of students' knowledge of STEM-related careers (group B) and their personal beliefs, values and self-perceived abilities to accomplish education and career-related tasks (group C) (Fouad 2007). Finally, social views and popular stereotypes of STEM industries and careers (group D) are also acknowledged as influential, especially with regard to a damaging role of negative stereotypes (Sjøberg and Schreiner 2010).

I. Kudenko
National STEM Learning Centre, York, UK

C. Simarro (✉) • R. Pintó
Universitat Autònoma de Barcelona, Barcelona, Spain
e-mail: cristina.simarro.rodriguez@uab.cat

© Springer International Publishing AG 2017   323
K. Hahl et al. (eds.), *Cognitive and Affective Aspects in Science Education
Research*, Contributions from Science Education Research 3,
DOI 10.1007/978-3-319-58685-4_24

Applying this framework to the analysis of educational initiatives and policies helps identify which groups of factors these interventions try to influence and how they do it. Here, the cooperation between education and social partners (such as representatives of STEM industries) is justifiably acknowledged as a potentially good approach for updating, enriching and contextualising science education, as well as providing valuable expertise and resources (see Wright 1990 for a comprehensive account of the origins of these relationships). There is evidence suggesting that STEM employers' involvement in education creates a multidimensional positive impact on students, affecting their academic outcomes, work and personal skills, career awareness and preferences (Burge et al. 2012). Specifically, it has been argued that a closer cooperation of school with STEM-related industries could help in raising students' awareness of STEM careers, bringing the world of work closer to education (ERT 2009). However, many questions about the process, forms and effectiveness of such cooperation remain under-researched (Andrée and Hansson 2015; Mann 2012).

There have been some advances in researching limitations and gaps in school-industry partnerships, which, it has been argued, can seriously jeopardise the effectiveness and actual impact of such arrangements (BSCR 2011). Most of these partnerships have a voluntary nature (Marriott and Goyder 2009), which can be positive in terms of flexibility and innovation, but may also entail blurred objectives and lack of commitment and sufficient expertise in the education field. Evidence suggests that this lack of clarity about the main objectives of employer-school interactions as well as a lack of comprehensive evaluation may have a negative impact on the success of such programmes (Burge et al. 2012; NCSR 2008). Additionally, most of the existing school-industry partnerships are isolated and short term.

It is not surprising, therefore, that those in charge of linking industries with education feel concerned by the lack of clear guidance on how to make school-industry collaboration as worthwhile as it should be (CBI 2012). This, however, requires a good understanding of what makes collaborations effective in promoting STEM education and careers and how to maximise their impact on students' aspirations.

There is an equal need to study and learn from the existing initiatives implemented within the framework of school-industry partnership. This research was set up to address some of these concerns and, more specifically, intended to answer the following questions:

- What are the common characteristics of European school-industry partnerships in STEM education, and how well do they match theories related to STEM aspirations promotion?
- What are the barriers and gaps in school-industry collaboration across Europe, and what could be done to resolve them and, consequently, to increase their effectiveness?
- What can we learn from evaluating the impact (on teachers' engagement and students' STEM careers aspirations) of school-industry collaborations implemented under the auspices of a common Europe-wide initiative?

Context and Methodology

The data for this paper comes from the European project ECB-inGenious, a multi-stakeholder FP7 initiative which involved over 40 partner organisations representing European industry, policy makers and STEM educators. The overall objective of the project was to foster young people's aspirations towards STEM careers. To this aim, we facilitated schools' engagement with STEM education activities, resources and events designed and supported by industry partners. The project ran for 3 years (2011–2014) engaging over 1500 classrooms across Europe. Around 160 schools in each project year received additional support, professional development and resources from industry partners in exchange for their participation in rigorous evaluation of project activities. After each project year, these 'pilot schools' were able to reapply and remain in the project.

The project provided a stable and sustainable platform for an ongoing dialogue between schools, industry experts and educational specialists. Specifically, teachers had numerous opportunities to engage in professional learning and networking during face-to-face activities (three summer schools, three teacher academies and nine teacher workshops) and online (through webinars and chats with industry experts, themed professional communities and teacher forums). As the project developed, they also gained access to an online database of educational resources developed/supported by industry partners from various EU regions.

ECB-inGenious had multiple research and evaluation activities focusing on the role of school-industry partnerships in developing, promoting and supporting STEM enrichment activities in schools across Europe. Seeking a broader perspective on the research questions, we combined quantitative and qualitative research methods (Teddlie and Tashakkori 2009). It commenced with the examination of existing school-industry partnerships in Europe, reviewing their aims, methodologies and educational policies and sampling their STEM enrichment activities. This examination was based on the responses collected from STEM industries and industry networks to an online survey consisting of both open and closed-ended questions (ECB-InGenious 2011a).

A total of 153 different local, regional, national and European STEM enrichment initiatives developed by business and industry for schools were gathered from 17 countries. Due to a varying quality of information provided, only 79 of them, covering 14 EU and EU partner countries, were included in the next stage of detailed quantitative and qualitative examination (Fig. 1). We used descriptive statistical methodology to analyse closed-ended responses, while open-ended questions, containing detailed descriptions of initiatives, were analysed using a theory-driven content analysis (Namey et al. 2008) and categorised by two researchers working independently. The first researcher classified school-industry initiatives by types of involved activity (e.g. talk) and identified which factors known to influence students' career aspirations (A, B, C and D) were addressed by each initiative. The process was independently repeated by the second researcher, and then their results were compared and validated by computing Cohen's kappa (0.80 and 0.82, respectively), a standard measure for

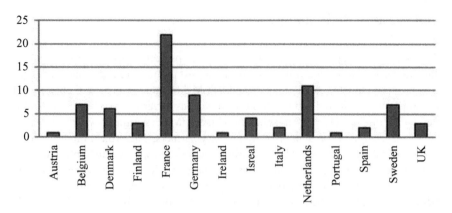

Fig. 1 Country origin and number of practices included in the analysis

inter-coder reliability (Neuman 1997). To double-check that no important information was missing from excluding the incomplete datasets, we applied the proposed
categorisation to all collected initiatives ($n = 153$). This exercise confirmed the robustness and validity of the instrument and showed that the subsample of 79 initiatives,
despite a varying number of activities sampled from each country, was largely representative of all collected activities.

In parallel, the project carried out a wider examination of school-industry collaboration in STEM education, seeking to identify educational and business needs,
and existing gaps and various barriers to school-industry partnerships. We organised
15 national and one European workshop consultations with various stakeholders
(teachers, industry representatives, policy makers, STEM education experts and
providers of professional development and resources). The outputs of these workshops were summarised in corresponding national 'needs analysis' reports and a
European white paper on the state of school-industry collaboration in STEM education in Europe (ECB-InGenious 2013). Again, content analysis was used, this time
based on data itself.

Finally, the project focused on testing existing and newly created STEM enrichment activities for schools and on assessing their impact on teaching of STEM disciplines, student views and career aspirations. In total, 35 school-industry initiatives
were prepared for 'testing' and subjected to rigorous evaluation: 14 were available
in year one of the project, 24 in the second and 29 in the final year. All of them were
initially available in English and at least one more European language, but, with
time, most of the activities were translated into five or more of the native languages
of project participants. Schools participating in evaluation were free to choose any
of these activities but were expected to implement at least three annually.

Student and teacher online questionnaires were administered before, during and
after each project year (Table 1) (ECB-InGenious 2011b, 2014b).

All questionnaires mainly consisted of close-ended multi-choice nominal and
scaled questions. There were both single-choice and multi-choice nominal questions, while a typical scaled question measured responses on a four-point Likert-type

Table 1 Evaluation questionnaires and the number of responses in each project year and overall

	Questionnaires	N° of responses			
		Year 1	Year 2	Year 3	Overall
Teachers	School baseline information (PA1)	168	170	179	517
	Teacher baseline views and expectations (PA2)	127	175	176	487
	Intended use of a project activity (NA1)	254	444	582	1280
	Activity evaluation and feedback (NA2)	206	387	524	1117
	End-of-the-year views and feedback (PO1)	104	118	192	414
Students	Baseline views (PA3)	3260	5816	5347	14,423
	Activity evaluation and feedback (NA3)	3198	6650	9811	19,659
	End-of-the-year views and feedback (PO2)	N/A	2952	5071	8023

scale: 'strongly agree', 'agree', 'disagree' and 'strongly disagree'. For reporting purposes this rating scale was often converted into an 'agree-disagree' dichotomy. When necessary, the four-point scale was expanded to include a fifth option, which accounted for either no opinion ('don't know') or inappropriateness of a question ('not applicable'). These answers were discarded when converting to a dichotomous scale. Occasionally we required a more sensitive instrument to measure a range of participants' views, so a ten-point semantic differential scale (Osgood 1964) was used. For example, to measure students' perception of project activities in comparison to their experience of everyday classroom work, we employed a scale that ranged from one ('much worse') to ten ('much better'). Then we used standard statistical methods to generate descriptive and inferential statistics and analyse quantitative variables and their relationships.

Finally, to capture qualitative, in-depth information and to give more opportunities for participants to respond in detail, a few open-response questions were added to each of the surveys. Answers to these questions were mainly used for illustrative purposes, but we also applied basic content analysis techniques to explore their range and prevalence of certain narratives.

At the start of each project year, we collected baseline data on schools, teachers and students participating in testing of the project activities. We asked about schools' history of collaboration with STEM industries, views on STEM learning, attitudes to STEM industries and student career inclinations. At the end of each project year, teachers and students completed final questionnaires answering similar attitudinal questions and providing feedback on the activities and their impact. In addition, teachers also filled in two special forms per each implemented activity: one form immediately before the activity to capture details of the intended use and one immediately after the trial to gather implementation details and evaluation feedback. A quick and simple survey was used to gather student reflections on the practice.

All questionnaires were translated into 15 European languages, which covered most of the project participants' countries (those not covered had to use an English version of the questionnaire). Due to various constrains, we could only collect student responses from a sample of benefitting students, which were nominated by

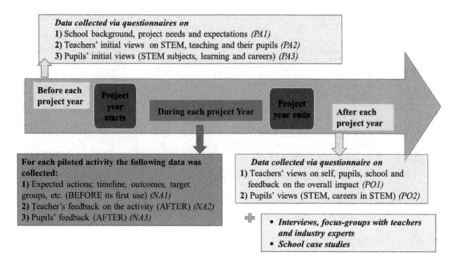

Fig. 2 Project evaluation timeline and instruments

their teachers. However, teachers had clear guidelines requesting that the same students complete every project questionnaire.

We also collected extensive qualitative data using case studies written by teachers, focus groups and interviews with teachers, students and head teachers, as well as other data sources. The evaluation process and instruments are detailed in Fig. 2.

Although data collection and analysis had certain methodological limitations and challenges (e.g. duplicate entries), they were mitigated by the volume, geographical spread and historical consistency of the collected responses (Ross 1992). We also triangulated (Patton 1999) student data with teachers' views on changes in students' perceptions of STEM, which showed a consistent picture of improving learning and career aspirations.

Results and Discussion

Common Characteristics of Sampled School-Industry Partnerships

The 79 educational initiatives sampled in the project confirmed the existence of gaps and limitations in school-industry collaboration identified in the literature. For instance, although some of the initiatives were applicable to more than one age group, the overwhelming majority of them were targeting secondary school students (90%) (Fig. 3).

The duration of the analysed initiatives is another characteristic worth a special note. Nearly two thirds of the reviewed initiatives were short-term/one-off activities

Fig. 3 Distribution of school-industry initiatives sampled in the project ($N = 79$) by the age of students they targeted (some initiatives were applicable to more than one age group)

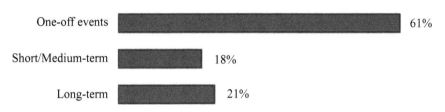

Fig. 4 Distribution of school-industry initiatives sampled in the project ($N = 77$) by the time scale of their application (Note: initiatives aimed only at teachers are not included)

(64%), whereas long-term activities, i.e. those carried out during 1 month or more and which, consequently, would have more opportunities for students to work in-depth and jointly with STEM professionals, represented a 21% of the gathered activities (Fig. 4).

Most of the existing practices were focused on the production of learning resources for STEM subjects (46%), with the involvement of industries being minimal (mostly reduced to funding and project management). Activities with a bigger involvement of industry representatives in STEM education, e.g. industry visits/talks or STEM professionals' involvement in school projects, were less frequent (24% and 15% correspondingly). Worryingly, only 6% of initiatives included teacher training and professional development (Fig. 5).

Of special interest to our research was the link between the reviewed initiatives and the four groups of factors, identified in the literature as influencing student aspirations towards STEM careers. This analysis has shown that most initiatives were focused on supporting STEM curriculum learning (factor group A, 62%), giving less support for career information (factor B, 41%) and often neglecting personal and social attitudinal issues (factors C and D, 25%) (Fig. 6).

Finally, the study also confirmed that impact evaluation, especially with regard to pupil outcomes, remains a 'missing part' in most of the cases: only 27% of the initiatives reported some sort of impact evaluation, and, even then, most of the evaluation activities were reduced to counting participants and conducting satisfaction surveys.

Overall, the analysis of sampled initiatives identified the following gaps and inefficiencies in school-industry partnerships. First, it showed the imbalance in addressing the factors that influence career aspirations of students. While research

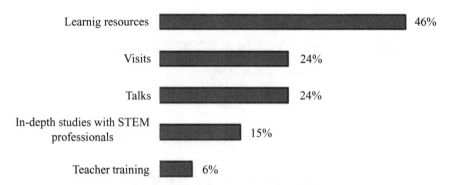

Fig. 5 School-industry initiatives by the type of involved activities (one school-industry initiative could include a few activity types (total sum is greater than 100%))

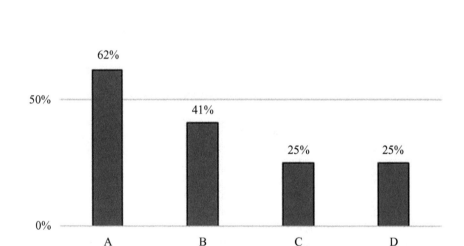

Fig. 6 School-industry initiatives by their focus on the group of factors, which influence students' career motivation. Most activities address more than one factor (total sum is greater than 100%)

evidence shows that effective initiatives have to address all (or most) factor groups outlined previously (factors A, B, C, D), in reality, career information (B) and psychological (C) and social (D) factors are often neglected. Secondly, it confirmed that industrialists prefer to design their initiatives for secondary school students. This bias in selecting the target audience also contradicts the research evidence, which shows that by the age of 14 students' attitudes and interests in STEM learning and careers are largely formed (Archer 2013).

Table 2 Obstacles and barriers identified by stakeholder representatives as hindering school-industry collaborations

Barriers		Type of barrier	
Lack of resources (economic/human/time)	23%	Structural	35%
Lack of support	10%		
Geographical closeness between schools and companies	2%		
Goals of the collaboration	18%	Motivational	33%
Lack of partners interested in collaborating	9%		
Lack of continuity/commitment between partners	6%		
Communication between partners	13%	Procedural	17%
Existing regulations	4%		
Different realities of the worlds of industry and education	13%	Cultural	16%
Matching of schedules	2%		
Negative stereotypes of industry in school	1%		

Many STEM industries do not engage with schools on continuous basis and tend to offer one-off events and short and simple activities. This is despite a clear warning coming from academic research, which demonstrates that one-off interventions are significantly less likely to have a long-term impact (Savickas et al. 2009). Finally, absence of considerations for impact evaluation undermines their potential for self-improvement, which further reduces their effectiveness in raising student career aspirations (Burge et al. 2012; Marriott and Goyder 2009; NCSR 2008).

Current Barriers and Gaps in School-Industry Collaboration

The 'needs analysis' data collected within the project proved very useful in interpreting and understanding the origin of some of the gaps described in a previous section. Specifically, national reports (ECB-InGenious 2013, 2014a) collecting the opinion of representatives from the educational, industrial and political sectors pointed to the existence of the following four major groups of obstacles:

1. Structural obstacles (related to partners' limited availability of resources, support and infrastructure).
2. Motivational obstacles (related to disjointed and even contradictory interests, goals and motivations of the parts involved).
3. Procedural obstacles, (related to the way school-industry links are managed, e.g. lack of stable organisational and networking structures).
4. Cultural obstacles (related to different ways in which school and industry approach STEM education as well as mistrust/misunderstanding/lack of knowledge of the aims and cultural settings of the other side in a partnership). Table 2 shows these types of barriers and the frequency of their appearance in the reports.

As we can see from Table 2, most of the identified barriers involve structural (35%) and motivational (33%) obstacles. Procedural (17%) and cultural (16%) obstacles were less likely to be explicitly mentioned. In reality, however, all groups of obstacles are interrelated (e.g. procedural obstacles may cause or contribute to the creation of structural barriers, and vice versa). Hence, a sensible approach to address any of these obstacles will require a complex intervention strategy, especially if one takes into consideration the different levels of influence that it would involve (strategic, tactical and operational) (ECB-InGenious 2013).

Project Impact on Schools

Evaluation data collected from teachers and students before their involvement in project activities confirmed the existence of serious 'gaps' in school-industry partnership and demonstrated how the absence of such collaboration negatively affects STEM education and students' career aspirations.

More than half of teachers (60%) were of the opinion that their students were unable to connect school lessons in STEM subjects with their everyday lives, while the number of teachers who thought their students had a good understanding of a variety of STEM-related careers was even smaller (35%). Moreover, four out of ten students said that they were not learning about STEM jobs in school, and nearly just as much (34%) did not see any practical use for the knowledge they gain in science lessons. At the same time, students were well aware that STEM industries are very important in the society (84%) and that their achievements in STEM subjects are important for their personal future (78%).

Not surprisingly, teachers felt that their current ability to use contextual examples and career information was limited and that they needed more support in doing this. Fifty-one percent of them thought it was 'hard to provide real-life illustrations and present-day industry examples in their lessons'. This conclusion is consistent with teachers' responses to another statement in the baseline survey 'I often use resources from modern industries in my teaching': only 15% of participants stated their strong agreement with this proposition. Interviews with teachers also showed that, with a few exceptions, schools' engagement with industries before the project was patchy and irregular. Hence, teachers were very keen to gain access to industry educational resources to help improve students' interest and knowledge of STEM subjects by illustrating real-life examples and 'cutting edge' industry applications of STEM knowledge. Yet, they were less certain of other areas for collaboration, e.g. learning about STEM careers (Fig. 7), or forms and activities this relationship should involve. At the onset of the project, some teachers questioned motives of industry partners' involvement in education and had reservations about letting them in schools.

Teachers needed additional support and more opportunities to learn about project industry partners, to understand the 'inner workings' of the world of business and learn about the fundamentals of school-industry collaboration. They also wanted

Q: Please state all the learning outcomes you would like to achieve with this activity

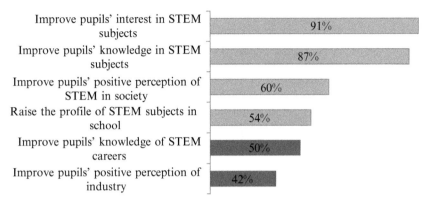

Fig. 7 Teachers' expectation of learning outcomes for students from implementing project activities (*N* = 1280)

more networking opportunities to learn from industry experts and other teacher participants and share 'best practice' in STEM enrichment activities.

> …The chance to talk with the facilitators face by face and fix all queries – personal contact is really important. Swapping experience with some of the participants, getting inspiration... (Teacher from Czech Republic, project year 1).

Using teacher and student data collected during the 3 years of the project, we were able to assess its longer-term impact on the quality of STEM teaching and learning in schools as well as on student perceptions of STEM subjects and careers. In year one, just over half (54%) of the teachers reported that there had been a high to medium increase in their use of industrial materials in their lessons. By year three, this was reported by 87% of participating teachers. Initially many teachers opted for simpler, less demanding activities, which were relatively easy to implement and which could be carried as 'stand-alone' interventions. They were equally very likely to follow a prescribed 'script' and refrained from modifying or changing the recommended activity.

However, the longer teachers remained in the project, the more confident and competent users of industry resources they became: they were implementing more activities and they were doing it in a more creative and complex way. In the first year of joining the project, teachers on average implemented fewer activities (*n* = 2.3) than in their consecutive years, this raising to 3.4 in the final year. By the end of the project, more than three quarters (77%) of teachers reported using inGenious activities in combination with other school-industry activities (both project-based and outside the project). Hence, teachers reported noticeable improvement in their ability to deliver STEM career learning activities (Fig. 8).

Moreover, the longer schools stayed in the project, the more they became engaged with different types of school-industry collaborative activities. Table 3 compares engagement in these activities as reported by both new and more experienced

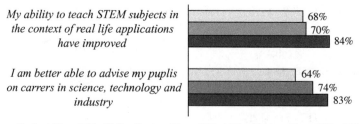

Fig. 8 Teachers' ability to provide enriched curriculum learning of STEM subjects as reported in different years of the project. The percentage of teachers agreeing/strongly agreeing with statements (Likert's four-point scale), $N = 375$

Table 3 Comparative effects of different length of project participation on school-industry collaboration

School-industry collaboration activities reported by schools with different years of project participation	Number of years spent in the project		
	New participants ($n = 76$)	1 year ($n = 33$)	2 years ($n = 72$)
Hands-on activity with industry representative	37%	56%	65%
Running/supporting/engaging in STEM club	41%	53%	65%
Hands-on activity with teacher using industry resources	52%	89%	89%
Running competition in STEM subjects	55%	81%	85%
Resources from industry used in normal classroom activity	61%	92%	82%
Professional development of teaching staff	65%	81%	83%
Visiting speaker from industry	72%	81%	87%
Visit to industry/resource centre	85%	92%	92%

participants at the start of the final year of the project. The difference between newcomers and schools that were in the project from the start is between 20 and 30 percentage points on most of the activities, especially with regard to actual involvement of industry representatives in school.

This was as much about developing a culture of learning over time as it was using particular activities and being familiar with their outputs. This appears to be what for many teachers was a pivotal point:

> I have thoroughly enjoyed being involved with inGenious over the last 3 years. It has transformed my teaching and shown me the importance of collaborations both with industry and with international colleagues. (Teacher from UK, project year 3)

When teachers reflected on how the project was affecting their students, they reported a gradual but constant improvement in student learning of STEM subjects. More importantly, this was matched by a noticeable increase in their awareness and interest in STEM careers, with the highest impact being achieving by the project final year (Figs. 9 and 10).

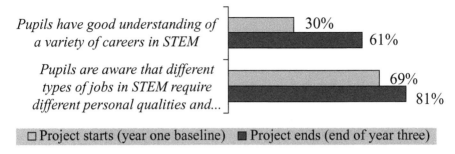

Fig. 9 Students' awareness of STEM careers reported by teachers before and after the project. The percentage of teachers agreeing/strongly agreeing with each statement (Likert's four-point scale), $N = 375$

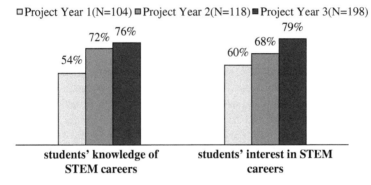

Fig. 10 Percentage of teachers in each project year who reported high and medium impact on student career perceptions (impact was measured on a four-point scale)

In the first project year, slightly more than a half of teachers defined the impact on student career learning and aspirations as high or medium, but by the end of its third year, this number was more than three quarters. A Finnish primary teacher, who participated in all 3 years of the project, described the final impact as follows:

> Thanks to this project, I think that I opened a little door to the corporate world and it aided their (i.e. students') interest in profession in general and in relation to their future.

This perception is confirmed by students' data, which showed a statistically significant positive change in young peoples' inclinations towards STEM-related careers (Fig. 11), both for primary and secondary school students.

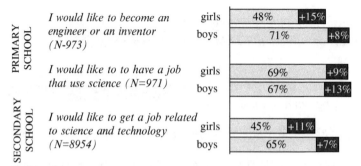

□ Project starts (year one baseline) ■ Change by the project's end (end of year three)

Fig. 11 Improvements in students' STEM career aspirations. The percentage of students who agreed or strongly agreed with each statement (Likert's four-point scale)

Conclusions

Considering the complexity and multidimensionality of factors that influence STEM career choices, our research has demonstrated that school-industry partnerships can provide for this complexity and have a positive role to play. However, our study has confirmed that some gaps and barriers present strong challenges to the establishment, effective implementation and sustainability of such kinds of partnership. Our analysis of a sample of European school-industry initiatives has demonstrated that present-day partnership arrangements can be of diverse nature and have considerable differences in terms of their aims, target audiences, duration, type of activities and the level of interaction between students and STEM professionals. Importantly, they also differ in their capacity to address different groups of factors known to impact young people's interest and career aspirations in STEM-related subjects (i.e. factor groups A, B, C, D).

At the same time, we have identified the prevalence of certain biases in the industry offer of educational enrichment activities, which could have a negative effect on school-industry collaborative work and create additional gaps in the provision of STEM enrichment activities. For example, industry initiatives tend to focus on secondary students at the expense of primary school groups and often overlook the need for rigorous evaluation of outcomes and impacts. Moreover, we also evidenced the existence of cultural barriers, which separate the world of education and industry and create misconceptions and suspicions about the motivation, needs and requirements of potential partners in school-industry collaborations. This has been shown as an additional negative factor that impairs collaboration and reinforces the existing gaps.

We have also shown that for a partnership to be effective, it has to be sustainable and has a long-term commitment from both sides. But such partnerships are not easy to develop and require additional structural and organisational support and guidance on how to make collaborative work experiences as worthwhile as they should (CBI 2012). In this sense, ECB-inGenious has become a successful European platform bringing schools and industry stakeholders together and creating opportunities for networking, professional learning, understanding of each other's needs, sharing of good practice and facilitation of school-industry collaboration. The project provided a stable sustainable platform for an ongoing dialogue between schools with industry experts and educational specialists, which helped to overcome many structural, procedural and cultural constraints identified in the needs analysis.

A very important enabler of this success was extensive professional development and support offered to teachers within the project. Teachers have noticeably improved their use of industry enrichment activities and STEM learning resources. However, this was a gradual process. The longer teachers stayed in the project, the more confident they became in addressing students' STEM career aspirations. This allowed them to be more creative and experimental in the classroom, engaging students in elaborate activities and achieving higher impact on student career aspirations. The evidence gathered in the project showed that it produced wider benefits, positively impacting schools' ability to maintain their current industry partnerships and develop new collaborations and links to the world of STEM industries.

Finally, we see the need for further research to identify other actions and policies that can tackle structural, motivational, procedural and cultural barriers, helping teachers effectively integrate industry support in STEM education and enabling schools to open up to real-life challenges (EC 2015b).

Acknowledgements ECB has been funded with support from the European Commission under Seventh Framework Programme, Grant agreement no: 266622. This publication reflects the views only of the authors. The Commission cannot be held responsible for any use which may be made of the information contained therein.

References

Andrée, M., & Hansson, L. (2015). Recruiting the next generation scientists and industrial engineers: How industrial actors engage in and motivate engagement in STEM initiatives. *Procedia - Social and Behavioral Sciences, 167*, 75–78.

Archer. (2013). Young people's science and career aspirations, age 10–14. ASPIRES final report. Retrieved from https://www.kcl.ac.uk/sspp/departments/education/research/aspires/ASPIRES-final-report-December-2013.pdf

Burge, B., Wilson, R., & Smith-Crallan, K. (2012). *Employer involvement in schools: A rapid review of UK and international evidence.* (NFER Research programme: From education to employment). Slough: NFER.

Business-School Connections Roundtable. (BSCR). (2011). *Realising potential: Business helping schools to develop Australia's future.* https://docs.education.gov.au/system/files/doc/other/australian_government_response_realising_potential.pdf. Accessed 23 Dec 2016.

Confederation of British Industry (CBI). (2012). *Learning to grow: What employers need from education and skills. Education and skills survey 2012.* http://www.ucml.ac.uk/sites/default/files/shapingthefuture/101/cbi_education_and_skills_survey_2012.pdf. Accessed 23 Dec 2016.

DeWitt, J., Archer, L., & Osborne, J. (2014). Science-related aspirations across the primary–secondary divide: Evidence from two surveys in England. *International Journal of Science Education, 36*(10), 1609–1629.

ECB-InGenious. (2011a). *InGenious project. Observatory Methodology. Report to the European Commission.*

ECB-InGenious. (2011b). *InGenious project. Protocol of experimentation for the leading pilot schools. Report to the European Commission.*

ECB-InGenious. (2013). *InGenious project. Needs analysis for school-industry partnerships. Report to the European Commission.*

ECB-InGenious. (2014a). *InGenious project. European Synthesis Report. Report to the European Commission.*

ECB-InGenious. (2014b). *InGenious project. Final evaluation report. Report to the European Commission.*

European Commission. (EC). (2013). *PISA 2012: EU performance and first inferences regarding education and training policies in Europe.* http://ec.europa.eu/dgs/education_culture/repository/education/policy/strategic-framework/doc/pisa2012_en.pdf. Accessed 23 Dec 2016.

European Commission. (EC). (2015a). *EU skills Panorama 2014. Analytical highlight. Focus on STEM skills.*

European Commission. (EC). (2015b). *Science education for responsible citizenship.*

European Round Table. (ERT). (2009). *Mathematics, science & technology education report. The case for a European coordinating body.* http://www.ert.eu/sites/ert/files/generated/files/document/mst_report_final.pdf. Accessed 23 Dec 2016.

Fouad, N. (2007). Work and vocational psychology: Theory, research, and applications. *Annual Review of Psychology, 58,* 543–564.

Mann, A. (2012). *It's who you meet: Why employer contacts at school make a difference to the employment.* http://www.educationandemployers.org/wp-content/uploads/2014/06/its_who_you_meet_final_26_06_12.pdf. Accessed 23 Dec 2016.

Marriott, N., & Goyder, H. (2009). Manual for monitoring and evaluating education partnerships. http://unesdoc.unesco.org/images/0018/001851/185117e.pdf. Accessed 23 Dec 2016.

Namey, E., Guest, G., Thairu, L., & Johnson, L. (2008). Data reduction techniques for large qualitative data sets. In *Handbook for team-based qualitative research* (pp. 137–163). Plymouth: Altamira Press.

National Centre for Social Research. (NCSR). (2008). The involvement of business in education: A rapid evidence assessment of the measurable impacts. http://www.educationandemployers.org/wp-content/uploads/2014/06/involvement-of-business-in-education-dcsf1.pdf. Accessed 23 Dec 2016.

Neuman, W. L. (1997). *Social research methods: Qualitative and quantitative approaches* (3rd ed.). Boston: Allyn and Bacon.

Osgood, C. E. (1964). Semantic differential technique in the comparative study of cultures. *American Anthropologist, 66*(3), 171–200.

Patton, M. Q. (1999). Enhancing the quality and credibility of qualitative analysis. *Health Services Research, 34*(5), 1189–1208.

Ross, K. N. (1992). Sample Design for International Studies of educational achievement. *Prospects, 22*(3), 305–316.

Savickas, M. L., Nota, L., Rossier, J., Dauwalder, J.-P., Duarte, M. E., Guichard, J., et al. (2009). Life designing: A paradigm for career construction in the 21st century. *Journal of Vocational Behavior, 75*(3), 239–250.

Sjøberg, S., & Schreiner, C. (2010). The ROSE project. An overview and key findings. Oslo.http://roseproject.no/network/countries/norway/eng/nor-Sjoberg-Schreiner-overview-2010.pdf. Accessed 23 Dec 2016.

Teddlie, C., & Tashakkori, A. (2009). *Foundations of mixed methods research: Integrating quantitative and qualitative approaches in the social and behavioral sciences.* Los Angeles: Sage.

The Royal Society. (2004). *Taking a leading role.* https://royalsociety.org/~/media/Royal_Society_Content/Education/2011-06-07-Taking_a_leading_role_guide.pdf. Accessed 23 Dec 2016.

Wright, P. W. G. (Ed.). (1990). *Industry and higher education: Collaboration to improve students learning and training. The society for research into higher education.* Bristol: Open University Press.

The Notion of Praxeology as a Tool to Analyze Educational Process in Science Museums

Juliana Bueno and Martha Marandino

Introduction

Museums are considered to be educational institutions, and since their origin they have changed their operational focus from caring for the collections to directing their attention to the public (Fayard 1999). The move toward the public has been acknowledged in the exhibitions as their production involves the transformation of scientific knowledge for education and dissemination purposes in order to make it comprehensible. In this chapter our goal is to analyze the knowledge presented at an exhibition, considering that it is a result of the transformation process from scientific knowledge to disseminated/exposed knowledge at museums (Mortensen 2010). The theoretical framework was finded in Anthropological Theory of the Didactic (ATD), especifically in the concept of praxeology (Chevallard 2007; Bosch and Gascón 2006), and it was able to reveal and to identify how the knowledge of biodiversity is presented in museum dioramas. We use praxeology as a tool to analyze the production of a diorama in order to answer our research question: how does the teaching process occur in an exhibition object – the diorama – in a science museum?

J. Bueno (✉)
Department of Education of the State of São Paulo, São Paulo, Brazil
e-mail: juliana.bueno@educacao.sp.gov.br

M. Marandino
University of São Paulo, São Paulo, Brazil
e-mail: marmaran@usp.br

© Springer International Publishing AG 2017   339
K. Hahl et al. (eds.), *Cognitive and Affective Aspects in Science Education Research*, Contributions from Science Education Research 3,
DOI 10.1007/978-3-319-58685-4_25

The ATD and the Educational Proposal of Exhibitions in Science Museums

The Anthropological Theory of the Didactic (ATD) studies the manipulation of knowledge with didactic purposes. It has become an important instrument to disclose the theoretical and practical framework of exhibition activities present in museums, because it enables the identification of the tasks (praxis) proposed for an exhibition's object, correlating them with a conceptual body of knowledge (logos) that supports its execution (Mortensen 2010). By identifying the tasks of the exhibition's objects, such as a diorama, their educational potentials can be revealed, which can contribute to the production processes of exhibitions in science museums.

It is important to emphasize that assembling an exhibition, which includes selecting its objects and texts, is intended to teach and communicate concepts and ideas from a specific format, that is, a museography. The teams that execute this selection have control, or vigilance, over the decisions oriented from an epistemological point of view (concerning the knowledge to be taught) as well as the museological point of view (concerning communication, educational, and artistic strategies that will be used). The idea of epistemological vigilance points out that if, on the one hand, the knowledge taught or divulged is maintained in the relationship of distance and proximity with the reference knowledge, it should, on the other hand, also correspond to a given reality and context and to different social practices (e.g., develop museum exhibits). The epistemological vigilance concept highlights that the objects of knowledge that will be taught are not misrepresented, replaced, but changed (Souza et al. 2012). By "change" we mean that knowledge goes through modifications that imply simplifications and adaptations in order to make it understandable to the general public. The work of actors who exercise epistemological vigilance is to control these changes so that knowledge does not present itself in the wrong way but is rather easier understood.

The decision on what will be chosen or not for an exhibition or for producing a diorama is associated with the development of an efficient means to communicate the selected content (Oliveira 2010). In this sense, we understand ATD as a framework that involves the theory (knowledge) and practice dimensions and which does not simply rely on understanding how the transposition of a certain knowledge into another occurred. It also allows objective descriptions in order to reveal the various steps of transposition as an epistemological reference model. To build this model, in order to bring relevant results, this theory proposes an anthropological study of the theory and the practice which goes beyond simply modeling and revealing the explicit and general forms of how scholarly knowledge turns into knowledge to be taught.

Chevallard (2006) developed the notion of praxeology, defined as the basic unit that analyzes human action, called praxeological organization (PO). A PO, according to the reference subject, can be modeled into two parts: one, the teaching part, associated with how to present a particular content – the didactic organization (DO) – and another, the part related to the body of knowledge socially produced by

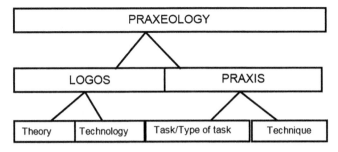

Fig. 1 Schema of a praxeological organization. Based on Machado (2011)

a group, which substantiates a theory and is justified in a technology, the mathematics organization or biological organization (BO). In the case of museums, we consider that the DO can also be called a museographic organization (MO), since it is through museography that the teaching strategies are expressed in a museum.

The study of praxeology in museums has been recently developed in order to investigate the learning environment of museum exhibitions, producing practical and theoretically grounded principles, with the conditions to be applied to the alignment between the design of exhibitions and the educational outcomes (Mortensen 2010). Such knowledge enables the identification of the *tasks* (praxis) proposed for the exhibition, correlating them with a body of conceptual knowledge that maintains its implementation (logos).

Figure 1 shows the identification of the elements that comprise the logos (*theory* and *technology*) and praxis (*task/type of task* and *technique*).

The study of praxeology in museums is relevant because when the *tasks* that involve the exhibition object are identified, it is possible to reveal the educational potentials of the diorama and therefore contribute to understanding the didactical intentions of the exhibitions at science museums, assuming that teaching and learning experiences occur in an informal education context.

Methodology

A long-term exhibition called "Zoology Research: biodiversity in the perspective of a zoologist" at the Museum of Zoology of the University of São Paulo, Brazil, was open from 2002 to 2011 and allowed the museum visitors the opportunity to get to know the aspects involved in the routine and work of zoologists, in order to highlight the importance of their research, to understand the origins of animal life, the natural processes that promoted morphological, genetic, and ecological diversification throughout the history of the planet, as well as the patterns that led to the current distribution between the different environments and continents. According to Marandino et al. (2009), this exhibition was divided into four modules in order to show the contents and objects that communicate aspects of ecology (with regard to the ecological relationships between the organisms and the physical world represented in the diorama) and biodiversity (with regard to the quantity and the diversity of animals exhibited). The modules are "Presentation and history of MZUSP;" "Origin of the species and large zoological groups;" "Evolution, diversity, and phylogeny – Activities of the zoologist;" and "Neotropical Fauna and Marine Environment." This last module is composed of representations of the Neotropical ecosystems and had five dioramas in the following environments: Atlantic Forest, Amazon Forest, Cerrado, Caatinga, and marine environment. Also in this module, there was a large map of the Neotropical region covering the floor, and taxidermy specimens of migratory birds representing a flock were suspended from the ceiling. In this chapter we analyze the diorama called "Amazon Forest" that was part of the "Neotropical Fauna and Marine Environment" module (Fig. 2).

The methodology of this research was based on a qualitative approach, and it includes data obtained by three collection tools:

(a) Documents: curatorial project and folders of the exhibition were analyzed looking for the concepts of ecology and biodiversity and also the educational and communicational purposes of the exhibition. Using the perspective of textual discursive analysis (Moraes and Galiazzi 2007), we selected and transcribed parts of the documents and classified them in two units: (1) parts of the text related to concepts of ecology and biodiversity and (2) parts of the text related to the objectives and the scientific and museographic conceptual proposal of the exhibition as a whole and about the Amazon Forest diorama. These units were used to characterize the logos – *theory* and *technology* – of the Amazon Forest diorama (as indicated in Table 1).

Fig. 2 Module "Neotropical Fauna and Marine Environment" of the MZUSP exhibition

Table 1 Intended praxeology of the "Amazon Forest" exhibition set – MZUSP – logos characterization

Theory Θ	Diversity of species and ecosystems of the Amazon Forest
Technology θ	An ecosystem composed of different environments and with great plant and animal diversity and also different ecology

(b) Interviews: conducted with two exhibition designers, following a semi-structured questionnaire to identify the intentions related to the understanding of ecology and biodiversity and the educational and communicational purposes of the exhibition. The interviews were tape recorded and transcribed as a whole, and the analysis was conducted using textual discursive analysis (Moraes and Galiazzi 2007) where parts of the interviews were selected in relation to the two units mentioned before: (1) related to the concepts of ecology and biodiversity and (2) related to the objectives and the scientific and museographic conceptual proposal of the exhibition as a whole and about the Amazon Forest diorama. Data from documents and interviews were used to characterize the logos – *theory* and *technology* – of the Amazon Forest diorama, helping to build Table 1.

(c) Observation: the Amazon Forest diorama was observed, filmed, and photographed, and a detailed description was elaborated in order to identify ideas of ecology and biodiversity in their museographic elements, especially in the objects, text, and supporting images. The observation and description of the diorama were prepared using a "scanning" technique in order to identify and record each object and the ecological relationship in the scene. This procedure was based on the work of Oliveira (2010), who in turn based his work on Dean (1996) to develop the scanning technique, as this author discusses the production and analysis of exhibitions and studies how the public establishes relationships with these elements. Therefore, a description of the diorama Amazon Forest was performed, starting from right to left and from top to bottom, always starting from the back plane (painting in the bottom and/or on the side) to the front (the objects). Because there is a difference in planes between painting and objects and due to the size of the studied dioramas, this procedure was fragmented into smaller pieces, called quadrants, and each one was described following the procedure indicated (Fig. 3). The use of this technique was critical to the analysis, since it provided a detailed and accurate description of the elements exposed in the scene.

During the description process, a text was produced narrating in detail each element that composes the diorama, as the example below of the first quadrant description. The elements cited in the text and the forms that they represent in the scene were used to define the praxis or practical block of a praxeology: the *tasks*, the *type of tasks*, and the *techniques* in the diorama (as indicated in Table 2).

First quadrant (back to front, top to bottom and right side): In the upper back plane there are two trunks without the canopy. The one on the right is darker; the one on the left, lighter and thicker. These trunks descend to the ground and from them come vines that are entangled with each other with small leaves distributed on the surface. In the previous frame

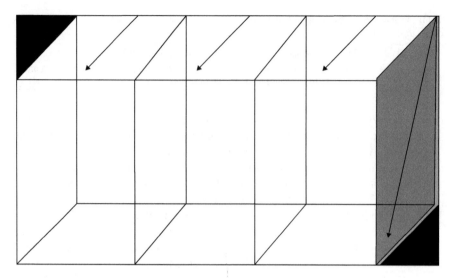

Fig. 3 Representation of the scanning technique developed by Oliveira (2010)

Table 2 Intended praxeology of the exhibition set of the "Amazon Forest" diorama – MZUSP – praxis characterization

Type of task (T)		
Tasks (t)	Support	Technique (τ)
T1: recognize ecological relationships	Panel	Identify organisms in the legend drawing; compare the organisms that are exposed in the diorama and raise assumptions about cause and effect
t: recognize the epiphytism		
T2: distinguish the species that compose the animal diversity	Diorama	Identify the species in the legend; observe, recognize, differentiate, and classify taxidermized and exposed species in the diorama or display case
	Display case	
t1: distinguish one iguana		
t2: distinguish the insect diversity		
T4: evidence the plant and animal richness	Panel	Read the informative text on the panel; identify organisms in the legend; observe, interpret, and verify observable aspects in the diorama, generalizing concepts
	Legend	
t: evidence the high plant and animal endemism	Diorama	

there are trunks of cut trees, covered with vines and a model of bromeliad with red flower. Next to the bromeliad, 1.5 m from the ground, is a specimen of sauin (Calliithix chrysolena) holding a small orange fruit facing the back plane of the diorama. About 10 cm below there is a specimen of another species of sauim (*Saguinus fuscicollis*), and below, about 50 cm from the ground, there is a specimen of a monkey (*Saimiri boliviensis*) holding a piece of brown food. Near the ground (10 cm) there are two specimens of squirrels (*Sciurus spadiceus*) resting on cut logs, one facing the public, holding a yellowish fruit, and another, with its back to the public. The soil is sparsely covered by shrubby vegetation 20–30 cm high, with some elevations, representing rocks or exposed roots of plants, and the presence of moss in some regions.

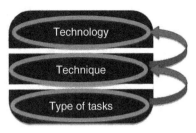

3. Interpretation of the *Technique: Technology*

2. Interaction/relation with a *Technique*

1. Identification of the *Task and/or Types of Tasks*

Fig. 4 Schema that indicates the construction of a praxeology (Achiam 2013)

The data obtained through the collection tools were analyzed by selecting and organizing them in order to construct the praxeology, understood here as a theoretical framework that at the same time works as a tool that is informed by the theoretical perspective of ATD. This procedure was done in a search to reveal the didactic potential of the Amazon Forest diorama. Achiam (2013) proposed a scheme that was used here as reference to build the praxeology (Fig. 4). Using data from the observation and description of each element of the diorama, we identify the *task* and the *type of tasks* and also, in relation, the *technique*, as suggested at the scheme. Together with the data from documents and interviews, we also interpret the *technology* and, in our case, the *theory* (Fig. 4).

For a characterization of the praxis, the data used were mainly the observation of the diorama and its expositive set. In order to identify the *task*, the *type of tasks*, and the *technique*, a focal question was used to help. The question defined was: "How can the visitor perceive ideas and concepts of biodiversity represented by Amazon Forest diorama and its exhibition set?" The identification of a *task* means determining the "action to be achieved" in relation to a particular item that is meant to be known and which involves human activity. In the case of the diorama, for example, it would be the action to recognize in a scene an Amazonian plant or distinguish an animal that lives in a given environment. After the *tasks* were identified, the next part involved the characterization of the *type of task* that groups the actions identified in the *task*. For example, actions such as "distinguish," in the diorama, an iguana or the insect diversity, comprise a specific *task* and can be grouped into a single *type of task*, for example, to "distinguish the species that compose the animal diversity." The identification of the *technique* represents "how to perform" a certain task and indicates the selection of a tool or museographic strategies to accomplish the *task*, that is, how to present a certain content. In our case, the *technique* in a diorama is given by expressions that indicate "how" the visitor accomplishes, for example, the *task* of recognizing the epiphytism, such as "identify" organisms in the legend drawing, "compare" the organisms that are exposed in the diorama, and "raise assumptions" about cause and effect.

To characterize the logos and the description of *theory* and *technology*, the data were obtained from interviews, documents, and observation of the exhibition set. The logical block of a praxeology comprises *technology* and *theory*, which are identified after listing the *tasks, type of tasks*, and the *technique* of the practical block of praxeology. To identify the *technology* means to relate the practical block with the logical block (Mortensen 2010). The *technology* emerges in order to justify the choices made; in other words, it requires looking at what is being placed into the practical block and theoretical block and can involve more than one exhibition set if we consider the MZUSP exhibition as a whole. In this chapter, the *theory* and *technology* were constructed based on the data identified for the diorama of the Amazon Forest and its supporting elements.

The Findings: Teaching with Dioramas in a Science Museum

The dioramas are specially assembled scenes that have been used since the nineteenth century in museums, in order to promote a realistic perception and therefore combine, in the case of natural sciences museums, scientific knowledge of plant and animal species, with reproduction techniques of scenes. These are built through light effects, painted backgrounds, and stuffed animals, that is, they exhibit a type of motionless theater, which places us in make-believe habitats that impress by their degree of realism (Almeida 2012). The history of these elements communicates the educational arguments for their inclusion in the exhibition: to contextualize the organisms and the environment and to facilitate the public's understanding of the information (Van Präet 1989).

The biological knowledge, in the case of dioramas, is usually connected to the observation, identification, and recognition of species of plants, animals, or fungi, their relationships with each other and with the environment, and also the identification of geophysical phenomena, such as rock formations, soil types, types of biome, and other topics.

Ash (2004) shows that a number of studies emphasize the potential of dioramas for such exhibitions as a means to promote the understanding of ecology, biodiversity, and conservation issues through contact with environments never before experienced by the public. In that view, dioramas are teaching objects, produced as a result of a museographic transformation which combine scientific, artistic, educational, and communication knowledge. From the perspective of ATD, it can be stated that dioramas have a PO, and this can demonstrate the contents and actions intended by their designers and producers.

In our research, the Amazon Forest exhibition set was composed of the diorama and of the supporting elements: display case and panel containing text, image, and layout with subtitles, as seen in Fig. 5.

According to Fig. 5, the diorama of the Amazon Forest has an "L" shape of about 3 m high, 4.5 m long, and 2.4 m wide at the larger part of the "L" and 1 m at the smaller part. The display case is open and has guardrails. Its illumination is at the front and it is directed toward the rear. It comprises stuffed animals and plant models with flowers and trees.

Fig. 5 Front view of the "Amazon Forest" exhibition set

There is a display case, at the lower part, 50 cm from the floor with three glass covered boxes with light in the background. Two of them include the invertebrates, and the other has a legend to identify some of the vertebrates.

Another element is the panel that includes a text, a schematic map of the area, and a scheme with the legend, which together display the characteristics of the Amazon Forest.

Based on the data obtained, we will now describe the logos, that is, the *theory* and *technology* identified in the Amazon Forest diorama. To do that, we selected two excerpts from the interviews with the exhibition designers of MZUSP. When asked about the role of the dioramas at MZUSP, the respondents indicated that:

> The purpose of the objects is to provide the contexts of the animals. How these different life forms live in the same habitat. (C1)

> [...] The diorama of the Amazon, we wanted to show the structural differences of ecosystems [...]. The diorama shows the ecosystem, the animals interacting, it shows the taxonomic biodiversity and the ecosystem biodiversity. (C2)

Considering the aspects cited and the data from the documents, the logos dimension of the praxeology of the "Amazon Forest" diorama is proposed as the following (Table 1):

To characterize the praxis of the diorama, we used the observation of the diorama and its exhibition set to answer the focal question: "How can the visitor perceive the ideas and concepts of ecology and biodiversity represented by the diorama of the Amazon Forest and its exhibition set?" This question directed us to a specific exercise, aiming to detail each element of the exhibition set, which in terms of praxeology meant to identify the *tasks*, *types of tasks*, and *techniques*.

The excerpt below provides part of the description of the diorama, developed using the scanning technique, which supported the construction of the praxis:

> In the 2nd quadrant, in the center, there is a cut tree trunk with ferns, vines, a pink orchid in tree trunks, a cuxui specimen (*Chiropotes satanas*) facing the back plane, standing on all four legs, and immediately below, 15 cm away, a night-monkey specimen (*Aotus sp*) inside a hollow trunk. In the front plane, sitting on a branch of a cut trunk, there is a dusky titi specimen (*coppery titi*) standing on all four legs and facing the public. At the bottom, on a branch of the cut trunk, there is an iguana *(Iguana iguana)* with its tail extending to quadrant 1.

The *tasks* identified were defined according to the role that each object plays in the museographic organization. For example, the *task* identified by "recognize the epiphytism" was determined by the observation and the description of the second quadrant of the diorama, shown in Fig. 6. To accomplish this *task*, the visitor should be able to perform the following *techniques*: observe the diorama; identify and recognize the plant's habitat, as the bromeliad and orchid were exposed in the diorama; and raise the assumption of cause and effect, i.e., the assumption that the scene reveals an example of epiphytism, since this is an ecological relationship that indicates the epiphytic way of life of an orchid growing on other plants.

Fig. 6 Inset of the diorama: tree trunk, ferns, and orchids

Similarly, with the presence of an iguana (*Iguana iguana*; Figs. 7 and 8) in the diorama, there is a *task* that reveals the intention to "distinguish species that compose the animal diversity of the Amazon Forest." For this *task* the visitor should perform the following *techniques*: read the legend; identify in the legend the popular and scientific names of the organisms; observe the diorama; relate the popular and scientific names of the stuffed organisms that are exhibited in the diorama; raise an assumption of cause and effect, for instance, assume that by exposing a wide variety of animals, the diorama reveals a characteristic of this biome, which is animal diversity.

Part of the exhibition set includes the preserved organisms presented in the display cases. Here, the identified *tasks* mobilize visitors to "distinguish the taxonomic diversity of the organisms exposed in the display cases," such as the organisms (invertebrates) in Fig. 9 and that infers the following *techniques*: read the legend; identify the phyletic group written in the text of the legend; observe the display case; recognize and relate a taxonomic group to the invertebrate specimens exposed in the display case.

Considering the aspects shown, we can express the praxeology in terms of *tasks, type of tasks*, and *technique*, like in the example above, for the Amazon Forest diorama as shown in Table 2.

The characterization of the logos and praxis of the intended praxeology of the Amazon Forest diorama allowed to indicate the intentions and choices made by

Fig. 7 Iguana specimen in the diorama

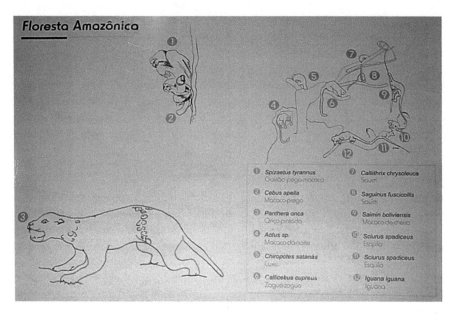

Fig. 8 Display case: legend and schema of each animal in the diorama

Fig. 9 Display case: invertebrate animals. Class: Insecta

designers in the production of this object, in addition to the concepts related to ecology and biodiversity expressed in it. It was also possible to identify the missing concepts and simplifications of ideas and concepts and discuss the potential and limitations of the teaching and learning process during the visits to the diorama. We point out these aspects in the following section.

Discussion

In this research, we show that the use of praxeology can be useful as a theoretical and methodological tool to highlight the different aspects of the educational process in museum exhibitions. The knowledge and the educational practices pass through in the museographic transposition from an established reference model and help to reveal "what is meant to be taught," "how knowledge is taught," and "what is actually taught." Revealing this process is part of the research program of the ATD, which seeks to model the institutional conditions under which a given knowledge and its teaching are built, articulated, and operated. Through the praxeology and the focal question, it was possible to identify each element of the exhibition set which has the potential to inform and teach scientific contents, using the ATD framework concepts as *tasks*, *type of tasks*, *and techniques* and hence write in a detailed manner how a diorama, as a teaching object, carries out its capacity to teach and disseminate aspects of the ecology and biodiversity of the Amazon Forest.

As pointed out by Winslow (2011), building praxeologies and defining epistemological reference models are not something that is natural or obvious a priori. It is a process to model a given didactic situation through description, experiments, and analyses, and its validity is determined according to its usefulness to explain the didactic phenomenon under study. In this sense, for Winslow, the ATD is an extremely descriptive research program, which does not mean it is neutral or disinterested, since it reveals flaws and inconsistencies that can inspire interventions or curricular reforms.

Achiam (2013) states that while often successful in prompting visitors to carry out intended actions, exhibits do not necessarily promote the intended interpretations of these actions among visitors. In her work she used the notion of praxeology as a model to remedy this shortcoming and suggested using it as a means to operationalize the link between exhibit features and visitor activities. In this work, we assume the same perspective, since through praxeology, it is possible to reveal the intentions, the presence, but also the absences and simplifications of the conceptual and didactic content in the production of a diorama, therefore contributing to inspire interventions and reforms in this didactic object in order to ensure that the desired discourse is closer to that understood by the public, as we shall see ahead.

The dioramas, in general, as well as other didactic objects, are human works and outcomes of didactic transposition processes and thus undergo simplifications and reductions but also add information or segregate them according to the teaching and learning goals. Without this, according to Chevallard (1991), as teachers (and more broadly as educators), we would be denying access to information and failing to exercise a fundamental activity of maintaining societies. In the museum context, transpositive processes occur for a broader socialization of knowledge, and in the case of museum exhibitions, this is done through objects, images, and texts.

As can be observed from the praxeology analysis, the Amazon Forest diorama and its exhibition set represent aspects of the biodiversity and ecology of a particular ecosystem. The production of this object involves selecting a conceptual content on this topic. However, it also fails to consider other conceptual contents that involve this subject. To analyze the presence and absence of ideas of the biodiversity and ecology in the diorama, we rely on the concept of epistemological vigilance proposed by Chevallard (1991).

Considering the diorama researched here, our study indicates that epistemological vigilance was present during its assembly, controlled by the agents of the noosphere. To Chevallard (1991), the *noosphere* is a central concept of this theory, and it is the space occupied by actors who are involved in the didactic work, and within it the actors face each other and seek to equate problems that emerge from the demands that come forth from society. Noosphere is the place where conflicts are developed, where negotiations are conducted, and where solutions mature. According to Brockington and Pietrocola (2005), the subjects of the noosphere proceed, mediating the needs of both society and the school system. In our case, the museum noosphere is formed by individuals who are involved in the production of the exhibition and who negotiate the objectives and purposes, that is, the "what" and "how" to exhibit the objects within the studied diorama. The actors of the museum noosphere had control of the knowledge defined to be presented through the objects, texts, and images. As the exhibition designers (C2) pointed out, they wanted to show the structural differences of ecosystems, using the diorama to show the ecosystem, the animals interacting, the taxonomic biodiversity, and the ecosystem biodiversity. However, it can be noticed that some ecological relationships were not shown in the exhibition set and some animal classes were also not represented.

The absence of some conceptual aspects in the exhibition set may give the impression that only those ecological relationships and species of plants and animals represented comprise the biodiversity of the Amazon Forest ecosystem. On the one hand, we understand that it is not possible to express all these elements through a single diorama and this is also pointless. On the other hand, these absences express the choices made by the designers, which can be justified both from an epistemological point of view and from a museological point of view. In this context, our study indicates that in addition to the epistemological control, there was also a museological control performed by the actors of the noosphere, based on the notion of how communication will be achieved in museums through dioramas. We call this museological control as museographic vigilance, which reveals that designers have intentions and constraints not only in terms of content but also on how to teach and disseminate in museums.

Our data allowed to identify how the choices made by the noosphere actors reveal the action of museographic vigilance on the production of the diorama, showing the regulatory capacity of this institutional sphere in the preparation of dioramas in science museums. As one of the exhibition designers (C1) stated, "The intention was to provide *a context for the animals* and the diorama was chosen as a museological object that best expressed this proposal," indicating the action of the designers, who selected the particular way for the exposition: through models of plants and animals conserved in a scene.

Oliveira (2010) states that the noosphere mediates the museographic transposition process through discussions between researchers, educators, curators, and technicians of the institution in the decision-making regarding what is more important to expose to the public. These areas of knowledge and their actors exert "pressure" that can influence how the content will be composed of the knowledge to be exposed. Therefore, the limitations of the exhibition space, the characteristics of the collection, and the spatial and temporal design of the scene determine "how" the dioramas will be constructed so that they can come close to or remain far from the reference knowledge.

The epistemological and museographic vigilance are complementary. As pointed out by Mortensen (2010: 323), "one key insight that may be drawn from this collective work is that exhibit content and exhibit form are not mutually independent." The choices of concepts and of objects and scenarios impose possibilities and restrictions on *what* and *how* to exhibit in science museums and then reveal the educational implications of the exhibits.

In that perspective, a diorama is an important topic to address in the educational activities of museums because of its ecological relationships and the organisms represented. A diorama helps to understand the limitations of the representations and how the teaching processes occur in the exhibition as selected by the actors and reflected from the choices and the power relations that occur in the production of objects in the exhibitions of museums (Marandino et al. 2015).

Another observation based on the identification of the praxeology of the diorama, from the point of view of epistemological and museographic vigilance, concerns the diversity of plants represented in the scene. Plants were not identified in the legends in the exhibition set. On the one hand, it is known that the research and the activity of the zoologist at MZUSP were related to animal diversity and it was also the focus of the exhibition. On the other hand, the Amazon Forest biome had many characteristics that defined it beyond the animals, and thus other organisms that composed it could also be identified, as was done with the invertebrates, for example. Furthermore, the phylogenetic identification of plants can be relevant to zoology research as it reveals aspects about behavior and ecology of a given organism, and therefore the presence of more explicit information could be significant. In that perspective, we consider that the plants in the scene could have been identified with labels in the legend, which would have helped the visitor characterize this ecosystem in a more intricate manner. However, showing the conceptual ideas about plants in detail was not the choice of the team that planned this exhibition, which reveals the options and the selections of the designers during the diorama production, as an evidence of the museographic transposition and the museographic vigilance process.

Dioramas are very interesting objects, with a great educational potential but with some conceptual and museographical limitations. Insley (2008) in his studies about long-term exhibitions points out that the conceptual content represented in dioramas that remain in place for a long period of time can become outdated, as they are designed to communicate information in a given period of time. The message then becomes outdated; therefore it would be useful to replace the objects with more contemporary ones. However, this can be a problem in terms of developing educational activities in museums because making changes in these objects represents high costs. According to Tunnicliffe and Scheersoi (2015), many long-term exhibitions of dioramas were disassembled and even demolished in the second half of the twentieth century, in part because they were considered old-fashioned. However, the researchers argued that we are experiencing the rebirth of these objects in the present day and that new dioramas are also being built with techniques to enhance the quality of reality, a typical characteristic of these objects.

We believe that the potential and limitations of dioramas should be elements for discussion and analysis by the public, which can be done during museum visits and in the teacher education programs, for example. In that perspective, museum educators and researchers can use praxeology as a tool to study and evaluate how the public reaches the logos and the praxis of dioramas and other type of exhibition (Mortensen 2010). In this context, Bueno et al. (2015) developed a guide to help teachers to analyze exhibitions in science museums and to help them to identify and reflect on the possibilities and limitations of these expositive objects. Bueno et al. offer a tool to analyze dioramas based on praxeology, which shows the great potential of this concept for the development and study of teaching and learning processes in museums.

The exhibition of the Museum of Zoology, which contained the Amazon Forest diorama, was discontinued in 2012 and reopened in August 2015 with another focus and a new title: "Biodiversity: know to preserve." This exhibition contains renovated dioramas conceived in a more modern museographic design. To do that, the organizing team of the museum proposed a new conceptual and museographic approach, which implied new choices related to content and objects, as well as new challenges regarding the "what" and "how" to teach and disseminate the conservation of biodiversity. Certainly, this new exhibition involves the work of the noosphere, revealing the fascinating potential of the dioramas as teaching objects in science museums.

References

Achiam, M. (2013). A content-oriented model for science exhibit engineering. *International Journal of Science Education, Part B, 3*(3), 214–232.

Almeida, A. P. (2012). Realismo e Fotografia: Dioramas de Hiroshi Sugimoto do Museu de História natural de Nova Iorque [Realism and photography: Hiroshi Sugimoto Dioramas of the New York museum of natural history]. *E-journal of Museologia & Interdisciplinaridade, 1*(2), 114–133. Retrived from http://periodicos.unb.br/index.php/museologia/issue/view/774.

Ash, D. (2004). How families use questions at dioramas: Ideas for exhibit design. *Curator, 47*(1), 84–100.

Bosch, M., & Gascón, J. (2006). Twenty-five years of the didactic transpositions. *ICMI Bulletin, 58*, 51–63.

Brockington, G., Pietrocola, M. (2005). Serão as regras da transposição didática aplicáveis aos conceitos de física moderna? [Are the rules of didactic transposition applicable to the concepts of modern physics?]. In *Investigações em Ensino de Ciências*. Universidade Federal do Rio Grande do Sul, Rio Grande do Sul, 10(3), 1–17.

Bueno, J. P. P., & Oliveira, A. D., & Vidal, F. K. (2015). Identificando o potencial de objetos expositivos para ações educativas em museus de ciências [Identifying the potential of exhibition objects for educational activities in science museums]. In M. Marandino, & D. Contier (Eds.), *Educação Não Formal e Divulgação em Ciência: da produção do conhecimento a ações de formação* (pp. 37–44). São Paulo: Faculdade de Educação da USP.

Chevallard, Y. (1991). *La transposición didáctica: del saber sabio al saber enseñado* [The didactic transposition: from science knowledge to knowldge taught]. Buenos Aires, Brazil: Aique Grupo Editor S.A.

Chevallard, Y. (2006). La théorie anthropologique des faits didactiques devant l'enseignement de l'altérité culturelle et linguistique: Le point de vue d'un outsider. [The anthropological theory of didactic facts before the teaching of cultural and linguistic otherness: The point of view of an outsider]. Paper presented at the Conférence plénière donnée le 24 mars 2006 au colloque *Construction identitaire et altérité: Créations curriculaires et didactique des langues*, Université de Cergy-Pontoise, Cergy-Pontoise, France. Retrieved from http://yves.chevallard. free.fr/spip/spip/IMG/pdf/La_TAD_devant_l_alterite_culturelle_et_linguistique.pdf.

Chevallard, Y. (2007). Readjusting didactics to a changing epistemology. *European Educational Research Journal, 6*(2), 9–27.

Dean, D. (1996). *Museum exhibition: Theory and practice*. London: Routledge.

Fayard, P. (1999). La sorpresa da Copérnico: El conocimento gira alredor del público [Copernicus's surprise: Knowledge revolves around the public]. Alambique. *Didáctica de las Ciencias Experimentales* 21, 9–16.

Insley, J. (2008). Little landscapes: Dioramas in museum displays. *Endeavour, 32*(1), 27–31.

Machado, V. M. (2011). *Prática de Estudo de Ciências: Formação Inicial docente na Unidade Pedagógica sobre a Digestão Humana* [Practice of science study: Initial teacher training in the pedagogical unit on human digestion] (Unpublished doctoral dissertation). Universidade federal de Mato Grosso do Sul, Campo Grande, Brazil.

Marandino, M., Martins, L.C., Gruzman, C., Caffagni, C.W.A., Islaji, C., Campos, N. F., ... Bigatto, M. A. (2009). A abordagem qualitativa nas pesquisas em educação em museus [the qualitative approach in research in museum education]. In *VII Encontro Nacional de Pesquisa em Educação em Ciência*. ABRAPEC. 1, 1–12.

Marandino, M., Achiam, M., & Oliveira, A. (2015). The diorama as a means for biodiversity education. In S. D. Tunnicliffe & A. Scheersoi (Eds.), *Natural history diorama: History, construction and educational role* (pp. 251–266). Dordrecht: Springer.

Moraes, R., Galiazzi, M. C. (2007). *Análise Textual Discursiva* [Discursive textual analysis]. Ijuí: Unijuí.

Mortensen, M. (2010). Exhibit engineering: A new research perspective. Unpublished doctoral dissertation. University of Copenhagen, Copenhagen, Denmark.

Oliveira, A. (2010). *Biodiversidade e museus de ciências: um estudo sobre transposição museográfica nos dioramas* [Biodiversity and science museums: a study on museographic transposition in dioramas]. Unpublished master's thesis. Universidade de São Paulo, São Paulo, Brazil.

Souza, W. B., Ricardo, E. C., Paiva, J. R., Neto, P. A.; Corrêa, R. W. (2012). A vigilância epistemológica de Chevallard aplicada ao espalhamento das partículas alfa [The epistemological vigilance from Chevallard applied to the scattering of alpha particles]. In *XIV Encontro de Pesquisa em Ensino de Física*. Maresias, São Paulo, Brazil, 1-9.

Tunnicliffe, S. D., & Scheersoi, A. (2015). Diorama as important tools in biological education. In S. D. Tunnicliffe & A. Scheersoi (Eds.), *Natural history diorama: History, construction and educational role* (pp. 133–143). Dordrecht: Springer.

Van-Praët, M. (1989). Contradictions des musées d'histoire naturelle et évolution de leurs expositions [Contradictions of natural history museums and their changing exhibitions]. In B. Schiele (Ed.), *Faire voir, faire savoir – la muséologie scientifique au présent* (pp. 25–34). Québec: Musée de la Civilisation.

Winslow, C. (2011). Anthropological theory of didactics phenomena: Some examples and principles of its use in the study of mathematics education. *Un panorama de TAD. CRM Documents, 10*, 117–138.

Expanding the Scope of Science Education: An Activity-Theoretical Perspective

Yrjö Engeström

Introduction

Two major forces demand an expansion of science education beyond the traditional focus on well-defined concepts codified in textbooks and transmitted within the classroom. The first force is students' increasing involvement in knowledge-related practices outside the school, in social media, and various life activities that question the authority of school knowledge and challenge the motivation of students. The second force is the increasingly problematic and societally contested character of natural phenomena, such as climate change, biodiversity, and sustainability of natural resources. This second force also puts authoritative scientific knowledge into problematic light as researchers and politicians heatedly argue on what is true and what is not.

Going beyond the classroom is usually framed as widening the context of learning. But how should we understand the notion of context? From the point of view of cultural-historical activity theory, human activity weaves its own context. Activity is in and for itself a relatively durable systemic and collective formation. The durability of activity stems from its object. The object of activity is not just a goal – it is the materially grounded long-term purpose that gives meaning and motive to the activity.

Activity theory suggests that learning is best analyzed and promoted as embedded in collective activity systems and their networks facing important challenges of change and renewal (Greeno and Engeström 2014). This means that learning and instruction are most effective and consequential when they are intertwined with community-level efforts of transformation.

Y. Engeström (✉)
University of Helsinki, Helsinki, Finland
e-mail: yrjo.engestrom@helsinki.fi

© Springer International Publishing AG 2017
K. Hahl et al. (eds.), *Cognitive and Affective Aspects in Science Education Research*, Contributions from Science Education Research 3,
DOI 10.1007/978-3-319-58685-4_26

357

Such transformation may be examined as processes of expansive learning (Engeström 2015). Expansive learning is understood as creation of new patterns of activity. There are three main dimensions in expansion, namely, the socio-spatial dimension, the temporal-historical dimension, and the ethical-political dimension. In expansive learning, the object of the activity is expanded so as to involve wider circles of participants and domains of influence, a longer time perspective, and qualitatively new demands of social and political responsibility.

In this chapter, I will not aim at a comprehensive overview of activity-theoretical research in science education (for such overviews, see Lee 2015; Plakitsi 2013a). Instead, I will draw on a range of activity-theoretical and conceptually closely related studies to discuss five layers of expanding the focus of science education. This endeavor is closely related to that of Penuel (2016: 1) who states: "I argue that education research is needed that focuses on how people use science and engineering in social practices as part of collective efforts to transform cultural and economic production."

In the following, the first layer of expansion of science education consists of bringing elements of societal practices into classroom science instruction. The second layer of expansion consists of taking an entire school as an activity system and community of learning. The third layer consists of analyzing and promoting science learning in activity systems outside the school. The fourth layer consists of involving indigenous and other communities as funds of knowledge and alternative epistemologies with powerful potential for science learning. Finally, the fifth layer consists of taking local, regional, and global social movements as dynamic contexts of activist science learning. At the end of the chapter, I will discuss a set of theoretical and methodological openings for research in science education in its expanding contexts.

First Layer of Expansion: Bringing Elements of Societal Practices into Classroom Instruction

This mode of expanding the scope of science education is exemplified in the recent work of Prins et al. (2016). These authors point out that one of the challenges of science education is the transformation of authentic scientific practices into contexts for learning. Bringing merely isolated techniques of observation, measurement, and experimentation from scientific practices into classroom instruction can easily lead the students to ask: Why are we doing this?

For Prins et al. (2016), this limitation may be overcome by building on activity theory:

> Activity theory is aimed at understanding the totality of human work and praxis, that is, collective activity systems in society. This implies not only examining what kinds of activities people engage in, but also who is engaged in that activity, what their goals and intentions are, what objects or products result from the activity, the rules and norms that

circumscribe that activity, and the larger community in which the activity occurs. (Prins et al. 2016: 2)

The authors chose the chemical practice "modeling drinking water treatment" as the focus of their study. This practice sought to represent the complete water treatment process using a series of mathematical models that enable the prediction of the quality of drinking water after various treatments, given a certain raw water quality. The practice was carried out by experts on water treatment working at the National Institute of Public Health and the Environment in the Netherlands; civil servants from the Ministry of Infrastructure and the Environment and the Ministry of Health, Welfare and Sport; process engineers employed at universities involved in the production of drinking water; and technologists from drinking water companies. This practice was modeled and turned into a curriculum unit by a group of six chemistry teachers. The curriculum unit was taught to 36 students of grades 10 and 11.

The curriculum unit was carefully designed. However, as might be expected, transplanting a scientific practice of this complexity into a classroom is problematic. For example, the students were addressed as junior employees of the Institute for Public Health and the Environment and given the assignment of modeling turbidity removal by coagulation and flocculation. The students appreciated the clear assignment, but "the appointment of students as junior employees was regarded as artificial in their school environment" (Prins et al. 2016: 19). The authors also note having found indications that the students perceived reflection on their learning as not meaningful: "Students are aware that they will not encounter a similar modeling problem in their school career" (Prins et al. 2016: 27).

While attempts such as the one reported by Prins et al. (2016) are valuable and may lead to impressive learning outcomes, they are limited by the somewhat simplistic idea that school instruction should reproduce or imitate scientific practice outside school. School instruction and school learning are driven by different objects and motives from those of scientific work, for example, in the Institute for Public Health and the Environment. Also the community, the division of labor, and the rules are foundationally different in these activity systems.

Therefore, rather than reproducing or imitating, it might be more meaningful to build on the very *differences* between the activities. Teachers and students might compare and contrast, for example, standard textbook accounts of drinking water treatment and the practices actually conducted in the responsible agencies. As Marton (2015) points out, difference is the mother of all learning. In activity-theoretical terms, we might push this observation a step further and argue that a conflict of motives is the mother of all expansive learning.

This line of research within the first layer of expansion, focusing on differences and tensions as potentials for expansion, is exemplified in the work of Tomaz and David (2015) in the tradition of everyday mathematics. They analyzed classroom events following from the introduction into instruction of everyday mathematical situations related to water consumption. The authors point out that negotiation between formalized knowledge and common knowledge is not free from conflicts and tensions (Tomaz and David 2015):

In the case of the rule of three, when the procedures of the students and the teacher came into contact, this contradiction prompted an expansion of the meaning of this rule, both for the students and for the teacher. This expansion happened when the students started questioning the mechanical use of the rule of three and considered it to be one of the possible procedures to solve problems involving proportional reasoning. […] some students understood both the similarities and differences between school and everyday procedures, what we have considered as an expansion of meanings for school procedures. (Tomaz and David 2015: 490)

Second Layer of Expansion: The Entire School as an Activity System and Community of Learning

Studies of science education – including those inspired by activity theory – mostly focus on events inside the classroom. The classroom is commonly taken as the focal and exclusive microcosm of learning. Studies that encompass the whole school are surprisingly rare. Yet, as indicated above, the school is a peculiar activity system that has its own historically formed characteristics that constrain and shape what is possible within a classroom.

A step toward analyzing the whole school as a context of science education was recently taken by Barma et al. (2015) in a study of two science teachers' effort to teach climate change. The teachers were being asked to: (1) change their teaching style to engage students to respond to open-ended questions, (2) address the topic of environment as a controversial issue, and (3) meet the principal's and the parents' expectations. These different demands translated into clashing motives. Building on the acceptance of the possibility that a controversial environmental issue would be discussed at school, the teachers decided to engage in planning a new teaching sequence. In a nutshell, students would have to engage in a quest to answer: What is global warming about? The evaluation assessment would be centered on their ability to debate the question in front of their peers, not on the degree of appropriation of related scientific concepts. The teaching sequence would ultimately aim at promoting the students' autonomy in relation to societal debate.

This expansive resolution of conflicting motives by the two teachers led to new ones which involved the values and culture of the entire school. The values of the teachers were clashing with the parents' expectations. Both teachers were struggling with the fact that they did not feel supported by their institution, using the analogy of a "pierced umbrella" when talking about the protection they were getting from their institution. Although the teachers criticized the "knives being thrown," they also expressed the need "to fit in the school frame." As the authors point out, "perhaps it is inevitable that bold actions of transformative agency require the involvement of communities beyond individual practitioners" (Barma, et al. 2015: 36).

Science education will increasingly face contested issues that spill beyond the boundaries of "pure science." Such issues are best dealt with in collaboration across the boundaries between different school subjects. This requires active support and encouragement from principals and parents at the level of the entire school.

Third Layer of Expansion: Science Learning in Activity Systems Outside the School

When science education is extended beyond the boundaries of the school, the next step is typically instruction and learning in science museums and science centers (Plakitsi 2013b). Also other contexts have recently been analyzed. The work of Mattos and Tavares (2013) in a Brazilian children's hospital is an interesting example.

In the hospital, the spatial arrangements, temporal rhythms and artifactual resources differed radically from those of a regular classroom. Unlike regular schools, the hospital school had to conduct its instruction fully linked with students' needs. Teaching included continuous negotiation of objectives among all parties involved in the activity. Contradictions emerged when a teacher, student-patients, families, and doctors attributed different meanings to this informal but supervised teaching activity. The classes were part and parcel of the child's medical treatment. Hospital instruction became a novel activity developed around the hybrid subject *student-patient-child*.

More elaborate arrangements of expanding beyond the school include partnerships between schools and various communities of scientists and technologists (France and Compton 2012). The Dutch *Technasium* schools are a prominent example of such partnerships. In these schools, students study a core subject called 'Research and Design.' By collaborating with business communities and higher education institutions, Technasium schools aim at showing pupils how versatile and interesting science and technology-related studies and professions can be. Cooperation with companies and institutions for higher education is embedded in the Research and Design subject. Technasium students work as a team on tasks and projects that lead to concrete results.

Technasium schools create partnerships with companies. This brings in the tension between education for the common good and the quest for private profit, usually not problematized in enthusiastic accounts of such partnerships. This may be balanced with partnerships with communities and citizen groups. Citizen science and participatory mapping are two prominent examples of possibilities for schools to work with communities for the common good rather than for private profit.

Citizen science is scientific work undertaken by members of the general public, often in collaboration with or under the direction of professional scientists and scientific institutions. Citizen scientist, in the modern sense, is defined as a scientist whose work is characterized by a sense of responsibility to serve the best interests of the wider community. Recent activity-theoretical research on citizen science includes the studies of Benavides (2016) and Vallabh et al. (2016).

The Third Finnish Breeding Bird Atlas (Valkama et al. 2011) is an example of citizen science. It was produced on the basis of observations reported by over 5000 volunteers around the country. The aim of the third atlas is to examine present distributions of birds and compare them with those published in the previous atlases. The atlas data can be utilized together with other long-term bird monitoring and other environmental data to investigate changes in biodiversity.

Participatory mapping – also called community-based mapping – is a general term used to define a set of approaches and techniques that combines the tools of modern cartography with participatory methods to represent the spatial knowledge of local communities. It is based on the premise that local inhabitants possess expert knowledge of their local environments which can be expressed in a geographical framework which is easily understandable and universally recognized. Participatory maps often represent a socially or culturally distinct understanding of landscape and include information that is excluded from mainstream or official maps (Belay 2012; Chambers 2006; Literat 2013).

A film *Mapping for Change: The Experience of Farmers in Rural Oromiya, Ethiopia*, edited by Jess Phillimore (2011), documents a participatory mapping experience of Ethiopian farmers in the Oromia Region. In a relatively degraded environment where soil fertility plummeted after the clearance of the natural forest, villagers collectively constructed a large three-dimensional map of their area. As the map was constructed and completed, it was invested with meaning that showed the situation in a new light, with potential for expansive transformation. The map helped both the villagers and the outside experts to identify the temporal and spatial patterns of deforestation and soil degradation and possible focal areas in which a reversal of these processes could be successfully initiated. The very building of the map included the important learning actions of questioning, analyzing, and modeling. The map was not merely a cognitive tool. It was above all a means for volitional breaking out of a paralysis. It was a representational device that allowed the villagers to expand their vision beyond the here-and-now, both in space and in time, and to start building a model for the future.

In the industrialized north, participatory mapping will most likely take the form of counter-mapping (Taylor and Hall 2013; Villaseca 2014). This implies critically examining dominant "official" maps and generating alternative maps that make visible resources that may otherwise remain inaccessible to those in need. For example, school students may learn to identify new biking routes, skateboarding spots and scenes of street culture, as well as locations of potential danger such as hotspots of drug dealing. Such geography from the ground up may be effectively blended with powerful digital tools such as Google Maps.

Fourth Layer of Expansion: Indigenous and Other Communities as Funds of Knowledge and Alternative Epistemologies

In scientific communities, there is increasing recognition that indigenous communities have access to knowledge that may be of crucial importance for the understanding and sustainable management of ecological change, human health, and use of natural resources (Breidlid 2013). Dialogue and collaboration between scientists and indigenous communities may become a model for responsible scientific practice (Alexander et al. 2011). Science educators may examine and foster such dialogues and use them as means to engage students in socially responsible science learning (Botha 2012). More generally, communities of different ethnic and cultural composition are important potential funds of knowledge (González et al. 2006) that can significantly enrich, sometimes also challenge, the knowledge included in standard science curricula of schools.

The work of O'Donoghue (2015) in the Environmental Learning Research Centre at Rhodes University in South Africa offers an example of cross-fertilizing indigenous knowledge and western scientific knowledge in community development projects and in classrooms. Over a seven-week module, science student teachers worked to differentiate key propositions in the science curriculum and to align these with the everyday knowledge practices and prior knowledge of school learners. The students worked in groups to unpack the science curriculum in relation to daily life and the issues of the day, in this case, focusing on fermentation chemistry.

A patchwork of fermentation practices were described, with side notes on patterns of how things had changed in ways that shaped emerging environmental issues that are included as topics in the science curriculum. Evidence for the Nguni mastery of fermentation that preceded the colonial era was gathered. The learners discovered that Nguni knowledge practices were not static. Notable innovations were the making of amarewu with cooked maize meal (introduced by the Portuguese) and later with a wheat ferment starter (wheat was introduced by the British). A similar trajectory of innovation was tracked in the making of dumplings and steam bread. Isonka (oven bread) was mastered in Dutch and later Africana trekboer kitchens. Dutch oven bread was later used by amaHlubi groups excluded by apartheid from the mainstream economy. Evidence for this was found locally in amaHlubi oral histories in the Herschel District. The isiXhosa knowledge practices were demonstrated in the science classroom and were discussed in terms of the diverse cultural practices found among the participating student teachers. These activities provided a cultural capital of knowledge practices that allowed the commensurable integration of western chemistry and indigenous knowledge in the student teachers' own teaching practices.

The demonstration of Xhosa fermented foods was followed by brainstorming and Web search activities to map out ideas and open up information on health risk issues that had arisen with the advent of the modern diet. Obesity, diabetes, and gluten allergies were highlighted here, as it was noted that bread is now the staple diet of modern South Africa. The group then did a "back analysis" of changing food production practices to identify how changed fermentation practices were, over time, shaping modern health risks. These activities served to open up environmental matters of concern to be included in teaching and learning materials.

Another important source for this layer of expansion is the work of Bang and Medin on possibilities of building science education on the logics of the indigenous knowledge of Native American tribes in the United States (Medin and Bang 2014). The authors show convincingly that the indigenous conception of the relationship between nature and culture as a whole is deeply different from that of the dominant western logic. They invite us to study carefully the cosmologies and epistemologies of non-dominant groups as a source of questioning, rethinking, and hybridization of science education based on standard western notions (Bang et al. 2013). Similar work on mathematics among indigenous people is conducted in Brazil by Tomaz (2013).

It is true that indigenous communities may quite rarely be encountered in regular school-based science education in the industrialized north. However, the increasingly multicultural composition of societies calls attention to diverse funds of knowledge and alternative epistemologies that reside in various ethnic and religious groups as well as among subcultures and cultural traditions more broadly.

Fifth Layer of Expansion: Social Movements as Dynamic Contexts of Activist Science Learning

Social movements around the world are increasingly challenging the ways large corporations and governments use science to exploit natural resources and to eliminate traditional, often ecologically sound practices of production and consumption. A prominent example is the Zapatista movement in Chiapas, Mexico.

Zapatistas are an indigenous social movement and self-governing community in southern Mexico. They resist the intrusion of genetically modified corn into Mexico, "the heartland of maize diversity." Their efforts include a seed bank, a genetic testing program, and a seed distribution program. In this quest, they collaborate with scientists and NGOs around the world – the Zapatista corn is distributed and cultivated in different parts of the world as an act of solidarity and support.

Zapatista corn performs the biocultural link between Zapatistas' political project and their maize plants. By creating alternative networks for corn circulation, the project allows international recipients to participate in Zapatistas' political bioculture, that is, to relate to seeds as potential food or plants that are deeply infected with the values of promoting self-sufficiency and resisting governmental and economic dependence (Brandt 2014: 876).

My own research group is currently conducting a study of learning in productive social movements in different parts of the world. Movements focused on sustainable and equitable uses of land and other natural resources, as well as those focused on food sovereignty, agroecologically informed sustainable farming, and more generally locally produced healthy food, are growing around the globe. These movements typically need to acquire and implement in practice concepts and principles that are grounded both in solid scientific research and in an understanding an appreciation of traditional methods and practices that have proven to be ecologically sustainable.

Such learning challenges generate collaboration across various types of expertise as well as new models of horizontal peer-to-peer instruction (Holt-Giménez 2006). Science education in schools could benefit greatly from involvement in such learning practices in social movements. For example, in France, community-supported agriculture movement has rapidly spread in the last 10 years in response to the overuse of provisioning services by intensive agriculture, the deterioration of product quality, the fluctuation of prices, and the disconnect between producers and consumers (Pelenc et al. 2013). In Japan, school students and teachers joined a successful community movement to revitalize the cultivation of a traditional local vegetable (Yamazumi 2009). As Roth (2010) points out, activism needs to be understood as a key quality of learning in science education engaged in societal transformations.

Theoretical and Methodological Openings

The five layers of expansion of science education presented above are not to be seen as a hierarchy leading toward universally "better" or more desirable forms. The layers represent emerging opportunities and practices highly dependent on specific cultural and historical circumstances. What works in one place and time may not work in another one. Thus, it may be wise to take these layers as a repertoire of possibilities that may be combined and hybridized in various ways.

The layers invite the reader to imagine science education in which classrooms and whole schools actively draw on the funds of knowledge of local communities, engage in participatory mapping and educate their students as citizen scientists, and collaborate with social movements in bringing about changes toward sustainable use of natural resources. Such partnerships and coalitions can make education a real force of societal transformation.

The layers of expansion were worked out on the basis of published research on science education in novel contexts and connections, still emergent but already significant enough to warrant a fairly optimistic perspective. However, another dimension of expansion is much more in an embryonic phase. This is the expansion of theoretical and methodological understandings that guide research in science education. Because of the embryonic state, I will here only briefly point out some possible openings that may be taken to pursue this dimension of expansion.

The first opening is that of analyzing the emergence of *transformative agency* as a decisive quality and outcome of science learning. In activity theory, this opening is grounded in Vygotsky's (1997) principle of *double stimulation*. A problematic situation (1st stimulus) often involves a paralyzing conflict of motives. Human beings can break out of such a paralysis by constructing and using material artifacts (2nd stimuli) that become meaningful signs and allow them to expansively transform the situation (Sannino 2015; Sannino and Laitinen 2015). Sannino's (2015) model of the principle of double stimulation was used as a conceptual lens by Barma and her coauthors in the study discussed earlier in this chapter. Also, the collectively created map of the Ethiopian farmers in the Oromia Region discussed earlier may be regarded as an effective second stimulus that empowered the participants to take agentive actions toward transforming the ways in which they as a community were using the natural resources of their environment.

The second opening is that of analyzing science education as *concept formation in the wild* (Engeström and Sannino 2012; Greeno 2012). Concepts are increasingly polyvalent, open-ended, and contested. Complementarity and movement between different modes or levels of conceptualization is of crucial importance for the robustness and continuous development of concepts.

The functioning of a theoretical concept may be seen as stepwise movement from a diffuse concrete phenomenon to an abstract model and then to rich, expanding concreteness – a process called *ascending from the abstract to the concrete*. The earliest, smallest, and simplest unit of such a complex totality is called a *germ cell*. It carries in itself the foundational relationship and contradiction of the complex whole (Engeström et al. 2012). It is ubiquitous, so commonplace that it is often taken for granted and goes unnoticed. It opens up a perspective for multiple applications, extensions, and future developments. It can be represented in multiple modalities, including artistic images and symbols. The Zapatista corn, discussed earlier, in its many representations may be regarded as a germ cell of an expanding concept of food sovereignty in the Zapatista context.

The third opening is that of analyzing learning as involving boundary crossing and hybridization between multiple worlds or domains. A powerful lens of this is the notion of *third space*, put forward by Gutiérrez et al. (1995). The third space is understood as a meeting point and negotiation arena between radically different, potentially conflicting perspectives and understandings. Negotiations in a third space can lead to powerful new hybrid conceptualizations and practices. The way O'Donoghue and his students brought into critical dialogue indigenous and modern scientific understandings of fermentation, discussed earlier, may be seen as a powerful example of boundary crossing and, at least potentially, creation of third spaces. The work of Tomaz and David on bringing together everyday mathematics and formal school mathematics is another powerful example of boundary crossing and third spaces.

The fourth opening is methodological. A good amount of design-based research is currently conducted in science education. From the point of view of activity theory, much of design-based research is limited in that learners and teachers are basically recipients of instructional solutions designed by researchers. In spite of recent more promising developments, design-based research still largely follows the linear model of intervention in which desirable outcomes are determined ahead of time and the agency of learners is thus minimized (Engeström 2011). To go beyond such linear models of intervention, a *methodology of formative interventions* is being worked out (Engeström et al. 2014). The most well-known formative intervention method is the *Change Laboratory* (Sannino et al. 2016). A group of practitioners analyze and redesign their activity in successive meetings with the help of conceptual tools from activity theory. The process is designed to trigger and support an intensive cycle of expansive learning that generates new models of activity not known by the interventionists ahead of time: learning what is not yet there (Engeström 2016).

These four openings are opportunities for enriching the theoretical and methodological resources available for research in science education. The expansion of contexts of science education will inevitably call for such new theoretical and methodological concepts.

Conclusion

In this chapter, I have proposed an expansive approach to science education, stemming from the theory of expansive learning developed within cultural-historical activity theory (Engeström 2015). My argument is that we should not restrict research in science education to classrooms and other settings of formal education. I have suggested five layers of expansion, starting from the bringing of elements of societal practices into science classrooms and ending with social movements as dynamic contexts for activist science learning. I am not arguing that a wider context is somehow better or more advanced than a more compact context. I am arguing for movement, negotiation, and hybridization between the contextual layers. This kind of inter-contextual movement has the potential to revitalize also classroom learning in science education.

I have also suggested four theoretical-methodological openings that may facilitate and enhance inter-contextual movement in science learning. These openings – transformative agency, concept formation in the wild, boundary crossing and third spaces, and formative interventions – are all under way as initiatives to sharpen up the theoretical and methodological frameworks of empirical educational research. Cultural-historical activity theory and the theory of expansive learning are significant resources in the development of these initiatives. Drawing on dialectical epistemology, they call attention to the object-oriented, contextually embedded, internally contradictory, and historically changing character of science learning and instruction.

References

Alexander, C., Bynum, N., Johnson, E., King, U., Mustonen, T., Neofotis, P., & Vicarelli, M. (2011). Linking indigenous and scientific knowledge of climate change. *Bioscience, 61*(6), 477–484.

Bang, M., Warren, B., Rosebery, A. S., & Medin, D. (2013). Desettling expectations in science education. *Human Development, 55*(5–6), 302–318.

Barma, S., Lacasse, M., & Massé-Morneau, J. (2015). Engaging discussion about climate change in a Quebec secondary school: A challenge for science teachers. *Learning, Culture and Social Interaction, 4*, 28–36.

Belay, M. (2012). *Participatory mapping, learning and change in the context of biocultural diversity and resilience.* PhD thesis. Grahamstown: Rhodes University.

Benavides. (2016). *Meanings teachers make of teaching science outdoors as they EXPLORE citizen science.* PhD thesis. Greensboro: The University of North Carolina at Greensboro.

Botha, L. R. (2012). Using expansive learning to include indigenous knowledge. *International Journal of Inclusive Education, 16*(1), 57–70.

Brandt, M. (2014). Zapatista corn: A case study in biocultural innovation. *Social Studies of Science, 44*(6), 874–900.

Breidlid, A. (2013). *Education, indigenous knowledges, and development in the global South: Contesting knowledges for a sustainable future.* London: Routledge.

Chambers, R. (2006). Participatory mapping and geographic information systems: Whose map? Who is empowered and who disempowered? Who gains and who loses? *The Electronic Journal on Information Systems in Developing Countries, 25*(2), 1–11.

Engeström, Y. (2011). From design experiments to formative interventions. *Theory and Psychology, 21*(5), 598–628.

Engeström, Y. (2015). *Learning by expanding: An activity-theoretical approach to developmental research* (2nd ed.). Cambridge: Cambridge University Press.

Engeström, Y. (2016). *Studies in expansive learning: Learning what is not yet there.* Cambridge: Cambridge University Press.

Engeström, Y., & Sannino, A. (2012). Concept formation in the wild. *Mind, Culture, and Activity, 19*, 201–206.

Engeström, Y., Nummijoki, J., & Sannino, A. (2012). Embodied germ cell at work: Building an expansive concept of physical mobility in home care. *Mind, Culture, and Activity, 19*, 287–309.

Engeström, Y., Sannino, A., & Virkkunen, J. (2014). On the methodological demands of formative interventions. *Mind, Culture, and Activity, 21*(2), 118–128.

France, B., & Compton, V. (Eds.). (2012). *Bringing communities together: Connecting learners with scientists or technologists.* Rotterdam: Sense.

González, N., Moll, L. C., & Amanti, C. (Eds.). (2006). *Funds of knowledge: Theorizing practices in households, communities, and classrooms.* New York: Routledge.

Greeno, J. G. (2012). Concepts in activities and discourses. *Mind, Culture, and Activity, 19*(3), 310–313.

Greeno, J. G., & Engeström, Y. (2014). Learning in activity. In R. K. Sawyer (Ed.), *The Cambridge handbook of the learning sciences* (2nd ed., pp. 128–147). Cambridge: Cambridge University Press.

Gutiérrez, K. D., Rymes, B., & Larson, J. (1995). Script, counterscript, and underlife in the classroom – Brown, James versus Brown v. Board of Education. *Harvard Educational Review, 65*, 445–471.

Holt-Giménez, E. (2006). *Campesino a campesino: Voices from Latin America's farmer-to-farmer movement for sustainable agriculture.* Oakland: Food First Books.

Lee, Y.-J. (2015). Activity theory and science education. In R. Gunstone (Ed.), *Encyclopedia of science education* (pp. 12–18). Dordrecht: Springer.

Literat, I. (2013). Participatory mapping with urban youth: The visual elicitation of sociospatial research data. *Learning, Media and Technology, 38*, 198–216.

Marton, F. (2015). *Necessary conditions of learning*. New York: Routledge.

Mattos, C. R., & Tavares, L. B. (2013). The multiple senses of science teaching at a hospital school. In *Proceedings of the ESERA 2013 conference: Vol. 1. Science education research for evidence-based teaching and coherence in learning* (pp. 16–25). Nicosia: European science education research association.

Medin, D. L., & Bang, M. (2014). *Who's asking? Native science, western science, and science education*. Cambridge: The MIT Press.

O'Donoghue, R. (2015). Working with critical realist perspective and tools at the interface of indigenous and scientific knowledge in a science curriculum setting. In L. Price & H. Lotz-Sisitka (Eds.), *Critical realism, environmental learning and social-ecological change*. London: Routledge.

Pelenc, J., Lompo, M. K., Balle, J., & Dubois, J.-L. (2013). Sustainable human development and the capability approach: Integrating environment, responsibility and collective agency. *Journal of Human Development and Capabilities, 14*(1), 77–94.

Penuel, W. R. (2016). Studying science and engineering learning in practice. *Cultural Studies of Science Education, 11*(1), 89–104.

Phillimore, J. (2011). *Mapping for change: The experience of farmers in rural Oromiya, Ethiopia*. Film available at https://vimeo.com/22123738

Plakitsi, K. (2013a). *Activity theory in formal and informal science education*. Rotterdam: Sense.

Plakitsi, K. (2013b). Teaching science in science museums and science centers. In K. Plakitsi (Ed.), *Activity theory in formal and informal science education* (pp. 27–56). Rotterdam: Sense.

Prins, G. T., Bulte, A. M. W., & Pilot, A. (2016). An activity-based instructional framework for transforming authentic modeling practices into meaningful contexts for learning in science education. *Science Education*. doi:10.1002/sce.21247.

Roth, W. M. (2010). Activism: A category for theorizing learning. *Canadian Journal of Science, Mathematics and Technology Education, 10*(3), 278–291.

Sannino, A. (2015). The principle of double stimulation: A path to volitional action. *Learning, Culture, and Social Interaction, 6*, 1–15.

Sannino, A., & Laitinen, A. (2015). Double stimulation in the waiting experiment: Testing a Vygotskian model of the emergence of volitional action. *Learning, Culture, and Social Interaction, 4*, 4–18.

Sannino, A., Engeström, Y., & Lemos, M. (2016). Formative interventions for expansive learning and transformative agency. *Journal of the Learning Sciences, 25*, 599–633.

Taylor, K. H., & Hall, R. (2013). Counter-mapping the neighborhood on bicycles: Mobilizing youth to reimagine the city. *Technology, Knowledge and Learning, 18*(1–2), 65–93.

Tomaz, V. S. (2013). A study of magnitudes and measurement among Brazilian indigenous people: Crossing cultural boundaries. In A. M. Lindmeier & A. Heinze (Eds.), *Proceedings of the 37th conference of the International Group for the Psychology of mathematics education* (Vol. 4, pp. 281–288). Kiel: PME.

Tomaz, V. S., & David, M. M. (2015). How students' everyday situations modify classroom mathematical activity: The case of water consumption. *Journal for Research in Mathematics Education, 46*(4), 455–496.

Valkama, J., Vepsäläinen, V., & Lehikoinen, A. (2011). *The third Finnish breeding bird atlas*. Helsinki: Finnish Museum of Natural History and Ministry of Environment. Available at http://atlas3.lintuatlas.fi/english ISBN 978-952-10-7145-4.

Vallabh, P., Lotz-Sisitka, H., O'Donoghue, R., & Schudel, I. (2016). Mapping epistemic cultures and learning potential of participants in citizen science projects. *Conservation Biology*. doi:10.1111/cobi.12701.

Villaseca, S. L. (2014). The 15-M movement: Formed by and formative of counter-mapping and spatial activism. *Journal of Spanish Cultural Studies, 15*(1–2), 119–139.

Vygotsky, L. S. (1997). The history of development of higher mental functions. Chapter 12: Self-control. In A.S. Carton & R. W. Rieber (Eds.) The collected works of L. S. Vygotsky. Vol. 4. The history of the development of higher mental functions (pp. 207–219). New York: Plenum.
Yamazumi, K. (2009). Expansive agency in multi-activity collaboration. In A. Sannino, H. Daniels, & K. D. Gutiérrez (Eds.), *Learning and expanding with activity theory* (pp. 212–227). Cambridge: Cambridge University Press.